Nanodevices for Photonics and Electronics

Nanodevices for Photonics and Electronics

Advances and Applications

edited by

Paolo Bettotti

PAN STANFORD PUBLISHING

Published by

Pan Stanford Publishing Pte. Ltd.
Penthouse Level, Suntec Tower 3
8 Temasek Boulevard
Singapore 038988

Email: editorial@panstanford.com
Web: www.panstanford.com

British Library Cataloguing-in-Publication Data
A catalogue record for this book is available from the British Library.

Nanodevices for Photonics and Electronics: Advances and Applications

ISBN 978-981-4613-74-3 (Hardcover)
ISBN 978-981-4613-75-0 (eBook)

Printed in the USA

Contents

Preface

The mastering of matter at the nanoscale level has enabled a completely new paradigm of scientific development. The possibility to tailor a material's properties by mastering its structure at the near-molecular level has greatly improved the overall quality of synthesized materials. Nowadays the link between the properties of a material and its structure is not only a way to describe the material's behavior but also a powerful tool to design improved and innovative devices.

Electronics has heavily exploited this knowledge to push the level of integration near the physical limit. But in the last two decades photonics has also shown a fast evolution toward miniaturized systems. It is reasonable to assume the two fields will somehow fuse together in the near future and will complement each other to overcome their own limits.

The field of nanostructured devices for photonics and electronics is extremely broad and it cannot be reviewed within a single monograph; thus this book addresses specific topics of particular relevance for their scientific novelty and for their importance in future technological development.

Chapters 2 to 4 describe the properties of photonic systems as they move from perfectly periodic systems (as in the case of photonic crystals in Chapter 2) toward aperiodic systems (as for the bioinspired aperiodic spirals in Chapter 3) and, ultimately, toward purely random systems described in Chapter 4. Each chapter contains an introductory part describing the main physics behind each topic, as well as applications of the photonic systems. Chapter 5 covers the topic of detectors for the terahertz (THz) region: this spectral range has a number of possible uses but it lacks both compact sources and detectors. The use of nanowires to develop THz detectors is described here. Chapter 6 deals with the topic of optomechanics and demonstrates how optical and mechanical modes can exchange energy. In fact at the nanoscale the frequency of mechanical modes approaches the gigahertz and can be effectively coupled with optical modes. Nanodots are among the most investigated nanostructures and their peculiar properties are reviewed in Chapter 7, together with applications in optoelectronics and photonics. The lack of an efficient silicon-based light source has forced the development of hybrid integration of III–V materials in silicon photonics: this is the topic reviewed in Chapter 8. Chapter 9 describes one of most investigated photonics research area: optical biosensing using microresonators. Finally Chapter 10 describes some recent advances in the development of transistors based on organic semiconductors.

Writing a monograph treating such different types of correlated topics would be hardly realized by a single person: I am grateful to all the researchers who accepted the invitation to contribute to this monograph.

Paolo Bettotti
Povo

Chapter 1

Introduction

Paolo Bettotti

Department of Physics, University of Trento, via Sommarive 14,
38123 Trento, Italy
paolo.bettotti@unitn.it

The prefix *nano* is a Greek word meaning *dwarf*. Dwarfs were small beings, and contrary to the brute force attitude of giants, they were considered intelligent and sly creatures. Dwarfs were skilled jewelers, able to forge finely magnificent, cutted treasures and magic arms. Thus, accidentally, the use of the prefix *nano* proves to be the proper one to describe the creative way nanotechnology improves existing technologies and invents new ones.

Nanotechnology is an extremely broad science that aims at controlling matter and the interaction between matter and energy at the nanometer scale. There is no exact length scale to define the realm of what is *nano*: in its more formal definitions *nano* refers to a scale of about 10^{-9} m, but there are several topics where this classification breaks. In electronics, *nano* means the scale at which quantum effects become important and the behavior of the devices depends on such effects. A quantum box, typically, has dimensions of a few nanometers only. On the other hand, nanophotonics refers to devices able to confine the light within volumes comparable to the photon wavelength. Thus, the term "nano" acquires a relative

Nanodevices for Photonics and Electronics: Advances and Applications
Edited by Paolo Bettotti
Copyright © 2016 Pan Stanford Publishing Pte. Ltd.
ISBN 978-981-4613-74-3 (Hardcover), 978-981-4613-75-0 (eBook)
www.panstanford.com

meaning that depends on the energy of the photons and can be scales over orders of magnitude.

After the first intuition of Greek philosopher Demokrit about the possibility that matter was constituted by small elementary "bricks," the idea to exploit the possibilities given by controlling matter at the atomic scale was suggested by R. Feynman's 1959 famous lecture "There's Plenty of Room at the Bottom." Later in 1974 the term "nanotechnology" was first used by Prof. N. Taniguchi [1] and in 1981 Prof. E. Drexler published the first article on molecular nanotechnology and foresaw some of the possible applications that may benefit from it [2].

Following its first definition during the 1970s, nanotechnology was largely driven by the evolution of the electronic industry because of the striking advantages in terms of miniaturization, cheapness, and speed achievable to promote a widespread diffusion of electronic gadgets. The development of microcomputers and the massive infiltration into the consumer market created a positive feedback to the micro- (and later nano) electronics development. Moore's law is of worldwide renown and, despite its purely phenomenological origin, still defines a benchmark for the evolution of central processing units (CPUs). Moore's law contributed to the definition of the International Technology Semiconductor Roadmap (ITSR) [3] and to the development of the semiconductor industry. The expectations of the ITRS were accepted as a guideline from the main stakeholders of the semiconductor technology and the technological development was driven by the predictions defined within this document. In particular the ITRS clearly underlined the areas of knowledge gaps that may have slowed down the entire innovation cycle. The ITRS is a successful example of a worldwide-accepted tool to plan the development of technological processes that, in turn, enable the fabrication of innovative products and is has therefore defined a virtuous feedback.

The outstanding results obtained by the extreme miniaturization have permitted the exploitation of the small relevant length scales. For example, nanoelectromechanical devices are now in the lime-light because of their mechanical resonances and the possibilities of coupling them and having them interact with electromagnetic fields. Mechanical resonators with resonant frequencies beyond

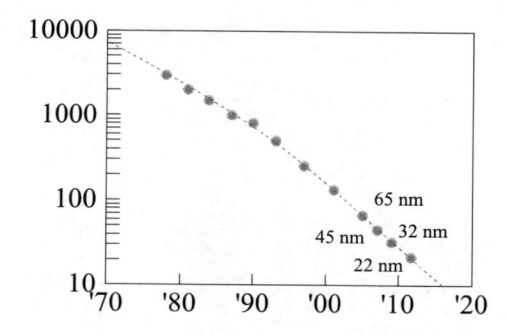

Figure 1.1 Moore's law expressed in terms of feature size. The trend is to decrease it by 0.7x every two years.

the gigahertz were demonstrated and optomechanics is nowadays an active research field with several possibilities in fields such as sensing and switching.

Despite the large achievements obtained by exploiting the nanoscale, the miniaturization of the devices cannot proceeds indefinitely and the ITRS itself predicts the need to lay the foundations of a post–complementary metal-oxide semiconductor (CMOS) era [4]. In fact, minimum length scales of CMOS elements are approaching the fundamental limit constituted by the single-atom-made device [5].

An interesting way to state Moore's law is the one based on the projection of the minimum feature size (Fig. 1.1). A simple linear extrapolation of the trend predicts that in a couple of decades the length of the minimum feature would reach the size of a handful of atoms. Such elementary units pose several questions at the border between science and philosophy: How many different functions may be implemented within a few- (or even a single-) atom device? What is the intrinsic stability of such elementary building blocks?

But even before reaching this science fiction limit, a number of bottlenecks have to be solved to go ahead with Moore's predictions: the size of integrated circuits is scaling at a fast pace and basic materials used in their fabrication are approaching fundamental limits (i.e., in terms of electrical conductivity and capacitance). There is a serious need to find suitable substitute materials or alternative device designs. An example is the use of copper instead of aluminum

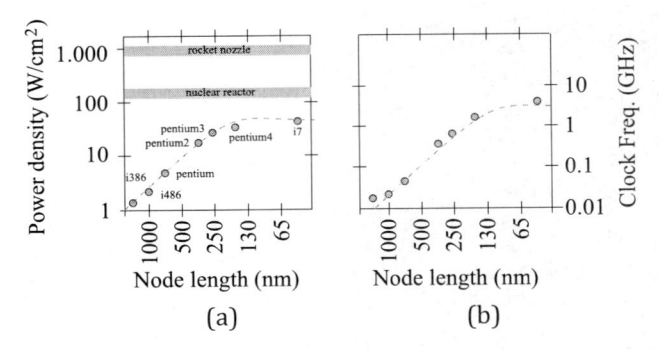

Figure 1.2 (a) Power dissipated by several models of CPUs versus gate length. (b) Clock frequency for the same models of CPUs versus gate length technology. The appearance of the multicore paradigm coincides with the inflection point on the frequency scaling graph.

in electrical interconnects and the use of a low-dielectric-constant material to replace silica [6, 7].

The huge effort spent in the last years to maintain Moore's predictions has ended with the introduction of new technological paradigms: multicore CPUs and parallelization models substitute high-frequency-clock single-core processors, after the fail of Dennard's scaling model [8], and true 3D interconnections became a reality with the introduction of 22 nm technology and of the first 3D transistor, which was put on the market by Intel (3-D Tri-Gate transistors) [9].

The approach to the physical limits of Moore's scheme can be inferred by looking at the evolution of the power consumed by modern CPUs, as shown in Fig. 1.2.

The ITRS community addresses the issue of the physical limit to miniaturization by increasing the complexity of the functions performed by devices. This idea is represented by Fig. 1.3: The performance of a device will not be defined solely in terms of speed (vertical axis), but also will depend on the complexity of the operation it will be able to execute (horizontal axis). Thus, the more-than-Moore approach not only focuses on the sole increase in the performance of a device (in terms of speed, integration level, and power consumption), but it tries to optimize the interaction of the device with the external environment.

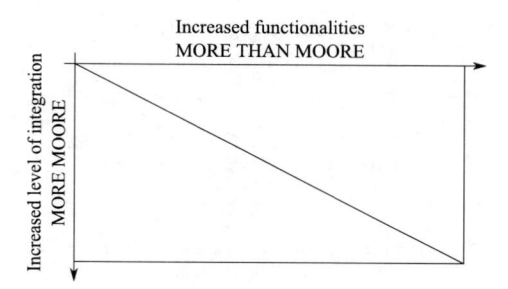

Figure 1.3 More-than-Moore approach couples the standard Moore approach based on increased performance with a parallel increase of the number of functions performed by the device. The vertical axis indicates the increase in the miniaturization level, which is also roughly proportional to the speed of the circuit. The horizontal axis depicts the complexity of the operation the device will be able to perform.

Considering the lesson learned from the roadmap-driven approach for the microelectronic industry, the more-than-Moore scheme wants to broaden the landscape of application of the next generation of optoelectronic devices by defining few thematic areas that may take advantage of dedicated roadmaps for their specific technologies.

A detailed discussion about the timing of this incoming revolution is out of the scope of this introduction and more in-depth analysis can be found in recent literature [10].

Among the different future scenarios that promise to overcome the approaching limits of the current nanoelectronic schemes, photonics is the most mature one and promises striking possibilities.

The large interest in photonics, and particular in integrated photonics, has been supported by the exponential development of the Internet and online services. At the same time several important advantages of photonics were quickly recognized:

- Photons are immune to electromagnetic interference (EMI) and ground loops. Thus, much robust connection links can be realized in industrial and harsh environments.
- Photons do not interact among each other (in the linear regime); thus, several different signals can be independently transmitted along a single optical cable without the

need to perform complex analysis (e.g., Fourier analysis) to recover the original signals.

- Photons do not dissipate heat and losses are reduced to extremely low values (~ 0.2 dB/km in optical fibers). Thus, optical signals have to be regenerated only after several tens of kilometers.

Since the initial development of optical communications, fibers proved to be the only system to support a large data transfer bandwidth and shortly substituted the electronic counterpart for large-distance communications. Nowadays, the limiting distance at which photonic communication systems compete with electronic communication systems is becoming shorter and shorter and the first optical cables for interboard connections recently reached the market.

In contrast to the electronic paradigm, photonics has been developed following several radically different implementations: a number of basic optical functionalities have been achieved using different materials and fabrication processes. This fact created a more fragmented technology constituted by different implementation schemes. Despite this fragmentation, silicon is recognized as a key enabling material also for photonics because of its compatibility with the already existing CMOS infrastructure—which is fundamental to achieving a smooth transition from electronics to optoelectronics and finally to photonic—and for its full compatibility with the transparency windows of the optical fibers. On the other hand, silicon suffers from fundamental limitations that prevent its ubiquitous use in photonics:

- Silicon is a poor light-emitting material. Thus, efficient light sources cannot be realized using it. Despite the large effort spent in the last decades to maximize the light emission from silicon, the integration of sources made of III/V semiconductors is at the moment the most efficient way to integrate a light source onto a silicon-based photonic integrated circuit (PIC).
- Silicon does not possess second-order optical nonlinearities. Some elementary building blocks of PIC—like modulators—exploit nonlinear phenomena to switch the

optical signal. Despite the demonstration of optoelectronics schemes to modulate the silicon refractive index using pn-junctions [11], other materials permit the realization of much faster and efficient devices and the full exploitation of the speed of photonics [12].

Despite these important limitations, silicon is still an actively investigated material for active photonics and devices with innovative design (such as modulators and switching) are continuously proposed in the literature [13, 14]. The goal is to maintain as much of the compatibility with the CMOS platform and, at the same time, improve the performance to fulfill the requirements of speed and energy consumption required by the next generation of PICs.

The energy consumption is a key parameter of PICs and cannot be underestimated. The analysis of the energy required by the data centers revealed that a nonnegligible amount of the power produced in developed countries is used to route Internet traffic only [15]. Considering that the large bandwidth enabled by photonics will further increase the bandwidth demand of high-performance computing in the years to come [16], PICs have to be much more efficient than their electronic counterparts in terms of energy consumed per operation. If highly efficient systems and architectures are not designed, the shift of the paradigm from electronics to photonics would bring only a limited advantage and the full potential of photonics will not be exploited. Some recent implementations of PICs require no more than few fJ/bit per switching operation, a value compatible with the scaling up to exa-scale computing [17, 18]. Despite these promising results, a real implementation of such advanced PICs will still require a few years.

During the last decades, nanotechnology has become a truly interdisciplinary science, with a growing level of interconnections among different branches. As a concrete example, we can mention the recently demonstrated experiments that mimic neural networks to perform complex functions (like pattern or speech recognition). This results from a network of semiconductor optical amplifiers that exploit a purely photonic approach and its intrinsic advantages in performing complex tasks compared to the pure binary scheme [19]. Another example is the use of biological mechanisms to arrange molecular-scale devices able to perform specific functions [20]. It is to be noted that the more-than-Moore approach mentioned above goes in a similar direction: the possibility to handle complex systems with highly dynamic behavior paves the way to the realization of functional architectures capable of executing complex tasks within a single instruction/operation.

This monograph describes topical examples about how material nanostructuring is used to modify material properties and to induce the desired functionalities.

Chapter 2 discusses both the basics as well as the applications of photonic crystals. This broad class of materials enables tight miniaturization of photonic devices and will probably find widespread use in the years to come for the realization of highly integrated photonic circuits. Chapter 3 introduces the broad class of aperiodic photonic structures. Their optical properties are considered and case studies are used to explain their peculiarities (such as the orbital angular momentum of light) as well as their applications. Chapter 4 discusses the basics of light localization and demonstrates how randomly structured materials can be engineered to control the propagation of light and to enhance its absorption to increase the efficiency of photovoltaics systems. Nanowire-based devices are described in Chapter 5. The basic electronic transport properties of these nanostructures are reviewed and their use as efficient detectors in the terahertz regime is examined in detail. As mentioned above the nanoscaling of materials shifts their natural resonance frequencies to rather large values; thus, the direct coupling between an electromagnetic field and the mechanical resonances becomes efficient and can be used to dynamically modulate their optical properties. Chapter 6 is about the field of optomechanics and presents some examples of dynamically actuated optical resonators. As the size of the nanostructures is further reduced, quantum mechanical effects become dominant. In Chapter 7, the use of nanocrystals as elementary functional elements of quantum photonics is addressed. The fabrication of nanostructure-based optoelectronic devices requires the integration of several different materials. To achieve these results, it is of fundamental importance to develop suitable technological processes that handle all these materials and permit a proper integration into highly controlled micro- and nanoscaled devices. Chapter 8 treats the topic of hybrid integration and reviews the main technologies and key achievements obtained in the last years. Chapter 9 describes the use of photonic microresonaters in the realization of optical biosensors. Finally Chapter 10 discusses the basic optoelectronic properties and use of small organic molecules. Organics are lighter, flexible, and cheaper than their inorganic counterparts. Thus the availability of reliable technologies based on them can revolutionize several optoelectronic sectors (from photovoltaics to lighting, from displays to sensors, to cite but a few).

References

1. Taniguchi, N. (1974). On the basic concept of "nano-technology," *Proc. Intl. Conf. Prod. Eng. Tokyo, II.*
2. Drexler, K.E. (1981). Molecular engineering: an approach to the development of general capabilities for molecular manipulation, *Proc. Natl. Acad. Sci.* **78**, pp. 5275–5258.
3. International Technology Roadmap for Semiconductors, http://www.itrs.net/.
4. Arden, W., Brillouët, M., Cogez, P., Graef, M., Huizing, B., and Mahnkopf, R. (2010). *"More-than-Moore" white paper*, http://www.itrs.net/Links/2010ITRS/IRC-ITRS-MtM-v2\%203.pdf.
5. Kuhn, K.J. (2010). CMOS transistor scaling past 32 nm and implications on variation, *Advanced Semiconductor Manufacturing Conference (ASMC), 2010 IEEE/SEMI*, pp. 241–246, San Francisco, CA.
6. Baklanov, M.R. (2012). Nanoporous dielectric materials for advanced micro- and nanoelectronics. In *Nanodevices and Nanomaterials for Ecological Security, NATO Science for Peace and Security Series B: Physics and Biophysics*, pp. 3–18.
7. Ueno, K. (2010). Material and process challenges for interconnects in nanoelectronics era, *2010 International Symposium on VLSI Technology Systems and Applications (VLSI-TSA)*, Hsinchu, Taiwan.
8. Esmaeilzadehy, H., Blemz, E., St. Amantx, R., Karthikeyan, S., and Burger, D. (2011). Dark silicon and the end of multicore scaling, *38th Annual International Symposium on Computer Architecture (ISCA)*, pp. 365–376, San Jose, CA.

9. http://newsroom.intel.com/community/intel_newsroom/blog/2011/05/04/intel-reinvents-transistors-using-new-3-d-structure.

10. Hoefflinger, B. (ed.). (2012). *Chips 2020: A Guide to the Future of Nanoelectronics*, Springer-Verlag, Berlin, Heidelberg.

11. Sacher, W.D., Green, W.M.J., Assefa, S., Barwicz, T., Pan, H., Shank, S.M., Vlasov, Y.A., and Poon, J.K.S. (2013). Coupling modulation of microrings at rates beyond the linewidth limit, *Opt. Express*, **21**, 8, pp. 9722–9733.

12. Petousi, D., Zimmermann, L., Voigt, K., and Petermann, K. (2013). Performance limits of depletion-type silicon Mach-Zehnder modulators for telecom applications, *J. Lighwave Technol.*, **31**, 22, pp. 3556–3562.

13. Shainline, J.M., Orcutt, J.S., Wade, M.T., Nammari, K., Moss, B., Georgas, M., Sun, C., Ram, R.J., Stojanović, V., and Popović, M.A. (2013). Depletion-mode carrier-plasma optical modulator in zero-change advanced CMOS, *Opt Lett.*, **38**, 15, pp. 2657–2659.

14. Mancinelli, M., Guider, R., Bettotti, P., Masi, M., Vanacharla, M.R., and Pavesi, L. (2011). Coupled-resonator-induced-transparency concept for wavelength routing applications, *Opt. Express*, **19**, 13, pp. 12227–12240.

15. Koomey, J.G. (2011). *Growth in Data Center Electricity Use 2005 to 2010*, Analytics Press.

16. Manipatruni, S., Lipson, M., and Young, I.A. (2013). Device scaling considerations for nanophotonic CMOS global interconnects, *IEEE J. Sel. Top. Quant. Electron.*, **19**, 2. doi:10.1109/JSTQE.2013.2239262

17. Notomi, M., Nozaki, K., Matsuo, S., Shinya, A., Sato, T., and Taniyama, H. (2011). fJ/bit integrated nanophotonics based on photonic crystals, in *37th European Conference and Exhibition on Optical Communication (ECOC)*, pp. 1–3, Geneva, Switzerland.

18. Notomi, M., (2013). Towards femtojoule-per-bit optical communication in a chip. In *15th International Conference on Transparent Optical Networks (ICTON), 2013*, pp. 1–4, Cartagena, Spain.

19. Larger, L., Soriano, M.C., Brunner, D., Appeltant, L., Gutierrez, J.M., Pesquera, L., Mirasso, C.R., and Fischer, I. (2012). Photonic information processing beyond turing: an optoelectronic implementation of reservoir computing, *Opt. Express*, **20**, 3, pp. 3241–3249.

20. Acuna, G.P., Bucher, M., Stein, I.H., Steinhauer, C., Kuzyk, A., Holzmeister, P., Schreiber, R., Moroz, A., Stefani, F.D., Liedl, T., Simmel, F.C., and Tinnefeld, P. (2012). Distance dependence of single-fluorophore quenching by gold nanoparticles studied on DNA origami, *ACS Nano*, **6**, 4, pp. 3189–3195.

Chapter 2

Photonic Crystals

Paolo Bettotti

Department of Physics, University of Trento, via Sommarive 14, 38123 Trento, Italy
paolo.bettotti@unitn.it

2.1 Introduction

Photonic crystals (PCs) are a broad class of materials with peculiar diffraction and light-guiding properties that are induced by their structuring on a length scale comparable to that of the wavelengths they have to interact with. The core capability of PCs is the possibility to fine-tune the dispersion properties of the bulk host dielectric through a proper micro-/nanostructuring. That is, the spectral properties of the material can be modified at will over a broad range of possibilities in order to attain the desired interaction among the propagating light and the dielectric host material.

The simpler case of PCs corresponds to a multilayered structure (also known as dielectric mirror or distributed Bragg mirror, or DBR [1]). Their story can traced back to the late 1800s [2], when the studies about interference raised the possibility to realize antireflection coating (ARC) using multilayered materials. ARCs

Nanodevices for Photonics and Electronics: Advances and Applications
Edited by Paolo Bettotti
Copyright © 2016 Pan Stanford Publishing Pte. Ltd.
ISBN 978-981-4613-74-3 (Hardcover), 978-981-4613-75-0 (eBook)
www.panstanford.com

where studied for an entire century before their generalization to higher dimensions were proposed in the two seminal papers published in 1987 by Yablonovitch and Pendry, which described 2D periodic structures with particular optical properties [3, 4]. It is only after these two publications that a suitable theory and formalism were developed to fully describe PCs and exploit their rich physics.

After the (nearly contemporary) publication of the first two seminal papers describing the basic principles of PCs, researchers realized the outstanding possibilities given by these materials and a huge interest regarding these micro- and nanostructures developed. It was immediately recognized that PCs can be used to tune in a very precise way the interactions between a propagating light beam and the materials with which it has to interact. This is a fundamental step in developing photonic devices with tailored properties. In PCs the tailoring of light–matter interactions can be pushed to such extreme levels that new optical effects, not achievable in homogeneous materials, appear. Some examples of these effects are superprism [5], slow light propagation regime [6], and negative refraction [7].

The great possibilities given by PCs can be understood by looking at the basic interactions that take place when an electromagnetic wave impinges on a material. As a photon enters the dielectric, its electric field polarizes the material itself. In turn this field of polarization slows down the propagating wave by a factor that is inversely proportional to the material refractive index (which, in turn, is bound to the atomic polarizability through the Clausius–Mossotti (or Lorentz–Lorenz) equation [8]):

$$\frac{4\pi}{3} N\alpha = \frac{\epsilon - 1}{\epsilon + 2}, \tag{2.1}$$

where N is the atomic density and ϵ is the dielectric constant.

This basic effect is well known in both bulk and integrated optics and it is exploited in the fabrication of elementary optical building blocks (e.g., wave retarders and wave plates).

A proper microstructuring of the dielectric permits one to heavily modify the behavior found in homogeneous media and to carefully tune the way in which those two fields interact with each other and affects the way the light propagates through the material.

While the electronic polarization acts on a length scale comparable to the atomic scale, the diffraction/interference effect requires an interaction length comparable to the wavelength of the photon. In other words, the final properties of PCs depend both on the material that compose them and by their nanostructuring (where the prefix "nano" refers to a subwavelength scale). In fact the nanostructure (e.g., both the symmetry and the periodicity) creates a dielectric lattice composed of elements having different refractive indexes. Each of these elements induces a controlled dephasing of the propagating field, while reflections at the interfaces produce interference that modulates the intensity of the field. The larger the refractive index contrast ($\Delta\varepsilon$) between the materials that compose the PC, the greater the effects on the propagating photons.

Most often PCs are periodic structures and they are naturally classified depending on their dimensionality. A multilayered material forms a 1D PC because it repeats itself along a single direction, while it is homogeneous (i.e., much larger than the wavelengths of interest) along the other two axes. A material structured on two different axes (i.e., like a chessboard) is a 2D PC, and finally, a dielectric that shows periodicities along all the three axes (like opal gems) is a 3D PC. Figure 2.1 shows a cartoon of these three possible structures.

Thanks to their relatively simple fabrication technology, 1D PCs were the first PCs to be deeply investigated. In particular porous silicon (PSi) enabled the possibilities to fabricate very complex lattices composed of hundreds of layers and with rather high refractive index contrast (of up to about 1.0) [9, 10]. Multilayers obtained by deposition methods (like chemical of physical vapor deposition) also raised a large interest [11, 12].

Because the PC effects are maximized only for light propagating along the direction perpendicular to the periodicity, their implementation in photonic integrated circuits (PICs) is a rather complex task

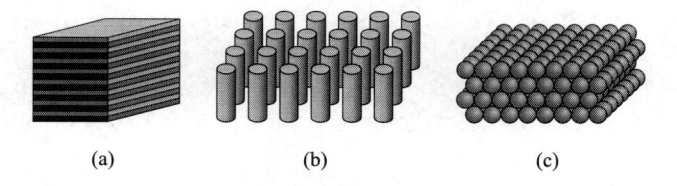

| (a) | (b) | (c) |

Figure 2.1   The typical arrangements of (a) 1D, (b) 2D, and (c) 3D PCs.

and they are used mainly as grating elements for coupling purposes. They are of interest in the fabrication of vertical cavity surface emitting lasers (VCSEL), where two DBRs surround the active layer and form a columnar-shaped laser cavity [13–15].

Two-dimensional PCs were initially studied in the microwave range using metals and millimeter-sized structures [16]. Later the advancements in micro- and nanofabrication allowed a reduction of the length scale of the lattice; thus working frequencies were shifted accordingly toward the near-infrared range. Two-dimensional PCs are ideal models of limited usefulness because they do not confine the light in the plane perpendicular to the periodicity; thus their in-plane characterization requires very thick samples. So far only a few systems were fabricated with high enough quality to exhibit optical properties typical of the pure 2D case [17].

The so-called 2.5-dimensional PCs are among the most investigated systems: a classical waveguiding mechanism is used to confine the light in the plane of a dielectric membrane, while the PC effects are exploited by a proper microstructuring of the membrane itself [18]. Group III–V materials (GaAs, AlGaAs) were largely used to fabricate 2D PCs because of their large refractive index, their efficient light emission, and the high-quality structures that were easily realized on them [19–21].

Three-dimensional PCs are divided into two main categories, self-ordered materials (like opals) and nanostructured 3D materials (like woodpile PCs [22]) that are fabricated using a bottom-up approach. The former have limited optical quality because of the large number of defects that are inevitably associated with macroscopic systems [23]. Opals were studied mainly as matrices to investigate these modifications induced to the spontaneous emission rates of an emitter embedded within an environment with a modified local density of states and as low-cost systems to develop dynamical PCs (such as sensors and large-area displaying devices) [24]. Three-dimensional PCs fabricated with a bottom-up approach have very high optical quality but their fabrication is a complex task and their use is limited to proof-of-principle demonstration of their basic optical properties [25, 26].

2.2 One-Dimensional PCs and Their Dispersion Properties

In this section we approach the design of a 1D PCs by firstly describing the basics of multilayer optics using classical optics. Afterward we will introduce an analogy with quantum mechanics to analyze dispersion of systems with higher dimensionality exploiting a description based on band diagrams.

The simpler case of a 1D PC corresponds to a multilayered structure (also known as a DBR [1]). Optical properties of a 1D PC can be described using a classic interference picture, as typically found in textbooks. The extension to higher dimensionality requires the introduction of a different formalism that resembles the Schroedinger equation for electrons in a periodic lattice (and hence the name "photonic crystals").

A light beam that impinges onto a multilayer is partially reflected at each interface. The reflection coefficients are proportional to the refractive index difference among the two adjacent layers, as described by the well-known Fresnel coefficients:

$$r_p = \frac{n_2 \cos \vartheta - n_1 \cos \phi}{n_2 \cos \vartheta + n_1 \cos \phi} \qquad t_p = \frac{2n_1 \cos \vartheta}{n_2 \cos \vartheta + n_1 \cos \phi}$$

$$r_s = \frac{n_1 \cos \vartheta - n_2 \cos \phi}{n_1 \cos \vartheta + n_2 \cos \phi} \qquad t_s = \frac{2n_1 \cos \vartheta}{n_1 \cos \vartheta + n_2 \cos \phi},$$

where r and t are the reflection and transmission coefficients and p and s indicate, respectively, impinging waves with the electric field either parallel or orthogonal to the incidence plane.

For the case of normal incidence, these simple formulae clearly demonstrate that the greater the refractive index difference between the materials that compose the PC, the stronger the reflections at each interface, and, ultimately, the larger the modifications induced to a photon that propagates through the PC.

Fresnel coefficients are useful to describe simple systems composed of one or two interfaces. On the other hand the description of more complex systems composed of tens or hundreds of layers is more effectively performed by using the approach called the *transfer matrix method*. Several books contain the formal derivation of the transfer (or scattering) matrix (TM) approach [27]; here we summarize the main results only.

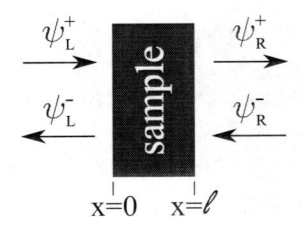

Figure 2.2 Sketch of the four components considered at each layer in a TM implementation. The subscript (L,R) indicates wherever the wave is either on the left or on the right side of the layer, while the superscript $(+,-)$ labels its propagation direction.

TM is a powerful method based on matrix multiplication and it is used to describe the transport properties of nonhomogeneous and nonperiodic materials. Each system's scatterer is modeled as a potential barrier and a corresponding TM is associated with it. Two descriptions are usually considered: in the TM case the matrix (**M**) expresses the right components impinging on the interface with the left components that leave the scattering element (see Eq. 2.2); on the other hand a scattering matrix (**S**) relates the outgoing terms with the ingoing terms (see Eq. 2.2). Figure 2.2 sketches symbols used for the four components (which are typically plane waves).

$$\begin{pmatrix} \Psi_R^+(x = \ell) \\ \Psi_R^-(x = \ell) \end{pmatrix} = \mathbf{M} \begin{pmatrix} \Psi_L^+(x = 0) \\ \Psi_L^-(x = 0) \end{pmatrix} \tag{2.2}$$

$$\begin{pmatrix} \Psi_L^-(x = 0) \\ \Psi_R^+(x = \ell) \end{pmatrix} = \mathbf{S} \begin{pmatrix} \Psi_L^+(x = 0) \\ \Psi_R^-(x = \ell) \end{pmatrix} \tag{2.3}$$

The two formalisms are correlated and it is possible to express the components of one description in terms of the other:

$$\mathbf{M} = \begin{pmatrix} M_{11} & M_{12} \\ M_{21} & M_{22} \end{pmatrix} = \begin{pmatrix} \dfrac{S_{22}S_{12} - S_{22}S_{11}}{S_{12}} & \dfrac{S_{22}}{S_{12}} \\ \dfrac{-S_{11}}{S_{12}} & \dfrac{1}{S_{12}} \end{pmatrix} \tag{2.4}$$

$$\mathbf{S} = \begin{pmatrix} S_{11} & S_{12} \\ S_{21} & S_{22} \end{pmatrix} = \begin{pmatrix} -\dfrac{M_{21}}{M_{22}} & \dfrac{1}{M_{22}} \\ \dfrac{M_{22}M_{11} - M_{12}M_{21}}{M_{22}} & \dfrac{M_{12}}{M_{22}} \end{pmatrix} \tag{2.5}$$

The physical meaning of the **S** components become clear if we assume that only one wave is impinging from the left ($\Psi_R^+ = 0$) and if we normalize its amplitude ($|\Psi_L^+| = 1$); then:

$$\Psi_R^+ = S_{21}\Psi_L^+ \tag{2.6}$$

$$\Psi_L^- = S_{11}\Psi_L^+ \tag{2.7}$$

$$\mathbf{S} = \begin{pmatrix} S_{11} & S_{12} \\ S_{21} & S_{22} \end{pmatrix} = \begin{pmatrix} r' & r \\ t' & t \end{pmatrix} \tag{2.8}$$

Thus the **S** matrix elements are the transmission and reflection amplitudes (those square values give the transmission and reflection coefficients, respectively).

Using the matrix approach the transfer function of a complex system composed of several types of scatterers is easily calculated. In fact the fields at each successive interface are determined if those at the previous interface are known. Thus the TM of a stack of films is given by

$$\mathbf{M_T} = \prod_{i=1}^{n} \mathbf{M_i} \tag{2.9}$$

$\mathbf{M_T}$ represents the TM of the entire stack of layers and the product runs over the n layers.

A particular type of multilayers is the lambda quarter thick layer. In this case the optical thickness of each layer is equal to

$$nd = \lambda_0/4, \tag{2.10}$$

where d id the physical thickness of the layer, λ_0 is the central wavelength of the BG, and n is the refractive index of the layer [28].

This geometry is one of the preferred schemes to fabricate dielectric mirrors, because it maximizes the width of the bandgap (BG) and thus the value of the reflection at λ_0.

2.3 Analogy with Quantum Mechanical Schroedinger Equation

The extension of the classical description to higher dimensionality cases, using the interference description, leads to a cumbersome mathematical treatment. Starting from the Maxwell equation it is possible to derive an eigenvalue formalism that closely resembles the Schroedinger description of electrons in an atomic potential [29]. Currently a wave equation is most used to describe the property of PCs.

Maxwell equations can be rephrased into the so called **master equation** (the complete derivation of the result shown here can be found in [30])

$$\nabla \times \left(\frac{1}{\epsilon(\mathbf{r})} \nabla \times H(\mathbf{r}) \right) = \left(\frac{\omega}{c} \right)^2 H(\mathbf{r}) \qquad (2.11)$$

This equation has the form of an eigenvalue problem with an operator that acts on the magnetic field pattern to produce an eigenvalue (which is proportional to the square of the mode energy).

From the master equation, it is possible to demonstrate a deep analogy with the case of electrons in a periodic potential. For photons, the atomic potential is substituted by the dielectric constant and the electronic wavefunction by the magnetic field.[a] A characteristic of the solution of the eigenvalue equation in a system with periodic potential is that the eigenmodes obey the Bloch theorem: If **a** is the periodic, discrete translational unit of the system, then

$$\epsilon(\mathbf{r}) = \epsilon(\mathbf{r} + m\mathbf{a}), \qquad (2.12)$$

with m an integer multiplier. Furthermore because the eigenfunctions must commute with the symmetry operators of the system it is possible to demonstrate that

$$\mathbf{H}_k(\mathbf{r}) = e^{ikx}\mathbf{u}_k(r + a) = e^{ikx}\mathbf{u}_k(r). \qquad (2.13)$$

Equation 2.13 is the known Bloch theorem rephrased for the photonic case.

The dielectric element of a PC is equivalent to the atomic sites of a crystalline lattice and photons behave as propagating electrons. Thus because **k** is a continuous function, photonic states are confined to continuous allowed bands. If the system is periodic the band diagram repeats itself every $2\pi/a$. Thus the extended band

---

[a]The magnetic field is considered because it removes the explicit dependence of the fields from the dielectric constant and considerably simplifies the solution of the equations; a generalized approach is otherwise required.

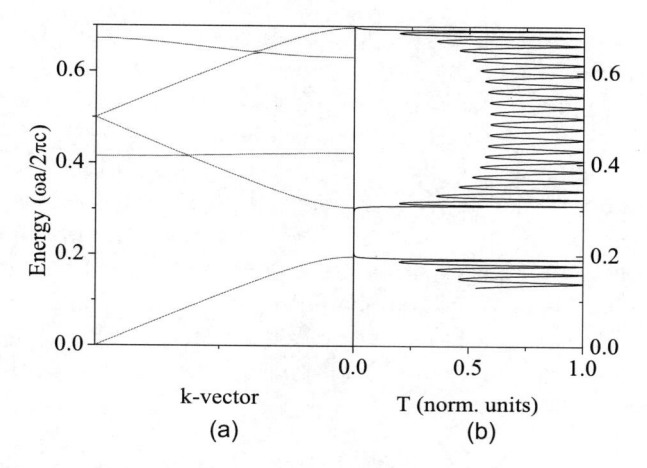

Figure 2.3 (a) Band dispersion of a lambda/4 multilayer. The layers have a refractive index of 3 and 1.5, respectively. (b) Simulated transmission spectra for the same structure.

representation is generally reduced to the band description within the first Brillouin zone (BZ) that contains all the information about the system and it is limited to $0 < \mathbf{k} < 2\pi/a$.

Figure 2.3a reports the band diagram of a $\lambda/4$ multilayer. Panel (b) shows the calculated transmission spectra for the same structure. It is clear that the transmission goes to zero for the energy range comprised within the PC gap.

The energy axis of Fig. 2.3 is normalized versus the period of the structure (in units of a/λ). Such normalization exploits the scale invariance of the master equation [30]. In fact:

- If the lattice period is scaled by a certain factor, $\mathbf{a} \rightarrow \mathbf{a}\ell$, then the eigenvalue is scaled accordingly: $\omega \rightarrow \omega/\ell$.
- The eigenvalues scale as the dielectric constant: If $\epsilon(\mathbf{r}) \rightarrow \ell\epsilon(\mathbf{r}) \Rightarrow \omega \rightarrow \omega/\ell$.

The length invariance implies that once a band diagram is computed for a lattice with a certain symmetry and for a given refractive index contrast, the result can be used to determine the optical properties of the PC working at different wavelengths ranges.

As for the electronic case, diffraction takes place at the band edges and BGs open. Within these energy intervals no photonic states are allowed and light propagation is forbidden. Such a BG corresponds to the stop band of Fig. 2.3: the region of low transmission originates because photons cannot propagate inside the PC if their energy falls within the BG.

By looking at the shape of the first band we note that the dispersion at small k follows a straight line, as it is typical for a homogeneous medium. In this region wavelengths are much larger than the typical length scale of the PC; thus the medium behaves as an effective material and photons do not feel the nanostructure. Thus the slope of the band resembles that of the light line (the linear dispersion law for photons that propagate through a homogeneous medium):

$$\omega(k) = \frac{ck}{n_{\text{eff}}},\tag{2.14}$$

where n_{eff} is an effective refractive index weighted over the PC microstructure.

At larger energy the photon wavelength becomes comparable with the period of the PC and dispersive effects appear. The derivative of the band slope is the group velocity ($v_g = \delta\omega/\delta k$); thus information about the speed at which photons travel across the PC can be extracted from the band diagram. At band edges, the slope tends to zero and the photon is at rest. Here two standing waves develop: They have the same wavelength (exactly two times the period), but because two solutions are permitted, they arrange differently the electromagnetic field within the PC layers; the two modes split in energy and a BG is formed. Figure 2.4 shows the time-integrated magnetic field for the first two bands in a 1D PC. Fields

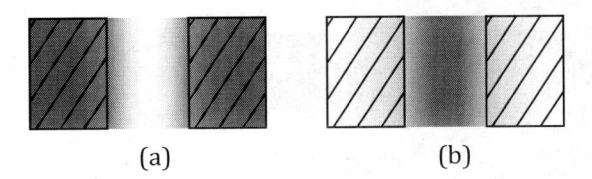

(a)   (b)

Figure 2.4 Time-integrated magnetic field in a 1D PC. The striped areas defined the regions of a high-index material. Both panels sketch the field distribution at the edge of the first Brillouin zone. (a) Field distribution of the first band (dielectric band) and (b) field distribution of the second band (air band).

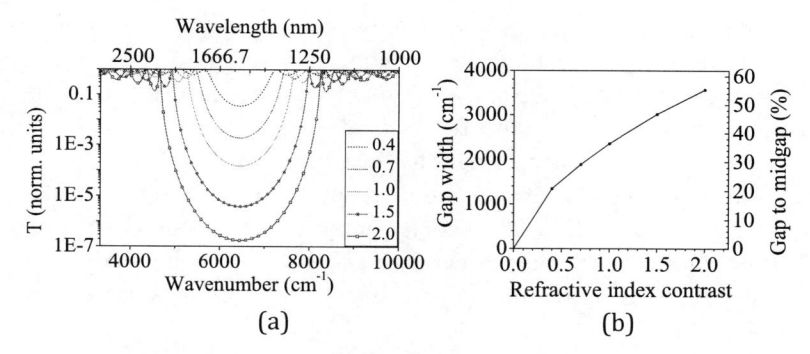

Figure 2.5 (a) Width and attenuation of a $\lambda/4$ DBR as a function of the refractive index contrast. The refractive index contrast values are defined in the graph legend. (b) Plot of the width of the BG versus the refractive index contrast; the nonlinear dependence between these two quantities is clear.

are plotted at the edge of the first Brillouin zone. The first band concentrates the **H** field within the high-index regions (the striped area defined the volume of the high dielectric constant), while the second band ejects the field into the low-index region.

The variational theorem states that photonic modes minimize their energy by concentrating the fields into the region of high refractive index. Thus the lower-energy mode maximizes the field within the regions of high refractive index (and is called dielectric mode), while the mode that has the maximum of the field in the low-refractive-index regions lies at higher energy (and is called air mode).

Figure 2.5 shows the effect on the width of the gap for a lambda quarter PC, for different refractive index contrasts: the BG width increases with the refractive index contrast. Figure 2.5a shows how the width and attenuation of DBRs that satisfy the $\lambda/4$ condition increase proportionally to the refractive index contrast of the materials used to fabricate the DBRs. A smaller refractive index contrast requires a larger number of periods to achieve a predetermined attenuation value (the number of period has been kept fixed for all the data shown in Fig. 2.5a). The bottom axis is given in wavenumbers to underline the fact that the center position of the gap does not shift if the $\lambda/4$ condition is fulfilled and to clearly show that the gap widens symmetrically to respect its midgap in energy. Figure 2.5b shows how the gap width increase proportionally to the refractive index contrast and it follows a sublinear relation. The *gap-to-midgap* quantity is often used to quantify the gap width normalized to its central frequency ($\Delta\omega/\omega$).

Regarding the dispersive effects, a band with zero dispersion implies that the photons are stopped within the dielectric (because $v_g \to 0$). This is obviously an ideal condition that cannot be realized in practice; nevertheless it is demonstrated that higher-order PCs effectively slow down the speed at which light propagates through them.

2.4 1D Defect Modes

An interesting question that arise is, How deep can we go exploiting the analogy among photons and electrons? Can we "dope" the photonic lattice—as in the semiconductor case—by introducing defect states?

If each period of a PC behaves as an atomic site (with the atomic potential replaced by a dielectric one), then by breaking the perfect periodicity of the lattice, we may introduce defect states within the photonic BG.

The easiest way to do it is to modify the thickness of a single layer within an otherwise periodic stack. Such a multilayer is called a microcavity (MC) and it is composed of two Bragg mirrors (often quarter wave stacks) that surround a central layer of different thickness (the cavity). Figure 2.6a reports the cross section of the structure. Figure 2.6b reports the profile of the field of a defect mode, that is, the only permitted photonic state within the BG of the $\lambda/4$ stacks (the green and white stripes define the two different refractive index layers). The plot in red on the left of the field pattern is the profile of the field and the connection between the modulation induced on the field and the dielectric constant is clear. Figure 2.6c compares the transmission spectra for a simple DBR (red dotted line) with that of the MC (black line). The spectrum of this structure is rather different form that of a single DBR; in fact while the width of the BG is nearly unaffected by the presence of the defect layer, a sharp peak with a high transmission value appears near the middle

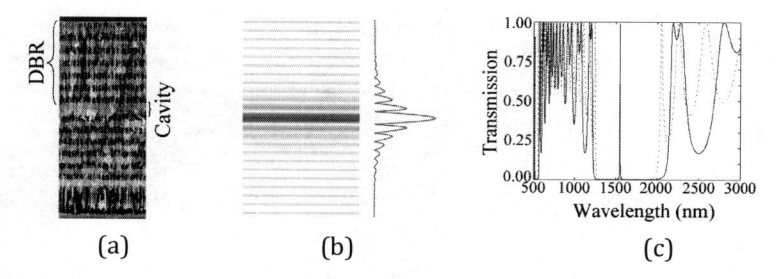

Figure 2.6 (a) Cross section of a microcavity fabricated in porous silicon. The lighter and darker bands are layers of a higher and a lower refractive index, respectively. Such modulation is achieved by inducing a periodic variation in the porosity of the material. (b) Snapshot of the E-field within the microcavity obtained from a finite-difference time domain (FDTD) simulation. It is clear how the field is highly localized within the cavity defect layer. (c) Transmission spectra of the cavity: the cavity peak appears within the stop band of the DBR.

of the gap. This peak is the defect state induced by breaking the symmetry of the periodic lattice.

We know from the variational theorem that photonic modes that occupy high-index regions tends to minimize their energy. Thus defects composed of a high-index dielectric *pull* the mode from the bottom of the conduction band within the BG. These so-called dielectric modes are equivalent to an electronic donor defect. On the other hand if the photonic defect is made of a low-index material it acts as an electronic acceptor level and it *moves up* inside the BG from the valence band.

There are a few important factors that describe the performance of MCs and permit one to compare different designs.

The quality factor is the ratio between the cavity resonant wavelength and its width:

$$Q = \frac{\omega}{\delta\omega} = 2\pi \frac{E_{\text{stored}}}{E_{\text{lost}}} \qquad (2.15)$$

The Q-factor is a measure of the rate at which a cavity loses optical energy because of its finite size (and imperfections in real samples). Q^{-1} is thus a measure of the energy lost at every round trip and is related to the lifetime of a photon that is trapped within the cavity: $\tau = Q/\omega = 1/\delta\omega$.

Photons are greatly slowed down while they travel across optical cavities (as shown by their flattened band dispersion). Thus, if the cavity is excited for a long enough time, it is possible to pile up a field intensity much larger than that of the incoming field. Assuming that the mirrors are the lossy elements of the cavity the field enhancement factor can be estimated from

$$\frac{I_{\text{cav}}}{I_{\text{pump}}} \approx \frac{Q}{2\pi\sqrt{R}} \qquad (2.16)$$

The field enhancement within the cavity is exploited to create and to study a huge number of optical phenomena, and it is of great importance in several technological applications (e.g., low-threshold lasers and wavelength conversion in nonlinear [NL] optical devices). The rich physics of the MCs arises from the heavily modified density of modes found within the cavities. In fact the mode density within a homogeneous material and within a cavity is described by the following equations:

$$\rho N_h = \frac{dN}{d\omega} = \frac{8\pi n^3 v^2}{c^3} \qquad \rho N_c = \frac{1}{\Delta v V} \qquad (2.17)$$

Thus their ratios scale proportionally to

$$\frac{\rho N_c}{\rho N_h} = \frac{1}{8\pi} \left(\frac{\lambda^3}{n}\right) \left(\frac{Q}{V}\right), \qquad (2.18)$$

where ρN_c is the mode density of the cavity and ρN_h that of the homogeneous medium. $dN/d\omega$ indicates the number of modes for the frequency interval; n is the refractive index, v the frequency, c the speed of light, and V the volume of the cavity mode. For a constant, however, this ratio corresponds to the Purcell factor and describes how the spontaneous emission of an emitter is weighted by the photon mode density of the volume in which it is inserted [31]. The Purcell factor tells us how much accelerated (or slowed down) the spontaneous emission of an emitter placed within a cavity is. The Purcell factor increases proportionally to the ratio Q/V; thus it is enhanced in cavities with narrow linewidths and small volumes. This factor is of great interest in laser technology because it can be used to reduce the lasing threshold. In fact the fraction of light emitted into a given laser mode can be approximated by the so-called *emission coupling factor*:

$$\beta = \frac{Pe}{1 + Pe}, \qquad (2.19)$$

where Pe is the Purcell enhancement, that is, the increase of the spontaneous emission of an exciton in a cavity, compared to its bulk value. If most of the light is emitted into a lasing mode, the losses due to spontaneous emission into nonlasing modes are greatly reduced as well as the laser threshold.

As for the electronic case the dispersion of an ideal defect is a flat line with an energy that falls within the BG region. Its energy position depends on both the material and the geometrical properties of the defect and can be tuned to the desired frequency by playing with these two parameters.

Defect modes are of interest not only for their stationary solution but also for their dynamic behavior. In fact photons are slowed down while traveling through a defect and pulses are subject to dispersion effects. Thus modeling of MCs is generally performed using simulation methods able to handle time-dependent events (like FDTD or the finite elements method) rather than by simply looking at the eigenstate (like using plane wave expansion codes).

In an ideal case dielectric and air defect modes have very similar spectral features. In real samples, however, the finite size of the PC introduces significant differences on these two types of defects that reflect into their final properties. In particular an air mode (which is a cavity formed by a material with a locally reduced refractive index) typically has a smaller Q-factor because of the greater losses associated with it. These are due to the combined effect of the lower local refractive index that reduces the light confinement and by the increased losses due to stronger scattering of the field outside the volume of the cavity. For these reasons dielectric bands are the most investigated in literature.

2.5 Two-Dimensional Photonic Crystals

Because of their limited dimensionality, 1D PCs cannot be used in the realization of a complex PIC and their use in optics is mainly limited to macroscopic systems (e.g., optical filters).

A much greater control over light propagation and mode confinement is achieved in dielectric media structured along two dimensions. Naively such systems can be thought as the superposition of more 1D PCs tilted by a certain angle. Because of the diffractive character of light propagation in 2D PCs and of the strong coupling that happens at each lattice point, optical properties of 2D PCs cannot be described as a linear combination of 1D lattices. Thus the classic description that makes use of the interference cannot handle efficiently this multidimensional problem and we have to revert to the wave equation description.

Compared to the 1D case, bidimensional PCs permit to choose among a much broader type of lattice geometries and topologies: their optical properties are heavily dependent on them and different structures can be used to enhance particular effects. The two most common types of geometry are square and triangular lattices, but several other geometries have been so far investigated, such as the Kagomé lattice [32], the hexagonal, and the quasicrystalline [33, 34], to cite but a few. Furthermore we can select between two different types of lattice topologies: either a connected high-index matrix with low-index scatterers embedded in it or a low-index connected medium with high-index scatterers (like the structure shown in see Fig. 2.1b). In a truly 2D configuration (of infinite height) these two dispositions have similar properties in terms of mode guiding and mode confinement, but in a real sample with finite height, the out-of-plane[a] coupling of the radiation plays an important role and determines the final losses of the PC. We will see in the following figure that the connected high-index matrix is the *de facto* solution because of its smaller losses.

Figure 2.7 shows the band dispersion for a square and a triangular lattice of air holes drilled in a silicon matrix ($\varepsilon = 12.25$). The radius of the holes is fixed to $0.5a$ (where a is the lattice period) in order to highlight the different band structures induced by the symmetry of the lattice for PCs having a similar filling fraction.

An important difference compared to the 1D case is that in multi-dimensional PCs the two polarizations are no longer degenerate and the gaps open at different energies and have different widths. This is a consequence of the different distribution of the fields in the high- and low-index regions. In fact light that propagates along the

---

[a]In a 2D PC the plane of propagation is the one perpendicular to the axis of the scatterers.

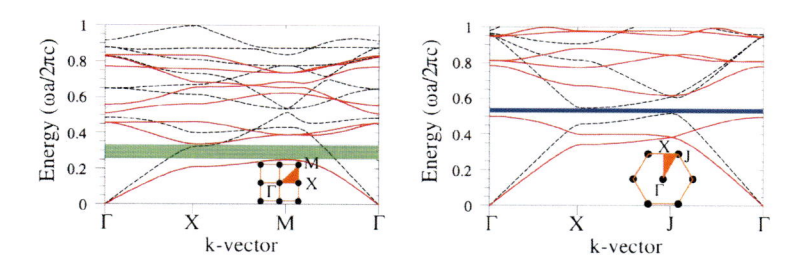

Figure 2.7 Band dispersion for a (a) square and a (b) triangular lattice of holes in a dielectric matrix ($\epsilon = 12.25$). The insets in the bottom of the graphs sketch the direct and the reciprocal lattices with black holes and polygons, respectively. High-symmetry points are defined with their usual labels. In both panels black dashed lines are for transverse electric (TE)-polarized bands, while red lines are for transverse magnetic (TM) polarization.

PC "sees" the microstructure and its reduced symmetry. In an ideal 1D PC the light propagates across a series of parallel interfaces that define the layers and that extend infinitively in the plane normal to the propagation direction. Thus both polarizations are invariants under any rotation operation and must have the same energy. On the other hand in a 2D PC, electromagnetic fields oscillating in different directions interact differently with the dielectric, for example, a field that oscillates parallel to the hole long axis *feels* a microstructure different from the one seen by a field that lies in the perpendicular plane.

It is possible to demonstrate that fields components in a 2D PC can be classified in terms of their symmetry under mirror operation to respect the plane perpendicular to the hole long axis [30]. These two polarizations confined either the **E** field or the **H** field in the *xy* plane. The former case is called **TE** and has E_x, E_y, H_z as components, while the other is called **TM** and its components are H_x, H_y, E_z.

The lifting of the degeneracy between the TE and TM polarizations has important consequences. The green rectangle in Fig. 2.7a depicts the BG that opens for the TM polarization (red lines). No gaps are present between the first two bands for the TE case (black lines). Thus a square lattice of holes cannot perfectly inhibit the propagation of unpolarized light. Figure 2.7b shows the band dispersion for the triangular lattice; here the blue region underlines the energy interval in which neither TE nor TM modes can propagate. In this energy range the PC behaves effectively as an ideal mirror that perfectly reflects any photons with energy comprised within the BG width. The two insets are cartoon views that overlap the direct and reciprocal lattices to underline the high-symmetry points considered in the band diagrams. The size of the scatteres in the direct lattice is not in scale with the simulated structures.

The width of both partial and complete BGs is defined by the symmetry, the refractive index contrast, and the filling factor (*f.f.*) of the structure. The BG map of Fig. 2.8 shows how the width of the gap of a triangular lattice of holes in silicon changes versus the structure *f.f.*, for both polarizations. The *f.f.* is defined as the ratio between the size of elementary scatterer (e.g., the radius of the holes) and the lattice period. The width of the BG depends heavily on the size of the scatterers, but because of the scalability of the Maxwell equations, the behavior of a given symmetry can be scaled to different lengths (that is why the y axis is given in a/λ units). The red area defines the region where a gap for the TE polarization opens, while the blue areas define the TM gaps. The complete gap opens up in a narrow region roughly comprised between 0.4 and 0.5 *f.f.*, where both TE and TM gaps overlap.

It is of interest to note that gap that falls at high energy and that are narrow in the *f.f.* coordinate are more sensible to the unavoidable imperfections generated during the fabrication. Thus they may not be properly reproduced in real samples. To note that opening of a complete BG requires refractive index contrast definitely larger than that required for the opening of a partial gap.[a] Each lattice geometry has its own BG map: the number of gaps sustained depends on the lattice symmetry, the refractive index contrast, and the *f.f.* Several examples of gap maps can be found in [30].

Soon after their discovery, there was a rush to find the geometry that maximizes the complete BG. It turned out that it is favored

---

[a]in general, the smaller the dimensionality of the PC, the smaller is the refractive index required to open a gap: in 1D PC gaps open even for $\Delta n \rightarrow 0$.

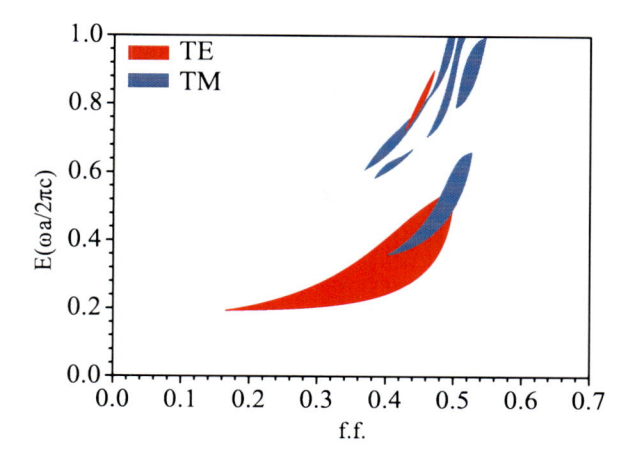

Figure 2.8 Map of gaps for a triangular lattice of holes. The connectivity of the triangular lattice favors the opening of large TE gaps (defined by the red areas) while TM gaps (in blue) are narrower. The complete gap opens up in the region of *f.f.* between 0.4 and 0.5.

in lattices with high symmetry (e.g., triangular, hexagonal, and quasicrystal-like), while lattices of reduced symmetry (e.g., square, rectangular) do not show it. This fact can be naively described considering that while TE gaps are wider in structures made of connected high-index regions, the width of the TM gaps is larger in lattices composed of nonconnected high-index regions. Thus lattices with high symmetry approximate those two limiting cases and, at the same time, maintain a relatively high degree of isotropy in the plane that permits the overlap between the gaps of the two different polarizations.

2.6 2D Defect Modes

Defects are probably the most important elements to provide functionality to any PIC. Compared to the 1D case, 2D systems have larger flexibility and defects can be grouped into two main categories, localized and extended.

A localized defect is formed by modifying the size of one or few neighbor scatterers. Optical cavities rely to this category: the defect

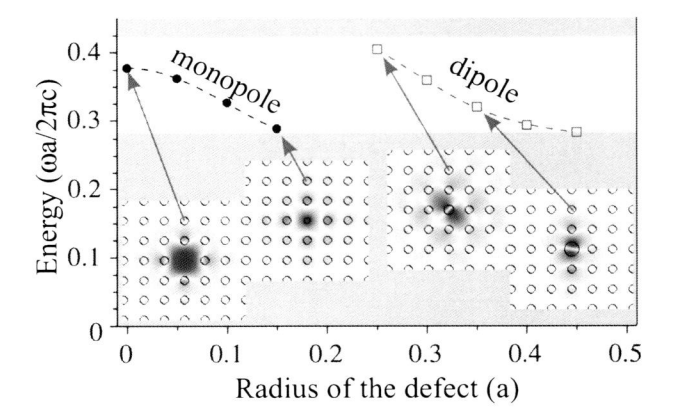

Figure 2.9 Evolution of cavity modes versus cavity geometry for an L1 defect state in a square lattice. Solid columns have an $\varepsilon = 12.25$, modes are TM polarized, and the inset represents the patterns of the E_z field component.

has a finite size and is used to localize and store energy within a small volume. A cavity can be formed by either an increase or a decrease of the volume of the perturbed scatterers. The energy position of the allowed states will move toward the dielectric band (if the local refractive index increases) or toward the air band (if the local refractive index decreases).

The simplest case of an MC is formed by removing a single hole from an otherwise perfect periodic lattice. As for the 1D case, the Q-factor of this cavity is defined by the refractive index contrast of the materials that composed the PC and by the number of periods surrounding the defect site. Figure 2.9 shows some examples of defects induced in a square lattice of dielectric columns by varying the radius of the defect site. The defect investigated is called L1 because it is formed by removing a single lattice site. The shaded gray areas define the BG. If the defect site contains a column with null radius (i.e., a single column is removed from the lattice) a defect state is introduced well inside the BG (the inset shown the corresponding E_z field component). As expected, by increasing the radius of the defect the energy of the defect decreases.

For the geometry considered here, if the radius of the defect column is smaller than that of the unperturbed lattice, monopole

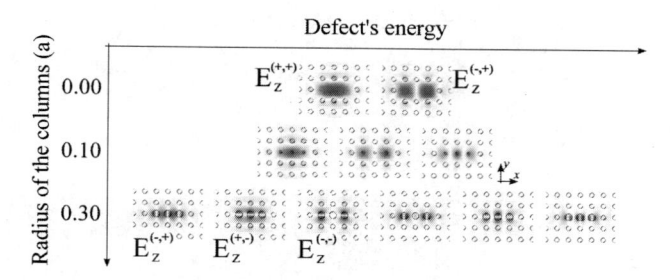

Figure 2.10 Some of the defect states supported by an L3 cavity. The labels describe how the E_z component transforms under mirror symmetry to respect the $x = 0$ and the $y = 0$ plane, respectively.

states only are sustained. A further increase of the defect size introduces several other states that differs for their symmetry (monopole, dipole, quadrupole, etc.) [35]. On the right part of Fig. 2.9 a few dipole states are shown. These defects are doubly degenerate; thus each of them is composed by two states that differ only for a 90° rotation of the filed pattern, that is, a symmetry operation spanned by the symmetry group of the square lattice (only one state is shown the figure).

A different, commonly investigated, defect is the L3, where three adjacent sites are perturbed. This geometry was the first one that permitted the fabrication of a PC MC with a Q-factor larger than 10^4 [36]. Figure 2.10 reports some of the resonant modes that are sustained in an L3 cavity. In an L3 cavity the larger volume of the resonator (compared to the L1 case) shows a higher density of localized states that are classified according to their energy and their symmetry. For a complete discussion about the modes in an L-*n*th cavity refer to [37].

More refined geometries for 2D cavities and their tuning are discussed in Section 2.8, where the out-of-plane losses from a thin slab are also taken into account.

An extended defect has a size that may be comparable with that of the PIC. For examples, waveguides (WGs) are extended defects and are created within a 2D lattice by modifying the size of neighbor scatterers along a continuous path; thus lines of *higher effective index* are created inside the lattice. Photons are forced to propagate along those paths because they are surrounded by dielectric mirrors

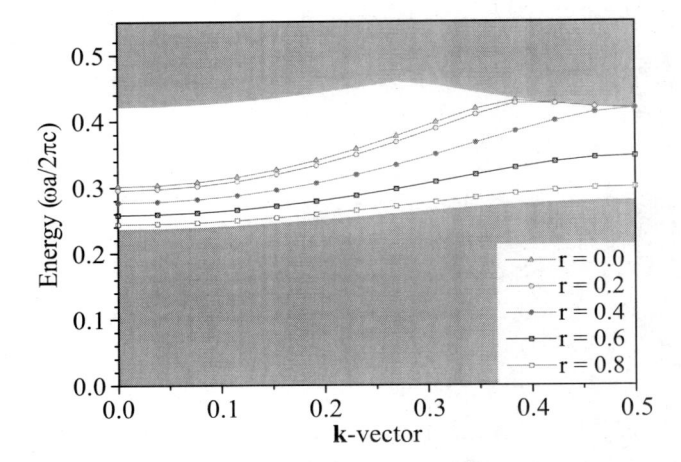

Figure 2.11 Simulated dispersions for different types of WGs in a square lattice. The WGs differ for the magnitude of the perturbation of the original lattice: the steeper the guided mode dispersion, the larger the difference between the size of the scatterers that compose the lattice and those that define the WG.

that reflect them irrespective of their incident direction (in case the 2D lattice has a complete BG).

Figure 2.11 reports the band structures for several types of WGs obtained by modifying the size of the scatterers along a single line in a square lattice of columns. The different WGs show large variation in their dispersion. The larger the difference between the size of the scatterers that define the WG and that of the unperturbed lattice, the steeper the dispersion of the guided modes. This fact can be qualitatively described, considering the role that the scattering events have in flattening out the band dispersion: a guided mode that propagates within a classical silicon WG resembles the case of a PC WG with no scatterers within the WG core and assumes a group velocity of about c/n_{eff}. On the other hand modes that propagate along WGs that contain large scatterers, *feel* a medium with a higher refractive index. In fact the scattering events that take place at each lattice site effectively slow down the light. The result is a reduced group velocity and more and more flattened band dispersion. In the limiting case in which the scatterers within the WG have the same size as those of the lattice, the guided mode must coincide with

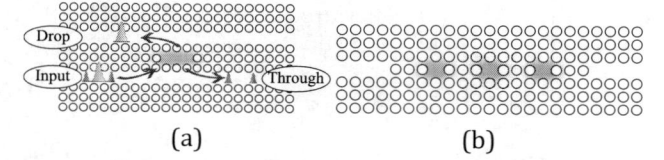

(a)   (b)

Figure 2.12 (a) Cartoon view of an add–drop filter. A pulse entering from the Input port interacts with the cavity site. The evanescent coupling with the defect routes the resonant frequencies out from the Drop port. All the other components exit from the Through port. (b) Scheme of a coupled-cavity waveguide. In this device the waveguide region is formed by a series of cavities. Photons propagate through a hopping mechanism.

the dielectric band (generally rather flat). An important difference between PC WGs and classical dielectric guides is the fact that PC structures can achieve an extremely small bending radius: 90° and even 120° turns were demonstrated with reasonable losses [38, 39]. Note that a significant part of the losses in sharp PC bends is due to reflections at the bending site and that these losses can be minimized by a proper perturbation of a few lattice sites localized around the bend itself [40].

WGs and MCs are very often coupled within the same PIC. WGs are the natural element to excite an MC and the basic structure composed of their interaction is called *add–drop filter* [41]. A scheme of an add–drop filter is shown in Fig. 2.12: A pulse propagates along the input WG and is coupled to the MC through its evanescent tail that extends to the cavity site. Once the MC is excited, only its resonant modes are coupled to the upper WG (named Drop channel) and routed along the Drop port. Add–drop filters are very common elements because they are used to demultiplex and to route the desired wavelengths along predetermined directions.

WGs and MCs can be combined together to obtain the so-called *coupled-cavity waveguides* (CCWGs). These structures combine several aspects of both cavities and WGs and permit one to carefully tune the dispersion of light pulses that propagate through them by exploiting the delays induced by the cavities placed along the WGs. Compared to a classical PC WG the CCWG permits one to obtain narrower bandwidths of the cavity band. This is because the photons propagate exploiting a hopping mechanism across the coupled cavities. A simple equation derived from a tight-binding model can be used to fit the band dispersion [42]

$$f(k) = f_0\left(1 + \frac{\Delta f}{f_0}\cos(kd)\right), \qquad (2.20)$$

where f is the mode frequency, f_0 is the frequency of an isolated, unperturbed cavity, Δf is the bandwidth of the guided modes, k is the wavevector, and d is the distance between adjacent cavities. The ratio $\Delta f/f_0$ quantifies the strength of the coupling among adjacent cavity sites (and is defined as the hopping parameter) [43].

2.7 Exploiting the 2D PC Dispersion Properties

The large optical anisotropy of PCs is a key element of their usefulness in that it can be tailored to achieve light dispersion effects that are otherwise not obtainable with optical homogeneous media.

To understand the basics of the light propagation within a 2D PC, we start by describing how light is diffracted at the interface between two homogeneous media. In this case the light dispersion is a linear function:

$$\omega^2(k) = \left(\frac{c}{n}\right)^2 (k_x^2 + k_y^2), \qquad (2.21)$$

where ω is the frequency of the light, c the speed of light, n the medium refractive index, and k_i the propagating wavevector along the i direction. In a homogeneous medium, if we consider all possible incident angles, we note that the isofrequency surface generates a cone (Fig. 2.13 shows a section of the cone along the incidence plane).

Light propagating across the boundary of two materials with different refractive indexes conserves the tangential components of the **k**-vector. Snell's law can be easily verified using this scheme, as shown in Fig. 2.13. The two circles define the isofrequency contours for monochromatic light that travels across the two materials, and the radius of the circles is proportional to the magnitude of the photon **k**-vector. The right half of the circumferences considers a general case of an incident photon (red arrow). By setting the continuity of the tangential components we find the directions for both refracted (cyan arrow) and reflected photons (green arrow). In

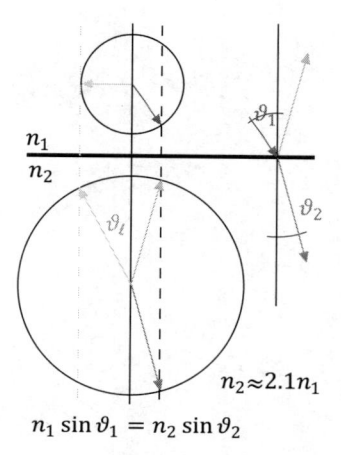

$$n_1 \sin \vartheta_1 = n_2 \sin \vartheta_2$$

Figure 2.13 Snell's law described using the isofrequency contour and a **k**-based description of propagating photons. The thick black line sketches the interface between the two materials. Red and cyan arrows are the incident and refracted beams, respectively. The green arrow on the right part of the figure is the reflected beam. The left half of the circles depicts the case of the limit angle.

the left half of the circumferences the case of the limit angle (gray arrows) is described.

The description of refraction on the basis of the **k**-vector permits the analysis of the complex band dispersion of the 2D PC in a simple way and to get a clear vision of the possible refracted beams and of their propagation directions. The general formula that describes the refraction at the interface with a 2D periodic structure is analogous to the law of diffraction grating:

$$\mathbf{k}_{\text{refl}, \parallel} - \mathbf{k}_{\text{inc}, \parallel} = m_1 \mathbf{G}_1 + m_2 \mathbf{G}_2 \tag{2.22}$$

The subscripts of the **k**-vectors refer to the reflected and incident beam, respectively, while m and $\mathbf{G}$ are the integer coefficient and the vectors of the periodic lattice, respectively. Energy conservation requires that all the **k**-vectors have to be of the same magnitude. Thus the possible solutions (reflected, refracted, and diffracted) must lie on a circumference of radius $n\omega/c$.

In the case of a PC lattice, the solutions do not form a continuous spectrum that lies on the circle but are restricted to a discrete set that represents the symmetry of the lattice itself [44, 45]. We

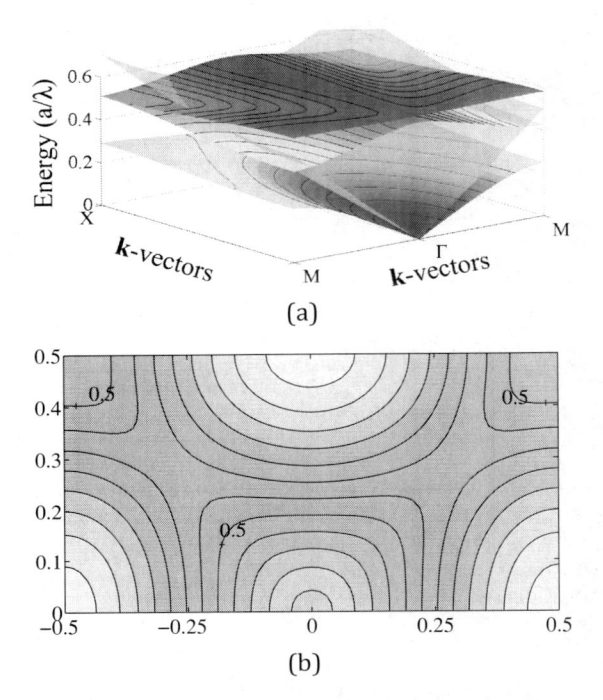

(a)

(b)

Figure 2.14   (a) Band dispersion for the first three bands of a square lattice of columns. The light-green surface represents the light cone. (b) Projection of the second band. The isofrequency lines have a square-like shape and may be used to achieve self-collimation or superprism regimes.

mentioned that the symmetry of the PCs defines the way they diffract the light. Here we show some examples based on a 2D lattice of cylindrical scatterers that span a square lattice.

Scatterers are supposed to be cylinders of radius $0.25a$ and having a refractive index of 3.5. Figure 2.14a reports a 3D view of the first three band dispersions for TM polarization. The lowest band (blue-/cyan-shaded surface) shows the typical linear behavior at a small $\mathbf{k}$, from the Γ point outward, and it stays below the light cone (depicted by the transparent green-colored cone). The photonic BG corresponds to the minimum energy separation between the first and the second band (in yellow/orange). The black contour lines draw the isofrequency lines for each band. In a homogeneous material they should appear as circles but the diffractive behavior of the PC force them to assume heavily distorted shapes. Figure 2.14b is the projection of the isofrequency contour for the second band. Two limiting cases of band dispersion are visible here, sharp vertexes and flat profiles. Considering the discussion above about the use of isofrequency lines to predict the diffraction of a light beam, the sharp vertexes are associated with a high angular dispersion for slight variation of the impinging photon direction. This fact has been exploited in the *superprism effect*, where photons of slightly different wavelengths are spatially separated, exploiting the different diffraction directions [46, 47]. Efficient and compact wavelength division multiplexers (WDMs) based on this effect were demonstrated.

On the other hand flat isofrequency contour lines are used to achieve the self-collimated regime. In this case the flat shape of the isocontour permits collimating of photons that impinge upon the PC interface with different angular directions into a beam that propagates normally to the isofrequency line. Figure 2.15 shows an example of a self-collimated beam. Figure 2.15a describes how the diffraction creates a collimated beam: waves emerging with different **k**-vectors are converted into a collimated beam exploiting

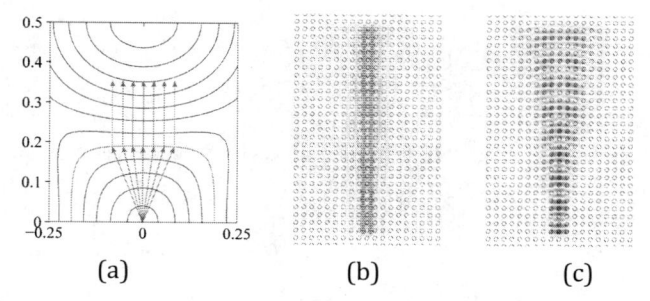

| | (a) | (b) | (c) |

Figure 2.15 FDTD simulation of a self-collimated beam propagating along a square lattice of dielectric columns. The source emits the E_x component. (a) Band diagram with the isofrequency contour considered in red. The green arrows depict the effect of the self-collimation: waves propagating with different **k**-vectors are diffracted into a collimated beam. (b) Self-collimated beam excited at a frequency of 0.5; this value corresponds to the flat dispersion region shown above. (c) Beam excited at a frequency of 0.225. This mode belongs to the first band and it clearly diverges after a short propagation distance.

the flat region of the band dispersion of a narrow energy interval. Figure 2.15b shows the propagation of a self-collimated beam: the beam propagates for several tens of lattice periods without showing appreciable beam divergence. In Fig. 2.15c a frequency smaller than the optimal one propagates with clear divergence. Rather complicate structures were so far proposed in the literature with mirrors and beam splitter elements integrated into a 2D lattice. These circuits are able to route the light beam without the need of explicitly defined WGs [48, 49].

2.8 2.5-Dimensional Photonic Crystals

The 2D systems discussed above assume an infinite length of the PC parallel to the longitudinal axis of the scatterers. Real samples may have a thickness of at most a few tens of wavelengths. This limitation is imposed by nearly all the fabrication techniques that rely either on deposition or on etching methods. Deposition methods are inherently slow, and generally, the higher the control over the deposited thickness, the smaller the deposition rate. Thus the fabrication of a 1D DBR of a few tens of layers may take several hours to complete. On the other hand etching methods have limited anisotropy and the shape of the scatterers is not perfectly controlled for high-aspect-ratio structures. Furthermore large roughness is often created onto all the surfaces of the PC.

Still, without effective confinement along the direction perpendicular to the PC plane, 2D structures will not be a useful system to fabricate real PICs.

The term "2.5D PC" is used to indicate systems that are structured as a 2D PC in the plane but exploit a guiding mechanism—typically based on total internal reflection—along the vertical direction. This is generally accomplished by etching the 2D lattice inside a thin film made of high-index material, surrounded by low-index claddings. The thickness of the thin film is such that it supports one or more guided modes.

This is an elegant solution that matches the capability of fine-tailoring the optical properties of the PC with a realistic fabrication process. In fact 2.5D PCs are fabricated in thin layers deposited with conventional deposition techniques (such as chemical or physical vapor deposition) on top of a low-index substrate (often silica). One of the advantages of this process is its compatibility with the standard complementary metal-oxide semiconductor (CMOS) platform that supplies high-quality materials (like the silicon-on-insulator substrates) and state-of-the-art fabrication technologies able to achieve spatial resolution of a few nanometers.

The main results discussed above regarding pure 2D systems can be transferred in the 2.5D case; the main difference is that in a thin slab of dielectric material the photonic modes have to fulfill both the PC and the total internal reflection conditions imposed by the finite height of the slab. This means that the supported solutions (guided modes) are reduced compared to the 2D case because all those modes that propagate with a **k**-vector projection above the corresponding limit angle couple with the continuum of the unbound radiation modes. Thus the band diagram of a 2.5D PC is confined below the light cones that is defined by the refractive index of the cladding(s) that surround the slab. Figure 2.16 reports the band diagram for a triangular lattice of holes drilled into a silicon slab. The simulation considered a symmetric slab ($\epsilon = 12.25$ surrounded by air on both sides) of thickness $0.5a$ having holes of radius $0.25a$. Red dotted lines depict the TE modes, while the black continuous lines are TM polarized. The continuum of radiation modes is defined by the gray area. In a 2D slab the radius of the holes is generally much smaller than the one that maximize the gap width

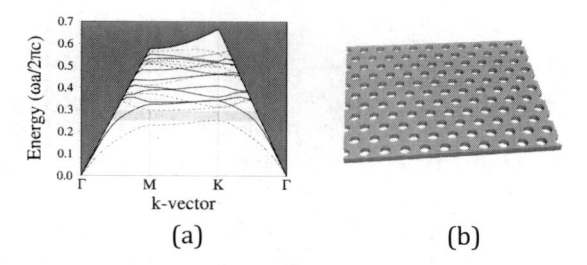

(a)   (b)

Figure 2.16 (a) Band diagram of a 2D slab. The gray area defines the continuum of radiation modes that are not guided within the slab. The guided modes are only those contained on the PC bands below the light cone. (b) 3D rendering of the simulated slab: the slab has a thickness of $0.5a$, the hole radius is $0.25a$, and the refractive index of the material is 12.25.

($r \approx 0.45$). In fact, a slab with very large gap in the plane would have unacceptably large losses due to the out of plane scattering that would be generated at the surface of the holes. Thus in order to keep a reasonable confinement of the mode in the vertical direction the *f.f.* of the PC have to be kept to rather low values.

The symmetric structure still allows for a separation of the two polarizations. In the more common case of a slab supported by a solid substrate, the separation of the TE and TM modes cannot formally applied but if the index of the substrate is much less than that of the slab (which is generally the case in order to have a good index confinement along the vertical direction), then the modes keep a high degree of parity and can still be classified looking at the symmetry of their field patterns.

The presence of the light cone produces a fundamental change in the confinement properties of the PC. In fact any MC formed within a slab will have components that couple with the radiation modes and this fact defines a limit on the maximum quality factor achievable for an MC in a slab. Despite this limitation an MC with an extremely high Q-factor can be realized in this thin dielectric layer.

As described by Eq. 2.18 the ratio Q/V is of paramount importance for several applications of MCs. In the case of rather weak confinement of the photonic mode, the Q-factor of the MC is generally proportional to its volume [37]. This fact can be phenomenologically explained, considering that the bigger the cavity, the longer the time required for a photon to travel across it and, eventually, escape. Moreover a large cavity may have a high confinement factor and then only a negligible part of the mode interacts with its boundaries. The disadvantages of large MCs are the limited integration level they can achieve and the large power they need to operate. On the other hand, MCs with a large Q/V ratio tend to have a Q-factor that decreases proportionally to their volume (this is the typical case for PC slabs). In this case the tightly confined cavity mode interacts strongly with the MC surface and the out-of-plane scattering becomes one of the main sources of losses that limits the achievable Q-factor. In a slab PC the Q-factor is often factorized into planar and vertical components:

$$\frac{1}{Q_{tot}} = \frac{1}{Q_{vert}} + \frac{1}{Q_{plan}}, \tag{2.23}$$

where Q_{tot} is the total cavity Q-factor, Q_{tvert} is the Q-factor in the direction perpendicular to the slab plane, and Q_{plan} is the one in the plane of the slab [50].

Thus the maximization of the confinement of photonic modes in tiny cavities is an active research topic with several implications both in fundamental and in applied science. Unfortunately it is also a formidable task that has to solve a reverse engineering problem and it does not admit a unique solution.

During the years several approaches were proposed to maximize the Q/V ratio. The general idea behind all the optimization methods is that of *tapering*. In its most generic form, tapering consists of the modification of a few, selected elements around the region of interest to permit an *adiabatic* transition of the optical modes across regions with different propagation properties.

The first attempts, to increase the Q-factor, deformed the holes around the defect site. With this trick the vertical Q-factor has been increased by an order of magnitude compared to simply shifting some holes from their lattice position [51].

The most promising methods to model high-Q-factor MCs are:

- **k**−space optimization.

  This method was initially proposed in [52]. It can be proved that the power radiated (lost) by modes that couple with the radiation modes depends only on **k**-vectors that lie within the light cone. Thus these modes fulfill the equation

  $$\vec{k}_{\parallel} < k = \frac{2\pi}{\lambda_0}, \qquad (2.24)$$

  where $\vec{k}_{\parallel}$ is the component of the **k**-vector lying in the plane of the slab and λ_0 is the free-space wavelength. Later the technique was formalized with a more in-depth description [53] and the estimated quality factor increases up to 10^5, while keeping the modal volume well below λ^3. More recently a significant increase in the measured total Q-factor was achieved by shifting the two holes that surround the defect site along its major axis [54]. The drawback of this approach is that the increase shows a sublinear dependence with the number of scatterers *moved* during the optimization and no general rules exist to determine which sites have to be perturbed and how they should be shifted from their original position. Thus an in-depth optimization requires a time-consuming trial-and-error approach because of the long simulation time.

- Genetic algorithm.
  Genetic (or evolutionary) algorithms are iterative methods that take some trial solutions as input and evolve them, following a certain scheme, to improve the fitness of the next-generation solutions, compared to a convergence criteria. These methods have been shown to be a powerful tool in that they can optimize specific parameters in order to achieve predefined optical properties and are not only related to the maximization of the Q-factor. Their main limitation is the long time required to achieve satisfactory designs [55, 56].

- Photonic heterostructures.
  A different approach was recently proposed. It is based on the formation of photonic heterostructures [57]. These are analogous of the electronic case; the main difference is the fact that the modulation of the potential is achieved by a slight modification of the lattice parameters along the WG length. Figure 2.17 represents the typical geometry of a photonic heterostructure [58]: the lattice period of a PC WG is symmetrically stretched around a central position (while it is unmodified along the perpendicular direction). The scale is greatly exaggerated in the figure (the modification of the period is of the order of percent). In the right part the band diagram of the two WG sections is qualitatively described: the bandwidths of the WG modes are shifted according to the different lattice periods. Thus along the WG section with the **b** lattice there exist modes with energy that falls within the BG region of the sections with period **a**. The energy of the mode of a heterostructure cavity is defined by the length of the WG and the profile of the electromagnetic field can be engineered to match the ideal Gaussian envelope profile, which is the one that minimizes the out-of-plane losses; thus, quality factors of these cavities go well beyond 10^6 [59].

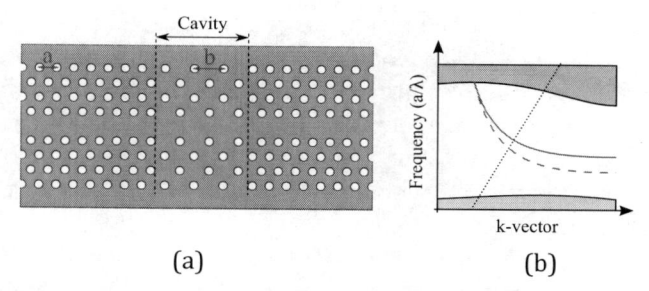

(a)   (b)

Figure 2.17   Schematic view of a photonic double heterostructure. (a) Top view of the 2D PC lattice. Three WG sections are placed sequentially: WG_a, WG_b, and WG_a. They differ in the lattice period, which has been slightly modified in these two types of WGs. The cavity is formed within the WG_b section. (b) Cartoon of the WG dispersion for the two different lattice periods. The section with the larger period (WG_b, dashed line) supports a propagating mode at lower energies. But these modes are within the gap of the two external sections WG_a (solid line) and are thus localized within the central region.

In all the aforementioned cases an important issue is the coupling among the PC-based devices with classic dielectric WGs or optical fibers. Mode matching has to be performed each time a propagating mode from a classical WG (or optical fiber) has to enter a PC region. Therefore both the mode profile (e.g., mode size needs to be adapted if the guiding sections have different sizes) and the propagating parameters have to change to satisfy the different propagation conditions relative to the different guiding mechanisms. In particular the coupling efficiency depends on which section, along the period of the PC, defines the interface between the classical WG (or fiber) and the PC section [60, 61]. Coupling is also important to maximize the amount of light emitted along predefined directions out of the slab plane [62] and to efficiently couple the mode with the slow light regime, where a group index in excess of hundreds can be achieved [63].

2.9 Three-Dimensional Photonic Crystals

Three-dimensional PCs are dielectrics structured periodically along all three directions. Ideally they have the most intriguing properties because the emission dynamics can be controlled to a great extent (e.g., an ideal 3D cavity would be able to store a photon for an indefinitely long time within a subwavelength volume). The problem is that their fabrication either is based on self-assembling processes that produce materials of limited quality or is an extremely complicated process still not compatible with the realization of real devices.

Two types of approaches are pursued for their fabrication, bottom up and top down. The bottom-up approach exploits self-assembling mechanisms and is generally used for the fabrication of opal-based 3D PCs. Solutions of monodispersed spheres (often silica, polystyrene, or titania) are left to arrange, under well-controlled drying conditions, to form face-centered-cubic (fcc) lattices (an fcc lattice does not shows a complete gap [64]). Large and macroscopic samples are formed with this technique but their quality is severely limited by the numerous defects that always form within them [65–67]. On the other hand the introduction of predefined defects at a specific site is a challenge still to be solved. Some tests demonstrated the possibility to introduce planar defects within the opal [68, 69] but the realization of a true 3D cavity has still to be demonstrated and this fact prevents the use of opals in 3D PICs.

Despite these limitations opals were actively investigated because of their simple fabrication method and the possibility to infiltrate the void between the sphere with dye-containing solutions and to study the behavior of emitters embedded within a 3D PC [70].

Their simple fabrication method allows the realization of large-area samples with enough high quality to be used in real market applications, as demonstrated by the several companies that supply opal-based optical filers and tunable displays.

An interesting structure is the inverted opals. This is the negative of the fcc mentioned above and is obtained by first fabricating a direct opal and then infiltrating the voids with a high-index material [71]. Compared to the direct type, the inverted structures shows a complete BG. Obviously the fabrication process is much more complex because the infiltration step has to be perfectly controlled in order to avoid the formation of partially filled volumes and also the complete removal of the spheres that define the direct lattice.

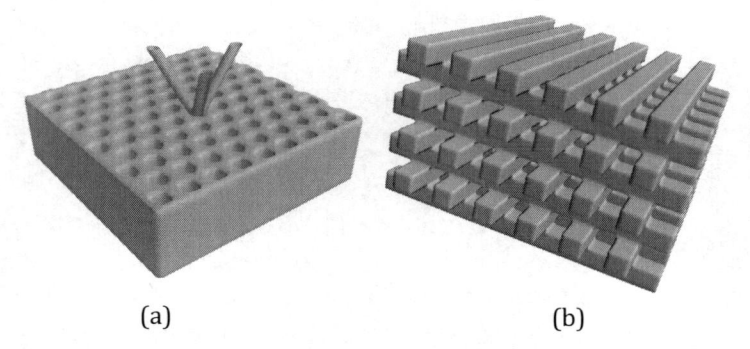

(a) (b)

Figure 2.18 (a) Scheme of yablonovite: three holes are drilled at every site of a 2D triangular lattice. (b) A woodpile PC is realized by overlapping orthogonally oriented arrays of sticks.

A much higher control over 3D PCs is obtained using top-down fabrication techniques. In this case the PC is realized by arranging/creating nearly every dielectric element. Two of the first investigated lattices were the yablonovite [72] and the woodpile [73] (shown in Fig. 2.18). The yablonovite has a complete gap in 3D and is fabricated by defining a triangular lattice on the surface of the sample and then by drilling, at each lattice site, three holes at 35° from the normal and rotated by 120° around the azimuth. The woodpile [73, 74] is based on layer-by-layer fabrication with arrays of rectangular sticks piled up after being rotated by 90° to respect the adjacent ones. Also this structure shows a complete gap in 3D if high-index materials are used for its realization. Compared to opals the introduction of defects in controlled positions can be achieved, but again, the fabrication process is extremely complicated and only proof-of-principle demonstrations of localized and extended defects were so far demonstrated [75, 76].

2.10 Nonlinear Optics

The noninteracting nature of photons is one of the main advantages of photonics compared to electronics. But at high field intensities the linear behavior of photons breaks up and light can be manipulated using light itself. A detailed description of NL optics can be found in several monographs [77, 78]; here we resume the main equations and how NL systems can effectively exploit the properties of PCs.

At high field intensities, the polarization (**P**) is described by a sum of NL terms:

$$P = \varepsilon_0[\chi^1\mathbf{E} + \chi^2\mathbf{EE} + \chi^3\mathbf{EEE} + \ldots] = \mathbf{P}^1 + \mathbf{P}^2 + \mathbf{P}^3 + \ldots \quad (2.25)$$

Thus photons with different energies are coupled by the NL terms and may interact among each other to exchange energy. The simplest cases are the generation of harmonics: a strong beam enters an NL material and generates harmonics at frequencies that are multiples of the pumping beam.

The NL terms are generated locally and they have to be phase-matched in order to interfere constructively and sustain a propagating beam. It is possible to demonstrate that a perfect phase matching among all the interacting beams maximizes the efficiency of the NL conversion:

$$\Delta k = 0 \quad (2.26)$$

In a homogeneous, bulk material, phase matching is achieved by proper shaping (cutting) of the NL crystal and careful alignment of the optical beams incident to it. But even in this way the linear dispersion of the refractive index often prevents the possibility to achieve interaction lengths long enough to effectively exchange power among the beams. To overcome these limitations NL systems use the so-called *quasi*-phase-matching condition that is achieved by a periodic reversal of the NL interaction in order to maintain the dephasing below 2π.

In the case of PCs, the periodicity of the lattice increases the probability to fulfill the phase matching: the reciprocal vectors of the lattice can be summed to those of the interacting beams to minimize their phase mismatch (as shown in Fig. 2.19). For example, in the case of the generation of a 2nd harmonic signal in a PC, the phase-matching condition is written as [79]

$$\mathbf{k}^{2\omega} - 2\mathbf{k}^{\omega} - \mathbf{G} = 0. \quad (2.27)$$

Furthermore the flattened band dispersion of the PCs permits one to increase the performance of NL devices by a factor roughly proportional to $(c/v_g)^2$ [80].

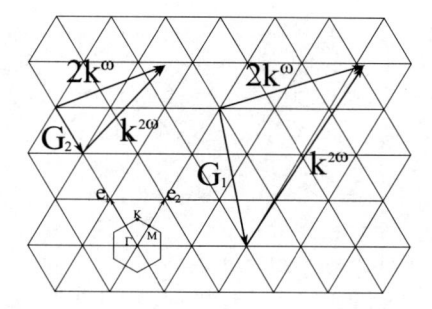

Figure 2.19 Example of phase matching of a 2nd harmonic generation process in a triangular lattice: the reciprocal lattice and its high symmetry points are shown.

Thus efficient NL effects have been demonstrated [81, 82] but despite these promising results several issues haven't been solved yet: severe losses come from both coupling [83, 84] and propagation of slow light modes [85, 86]. Furthermore group velocity dispersion has to be considered in order to keep high pulse intensity during the propagation along slow light WGs [87, 88].

2.11 Conclusions

PCs are still in the limelight of research on integrated photonics. Their great control over light dispersion permits the fabrication of artificial materials with highly controlled optical properties and allows the realization of complex PICs. This is a fundamental step toward an increased level of integration in optical circuits and maximization of the number of devices and thus the complexity of the operations that can be performed by PICs.

Currently only 1D systems have found real application as antireflective coatings and cavities in vertical cavity surface emitting lasers (VCSEL). This is mainly due to the high control obtained in the thin-layer deposition systems. PCs with higher dimensionality require further improvement of the nanofabrication technologies to achieve a level of reliability compatible with an industrial standard. Nevertheless the fast improvement of micro- and nanotechnologies will allow one to fabricate high-quality 2D PCs in the next few years.

The possibilities given by PCs of confining light within sub-wavelength volume in cavities with an extremely high Q-factor and of routing light pulses along sharp bends are key parameters to realize highly integrated PICs. Seminal papers about basic optical circuits in PC were published in [89–91]: WGs and MCs are the most important functional elements and from their combination several logic devices were demonstrated [92, 93]. During the last few years the advancements in the design of MCs in the $\mathbf{k}$−space and the implementation of PC heterostructures have greatly pushed toward the fabrication of extremely high-quality-factor resonators. All these results will be the foundation of the next era of integrated photonics.

References

1. Brovelli, L.R., and Keller, U. (1995). Simple analytical expression for the reflectivity and the penetration of a Bragg mirror between arbitrary media, *Opt. Commun.*, **116**, 4–6, pp. 343–350.

2. Rayleigh, L. (1887). On the maintenance of vibrations by forces of double frequency, and on the propagation of waves through a medium endowed with a periodic structure, *Philos. Mag.*, **24**, pp. 145–159.

3. Yablonovitch, E. (1987). Inhibited spontaneous emission in solid-state physics and electronics, *Phys. Rev. Lett.*, **58**, 20, pp. 2059–2062.

4. John, S. (1987). Strong localization of photons in certain disordered dielectric superlattices, *Phys. Rev. Lett.*, **58**, 23, pp. 2486–2489.

5. Kosaka, H., Kawashima, T., Tomita, A., Notomi, M., Tamamura, T., Sato, T., and Kawakami, S. (1998). Superprism phenomena in photonic crystals, *Phys. Rev. B*, **58**, 16, pp. 10096–10099.

6. Soljacic, M., Johnson, S.G., Fan, S.H., Ibanescu, S.H.M., Ippen, E., and Joannopoulos, J.D. (2002). Photonic-crystal slow-light enhancement of nonlinear phase sensitivity, *J. Opt. Soc. Am. B*, **19**, 9, pp. 2052–2059.

7. Notomi, M. (2002). Negative refraction in photonic crystals, *Opt. Quant. Electron.*, **34**, 1–3, pp. 133–143.

8. Feynman, R.P., Leighton, R.B., and Sands, M. (1989). Refractive index of dense materials, Sec. 3. In *Feynman Lectures on Physics.*, Vol. 2, Addison Wesley.

9. Ghulinyan, M., Oton, C.J., Gaburro, Z., Bettotti, P., and Pavesi, L. (2003). Porous silicon free-standing coupled microcavities, *Appl. Phys. Lett.*, **82**, 10, pp. 1550–1552.

10. Bertolotti, J., Gottardo, S., Wiersma, D.S., Ghulinyan, M., and Pavesi, L. (2005). Optical necklace states in Anderson localized 1D systems, *Phys. Rev. Lett.*, **94**, 11, p. 113903.

11. Patrini, M., Galli, M., Belotti, M., Andreani, L.C., Guizzetti, G., Pucker, G., Lui, A., Bellutti, P., and Pavesi, L. (2002). Optical response of one-dimensional $(Si/SiO_2)_m$ photonic crystals, *J. Appl. Phys.*, **92**, 4, pp. 1816–1820.

12. Hawkeye, M.M., Joseph, R., Sit, J.C., and Brett, M.J. (2010). Coupled defects in one-dimensional photonic crystal films fabricated with glancing angle deposition, *Opt. Express*, **18**, 12, pp. 13220–13226.

13. Dems, M., Chung, I.S., Peter, N., Bischoff, S., and Panajotov, K. (2010). Numerical methods for modeling photonic-crystal VCSELs, *Opt. Express*, **18**, 15, pp. 16042–16054.

14. Siriani, D.F., Tan, M.P., Kasten, A.M., Harren, A.C.L., Leisher, P.O., Sulkin, J.D., Raftery, J.J., Danner, A.J., Giannopoulos, A.V., and Choquette, K.D. (2009). Mode control in photonic crystal vertical-cavity surface-emitting lasers and coherent arrays, *IEEE J. Sel. Top. Quant. Electron.*, **15**, 3, pp. 909–917.

15. Kim, J.H., Chrostowski, L., Bisaillon, E., and Plant, D.V. (2007). DBR, Sub-wavelength grating, and photonic crystal slab Fabry-Perot cavity design using phase analysis by FDTD, *Opt. Express*, **15**, 16, pp. 10330–10339.

16. Özbay, E., Abeyta, A., Tuttle, G., Tringides, M., Biswas, R., Chan, C.T., Soukoulis, C.M., and Ho, K.M. (1994). Measurement of a three-dimensional photonic band gap in a crystal structure made of dielectric rods, *Phys. Rev. B*, **50**, pp. 1945–1948.

17. Bettotti, P. Dal Negro, L. Gaburro, Z., Pavesi, L., Lui, A., Galli, M., Patrini, M., and Marabelli, F. (2002). P-type macroporous silicon for two-dimensional photonic crystals, *J. Appl. Phys.*, **92**, 12, pp. 6966–6972.

18. Scherer, A., Painter, O., D'Urso, B., Lee, R., and Yariv, A. (1998). InGaAsP photonic band gap crystal membrane microresonators, *J. Vac. Sci. Technol. B*, **16**, 6, pp. 3906–3910.

19. Vahala, K.J., (2003). Optical microcavities, *Nature*, **424**, 6950, pp. 839–846.

20. Thon, S.M., Rakher, M.T., Kim, H., Gudat, J., Irvine, W.T.M., Petroff, P.M., Bouwmeester, D. (2009). Strong coupling through optical positioning of

a quantum dot in a photonic crystal cavity, *Appl. Phys. Lett.*, **94**, 11, p. 111115.

21. Malvezzi, A.M., Vecchi, G., Patrini, M., Guizzetti, G., Andreani, L.C., Romanato, F., Businaro, L., Di Fabrizio, E., Passaseo, A., De Vittorio, M. (2003). Resonant second-harmonic generation in a GaAs photonic crystal waveguide, *Phys. Rev. B*, **68**, 16, p. 161306.

22. Ogawa, S., Tomoda, K., and Noda, S. (2002). Effects of structural fluctuations on three-dimensional photonic crystals operating at near-infrared wavelengths, *J. Appl. Phys.*, **91**, 1, pp. 513–515.

23. Stein, A., Li, F., and Denny, N.R. (2008). Morphological control in colloidal crystal templating of inverse opals, hierarchical structures, and shaped particles, *Chem. Mater.*, **20**, 3, pp. 649–666.

24. Ozin, G.A., Arsenault, A.C. (2008). P-Ink and Elast-Ink from lab to market, *Mater. Today*, *11*, 7–8, pp. 44–51.

25. von Freymann, G., Ledermann, A., Thiel, M., Staude, I., Essig, S., Bush, K., and Wegener, M. (2010). Three-dimensional nanostructures for photonics, *Adv. Funct. Mater.*, **20**, 7, pp. 1038–1052.

26. Noda, S., (2006). Recent progresses and future prospects of two- and three-dimensional photonic crystals, *J. Lightwave Technol.*, **24**, 12, pp. 4554–4567.

27. Stenzel, O. (2005). *The Physics of Thin Film Optical Spectra, an Introduction*, Springer-Verlag, Berlin, Heidelberg.

28. Yeh, P. (1988). *Optical Waves in Layered Media*, Wiley.

29. Sakoda, K. (2001). *Optical Properties of Photonic Crystals*, Springer-Verlag.

30. Joannopoulos, J.D., Johnson, S.G., Winn, J.N., and Meade, R.D. (2008). *Molding the Flow of Light*, 2nd ed., Princeton University Press (PDF downloadable from http://ab-initio.mit.edu/book/).

31. Vahala K. (ed.). (2004). *Optical Microcavities*, World Scientific.

32. Takeda, H., Takashima, T., and Yoshino, K. (2004). Flat photonic bands in two-dimensional photonic crystals with kagome lattices, *J. Phys.-Condens. Matter*, **16**, 34, pp. 6317–6324.

33. Jin, C.J., Cheng, B.Y., Man, B.Y., Li, Z.L., Zhang, D.Z., Ban, S.Z., and Sun, B. (1999). Band gap and wave guiding effect in a quasiperiodic photonic crystal, *Appl. Phys. Lett.*, **75**, 13, pp. 1848–1850.

34. Man, W.N., Megens, M., Steinhardt, P.J., and Chaikin, P.M. (2005). Experimental measurement of the photonic properties of icosahedral quasicrystals, *Nature*, **436**, 7053, pp. 993–996.

35. Villeneuve, P.R., Fan, S., and Joannopoulos, J.D. (1996). Microcavities in photonic crystals: mode symmetry, tunability, and coupling efficiency, *Phys. Rev. B*, **54**, 11, pp. 7837–7842.

36. Akahane, Y., Asano, T., Song, B.S., and Noda, S. (2003). High-Q photonic nanocavity in a two-dimensional photonic crystal, *Nature*, **425**, pp. 944–947.

37. Okano, M., Yamada, T., Sugisaka, J., Yamamoto, N., Itoh, M., Sugaya, T., Komori, K., and Mori, M. (2010). Analysis of two-dimensional photonic crystal L-type cavities with low-refractive-index material cladding, *J. Opt.*, **12**, 075101.

38. Tokushima, M., Kosaka, H., Tomita, A., and Yamada, H. (2000). Lightwave propagation through a 120 degrees sharply bent single-line-defect photonic crystal waveguide, *Appl. Phys. Lett.*, **76**, 8, pp. 952–954.

39. Zhang, Y., and Li, B. (2007). Ultracompact waveguide bends with simple topology in two-dimensional photonic crystal slabs for optical communication wavelengths, *Opt. Lett.*, **32**, 7, pp. 787–789.

40. Jensen, J.S., and Sigmund, O. (2004). Systematic design of photonic crystal structures using topology optimization: low-loss waveguide bends, *Appl. Phys. Lett.*, **84**, 12, pp. 2022–2024.

41. Chutinan, A., Mochizuki, M., Imada, M., and Noda, S. (2001). Surface-emitting channel drop filters using single defects in two-dimensional photonic crystal slabs, *Appl. Phys. Lett.*, **79**, 17, pp. 2690–2692.

42. Yariv, A., Xu, Y., Lee, R.K., and Scherer, A. (1999). Coupled-resonator optical waveguide: a proposal and analysis, *Opt. Lett.*, **24**, 11, pp. 711–713.

43. Martinez, A., Martí, J., Bravo-Abad, J., and Sánchez-Dehesa, J. (2003). *Fiber and Integrated Optics*, Vol. 22, pp. 151–160.

44. Malkova, N., Scrymgeour, D.A., and Venkatraman G. (2005). Numerical study of light-beam propagation and superprism effect inside two-dimensional photonic crystals, *Phys. Rev. B*, **72**, p. 045144.

45. Witzens, J., Lončar, M., and Scherer, A. (2002). Self-collimation in planar photonic crystals, *IEEE J. Sel. Top. Quant. Electron.*, **8**, 6, pp. 1246–1257.

46. Baba, T., and Nakamura, M. (2002). Photonic crystal light deflection devices using the superprism effect, *IEEE J. Sel. Top. Quant. Electron.*, **38**, 7, pp. 909–914.

47. Wu, L., Mazilu, M., and Krauss, T.F. (2003). Beam steering in planar-photonic crystals: from superprism to supercollimator, *J. Lightwave Technol.*, **21**, 2, pp. 561–566.

48. Yamashita, T., and Summers, C.J. (2005). Evaluation of self-collimated beams in photonic crystals for optical interconnect, *IEEE J. Sel. Areas Commun.*, **23**, 7, pp. 1341–1347.

49. Lee, S.G., Oh, S.S., Kim, J.E., Park, H.Y., and Kee, C.S. (2005). Line-defect-induced bending and splitting of self-collimated beams in two-dimensional photonic crystals, *Appl. Phys. Lett.*, **87**, 18, p. 181106.

50. Lončar, M., Hochberg, M., Scherer, A., and Qiu, Y. (2004). High quality factors and room-temperature lasing in a modified single-defect photonic crystal cavity, *Opt. Lett.*, **29**, 7, p. 721.

51. Vučković, J., Lončar, M., Mabuchi, H., and Scherer, A. (2002). Design of photonic crystal microcavities for cavity QED, *Phys. Rev. E*, **65**, p. 016608.

52. Vučković, J., Lončar, M., Mabuchi, H., and Scherer, A. (2002). Optimization of the Q factor in photonic crystal microcavities, *IEEE J. Quant. Electron.*, **38**, 7, pp. 850–856.

53. Srinivasan, K., and Painter, O. (2002). Momentum space design of high-Q photonic crystal optical cavities, *Opt. Express*, **10**, 15, pp. 670–684.

54. Akahane, Y., Asano, T., Song, B.S., and Noda, S. (2003). High-Q photonic nanocavity in a two-dimensional photonic crystal, *Nature*, **425**, pp. 944–947.

55. Goh, J., Fushman, I., Englund, D., and Vučković, J. (2007). Genetic optimization of photonic bandgap structures, *Opt. Express*, **15**, 13, pp. 8218–8230.

56. Shen, L., Ye, Z., and He, S. (2003). Design of two-dimensional photonic crystals with large absolute band gaps using a genetic algorithm, *Phys. Rev. B*, **68**, 035109.

57. Istrate, E., and Sargent, E.H. (2006). Photonic crystal heterostructures and interfaces, *Rev. Mod. Phys.*, **78**, pp. 455–481.

58. Song, B.S., Noda, S., Asano, T., and Akahane, Y. (2005). Ultra-high-Q photonic double-heterostructure nanocavity, *Nat. Mater.*, **4**, 3, pp. 207–210.

59. Sekoguchi, H., Takahashi, Y., Asano, T., and Noda, S. (2014). Photonic crystal nanocavity with a Q-factor of ∼9 million, *Opt. Express*, **22**, 1, pp. 916–924.

60. Sanchis, P., Marti, J., Blasco, J., Martinez, A., and Garcia, A. (2002). Mode matching technique for highly efficient coupling between dielectric waveguides and planar photonic crystal circuits, *Opt. Express*, **10**, 24, pp. 1391–1397.

61. Sanchis, P., Bienstman, P., Luyssaert, B., Baets, R., and Marti, J. (2004). Analysis of butt coupling in photonic crystals, *IEEE J. Quant. Electron.*, **40**, 5, pp. 541–550.

62. Portalupi, S.L., Galli, M., Reardon, C., Krauss, T.F., O'Faolain, L., Andreani, L.C., and Gerace, D. (2010). Planar photonic crystal cavities with far-field optimization for high coupling efficiency and quality factor, *Opt. Express*, **18**, 15, pp. 16064–16073.

63. Vlasov, Y.A., and McNab, S.J. (2006). Coupling into the slow light mode in slab-type photonic crystal waveguides, *Opt. Express*, **31**, 1, pp. 50–52.

64. Galisteo-Lopez, J.F., Palacios-Lidon, E., Castillo-Martinez, E., and Lopez, C. (2003). Optical study of the pseudogap in thickness and orientation controlled artificial opals, *Phys. Rev. B*, **68**, 11, p. 115109.

65. Vlasov, Y.A., Astratov, V.N., Baryshev, A.V., Kaplyanskii, A.A., Karimov, O.Z.and Limonov, M.F. (2000). Manifestation of intrinsic defects in optical properties of self-organized opal photonic crystals, *Phys. Rev. B*, **61**, 5, pp. 5784–5793.

66. Wong, S., Kitaev, V., and Ozin, G.A. (2003). Colloidal crystal films: advances in universality and perfection, *J. Am. Chem. Soc.*, **125**, 50, pp. 15589–15598.

67. Lavrinenko, A.V, Wohlleben, W., and Leyrer, R.J. (2009). Influence of imperfections on the photonic insulating and guiding properties of finite Si-inverted opal crystals, *Opt. Express*, **17**, 2, pp. 747–760.

68. Yan, Q., Teh, L.K., Shao, Q., Wong, C.C., and Chiang, Y.M. (2008). Layer transfer approach to opaline hetero photonic crystals, *Langmuir*, **24**, 5, pp. 1796–1800.

69. Shi, LT., Jin, F., Zheng, M.L., Dong, X.Z., Chen, W.Q., Zhao, Z.S., and Duan, X.M. (2011). Threshold optimization of polymeric opal photonic crystal cavity as organic solid-state dye-doped laser, *Appl. Phys. Lett.*, **98**, 9, 093304.

70. Nikolaev, I.S., Lodahl, P., and Vos, W.L. (2008). Fluorescence lifetime of emitters with broad homogeneous linewidths modified in opal photonic crystals, *J. Phys. Chem. C*, **112**, 18, pp. 7250–7254.

71. Vlasov, Y.A., Bo, X.-Z., Sturm, J.C., and Norris, D.J. (2001). On-chip natural assembly of silicon photonic bandgap crystals, *Nature*, **414**, pp. 289–293.

72. Yablonovitch, E., Gmitter, T.J., and Leung, K.M. (1991). Photonic band structure: the face-centered-cubic case employing nonspherical atoms, *Phys. Rev. Lett.*, **67**, pp. 2295–2298.

73. Sözüer, H.S., and Dowling, J.P. (1994). Photonic band calculations for woodpile structures, *J. Mod. Opt.*, **41**, 2, pp. 231–239.

74. Ho, K.M., Chan, C.T., Soukoulis, C.M., Biswas, R., and Sigalas, M. (1994). Photonic band gaps in three dimensions: new layer-by-layer periodic structures, *Solid State Commun.*, **89**, pp. 413–416.

75. Kohli, P., Christensen, C., Muehlmeier, J., Biswas, R., Tuttle, G., and Ho, K.-M. (2006). Add-drop filters in three-dimensional layer-by-layer photonic crystals using waveguides and resonant cavities, *Appl. Phys. Lett.*, **89**, p. 231103.

76. Liu, R.-J., Li, Z.-Y. Feng, Z.-F. Cheng, B.-Y., and Zhang, D.-Z. (2008). Channel-drop filters in three-dimensional woodpile photonic crystals, *J. Appl. Phys.*, **103**, 094514.

77. Boyd, R.W. (2008). *Nonnlinear Optics*, Academic Press.

78. Sutherland, R.L. (2003). *Handbook of Nonlinear Optics*, Marcel Dekker.

79. Berger, V. (1998). Nonlinear photonic crystals, *Phys. Rev. Lett.*, **81**, 9, pp. 4136–4139.

80. Soljačić, M., and Joannopoulos, J.D. (2004). Enhancement of nonlinear effects using photonic crystals, *Nat. Mater.*, **3**, pp. 211–219.

81. Corcoran, B., Monat, C., Grillet, C., Moss, D.J., Eggleton, B. J., White, T. P., O'Faolain L., and Krauss, T. F. (2009). Green light emission in silicon through slow-light enhanced third-harmonic generation in photonic crystal waveguides, *Nat. Photonics*, **3**, pp. 206–210.

82. Rivoire, K., Lin, Z., Hatami, F., and Vučković, J. (2010). Sum-frequency generation in doubly resonant GaP photonic crystal nanocavities, *Appl. Phys. Lett.*, **97**, p. 043103.

83. Vlasov, Y.A., and McNab, J. (2006). Coupling into the slow light mode in slab-type photonic crystal waveguides, *Opt. Lett.*, **31**, 1, pp. 50–52.

84. Hugonin, J.P., Lalanne, P., White, T.P., Krauss, T.F. (2007). Coupling into slow-mode photonic crystal waveguides, *Opt. Lett.*, **32**, 18, pp. 2638–2640.

85. Hughes, S., Ramunno, L., Young J.F., Sipe, J.E. (2005). Extrinsic optical scattering loss in photonic crystal waveguides: role of fabrication disorder and photon group velocity, *Phys. Rev. Lett.*, **94**, p. 033903.

86. O'Faolain, L., White, T.P., O'Brien, D., Yuan, X., Settel, M.D., Krauss, T.F. (2007). Dependence of extrinsic loss on group velocity in photonic crystal waveguides, *Opt. Express*, **15**, 20, pp. 13129–131138.

87. Engelen, R.J.P., Sugimoto, Y., Watanabe, Y., Korterik, J.P., Ikeda, N., van Hulst, N.F., Asakawa, K., and Kuipers, L. (2006). The effect of

higher-order dispersion on slow light propagation in photonic crystal waveguides, *Opt. Express*, **14**, 4, pp. 1658–1672.

88. Inoue, K., Oda, H., Ikeda, N., Asakawa, K. (2009). Enhanced third-order nonlinear effects in slowlight photonic-crystal slab waveguides of linedefect, *Opt. Express*, **17**, 9, pp. 7206–7216.

89. Nakamura, H., Sugimoto, Y., Kanamoto, K., Ikeda, N., Tanaka, Y., Nakamura, Y., Ohkouchi, S., Watanabe, Y., Inoue, K., Ishikawa, H., and Asakawa, K. (2004). Ultra-fast photonic crystal/quantum dot all-optical switch for future photonic networks, *Opt. Express*, **12**, 26, pp. 6606–6614.

90. Notomi, M. (2011). fJ/bit photonic platform based on photonic crystals, *Proc. SPIE-Active Photonic Mater. IV*, **8095**, p. 809518.

91. Faraon, A., Majumdar, A., Englund, D., Kim, E., Bajcsy, M., and Vučković, J. (2011). Integrated quantum optical networks based on quantum dots and photonic crystals, *New J. Phys*, **13**, p. 055025.

92. Andalib, P., and Granpayeh, N. (2009). All-optical ultracompact photonic crystal AND gate based on nonlinear ring resonators, *J. Opt. Soc. Am.-B*, **26**, 1, pp. 10–16.

93. Zhang, Y.L., Zhang, Y., and Li, B.J. (2007). Optical switches and logic gates based on self-collimated beams in two-dimensional photonic crystals, *Opt. Express*, **15**, 15, pp. 9287–9292.

Chapter 3

Engineering Aperiodic Spiral Order in Nanophotonics: Fundamentals and Device Applications

Luca Dal Negro, Nate Lawrence, and Jacob Trevino

Electrical and Computer Engineering, Boston University, 8 Saint Mary's Street, Boston, MA 02139, USA

dalnegro@bu.edu

In this chapter, we present our work on the engineering of aperiodic spiral order for nanophotonic device applications. After introducing the guiding ideas behind aperiodic nanophotonics, we discuss the distinctive structural and optical properties of arrays of metal–dielectric nanoparticles with Vogel spiral geometry. This fascinating class of optical nanostructures offers unprecedented opportunities to manipulate light scattering and resonant phenomena over broad frequency spectra. In particular, photonic Vogel spirals support large photonic bandgaps (PBGs) as well as a high density of radially localized optical modes that cannot be found in regular photonic crystals or quasicrystal structures, and are ideally suited to boost polarization-insensitive light–matter coupling on planar substrates. In addition, optical Vogel spirals feature distinctive scattering resonances carrying orbital angular momentum (OAM) of light with

Nanodevices for Photonics and Electronics: Advances and Applications
Edited by Paolo Bettotti
Copyright © 2016 Pan Stanford Publishing Pte. Ltd.
ISBN 978-981-4613-74-3 (Hardcover), 978-981-4613-75-0 (eBook)
www.panstanford.com

a rich azimuthal spectrum, uniquely tailored by their distinctive aperiodic geometry. The direct generation and manipulation of *structured light* on optical chips using nanophotonic Vogel spirals provide exciting opportunities for engineering applications to broadband optical sources, secure communication, photovoltaics, and optical sensing.

3.1 Introduction to Aperiodic Photonic Structures

The distinguishing feature of periodic optical media is the formation of extended eigenmodes supporting continuous energy bands [1]. On the other hand, disordered optical media with randomly fluctuating optical constants, referred to as random media, can give rise to exponentially localized eigenmodes and support singular energy spectra with isolated δ-peaks [2]. A substantial amount of work has been devoted in the past years to better understand transport, localization, and wave phenomena in disordered random media [2–6]. These activities unveiled fascinating analogies between the behavior of electronic and optical excitations, such as disorder-induced Anderson light localization [6, 7], the photonic Hall effect [8], optical magnetoresistance [9], universal conductance fluctuations of light waves [10], and optical negative temperature coefficient resistance [5]. However, the engineering applications of random media are still very limited. In fact, random structures, while providing a convenient path to field localization, are not reproducible and lack simple design rules for deterministic optimization.

On the other hand, aperiodic optical media generated by mathematical rules, known as *deterministic aperiodic structures*, have recently attracted significant attention in the optics and electronics communities due to their simplicity of design, fabrication, and compatibility with current material deposition and device fabrication technologies [12–16]. Initial work, mostly confined to theoretical investigations of 1D aperiodic systems [17–24], has fully succeeded in stimulating broader experimental/theoretical studies of optical nanostructures that leverage deterministic aperiodic order as a comprehensive strategy to achieve new device functionalities. In addition to the numerous applications to optical sensing, light emission, photodetection, and nonlinear optics [13, 16], the study of deterministic aperiodic optical structures is a highly interdisciplinary research field, conceptually rooted in discrete geometry (i.e., tiling theory, point patterns theory, and mathematical crystallography [25–27]), symbolic dynamics [28–30], and number theory [31–33]. As a result, the optics of aperiodic media generates many exciting and cross-disciplinary opportunities that only recently began to be actively pursued [24]. The scope of this chapter is to provide a general background and to discuss the manipulation of light using plasmonic deterministic aperiodic nanostructures (DANS) with Vogel spiral geometry [16, 34]. As we will specifically discuss in this book chapter, DANS with isotropic Fourier space can be designed in the absence of Bragg peaks based on the concept of *aperiodic Vogel spiral geometry* [35]. It is our goal to show that Vogel spiral arrays of nanoparticles provide an appealing engineering approach for the control of polarization-insensitive planar diffraction phenomena, which result in the enhancement of linear and nonlinear *light–matter interactions in photonic–plasmonic nanostructures* [34, 36, 37].

3.2 Periodic and Aperiodic Order

One of the greatest intellectual triumphs of the twentieth century is the discovery of *aperiodic order* in the mathematical and physical sciences.

Periodic structures repeat a basic motif or building block in 3D space. In general, a vector function of position vector r satisfying the condition $\Phi(r + R_0) = \Phi(r)$ describes spatially periodic patterns because it is invariant under the set of translations generated by the vector R_0. As a result, we say that periodic structures display long-range order characterized by *translational invariance symmetry* along certain spatial directions.

The regularity of inorganic crystals best exemplifies periodic order since a certain atomic configuration, known as the *base*, repeats in space according to an underlying periodic lattice, thus defining a *crystal structure*. A fundamental property of the diffraction patterns of periodic structures is the presence of well-defined and sharp (i.e., δ-like) peaks indicating *long-range order*. As a result, the reciprocal Fourier space of periodic and multiperiodic systems is discrete (i.e., pure-point), with Bragg peaks positioned at rational multiples of the primitive reciprocal space vectors. This picture best exemplifies the notion of periodic arrangement of atoms that is at the origin of the traditional classification scheme of materials into the two broad categories of *crystalline* and *amorphous structures*.

The mathematical study of symmetry, planar tilings,[a] and discrete point patterns (i.e., Delone sets) paved the way for the discovery of aperiodic order in geometry. The study of tilings and point patterns has only recently been formalized using the advanced group-theoretic methods of mathematical crystallography and provides the natural framework to fully understand quasiperiodic and aperiodic structures. However, it was not until 1974 when the mathematician Roger Penrose discovered the existence of two simple polygonal shapes capable of tiling the *infinite* Euclidean plane *without spatial periodicity*. It was realized in the early 1980s that the diffraction patterns associated with such aperiodic point sets consist of *sharp diffraction peaks* with icosahedral point-group symmetry, which includes the pentagonal symmetry. It then became clear that the presence of sharp peaks in the diffraction spectra of materials does not necessarily imply structural periodicity. It was the pioneering experimental work of Dan Shechtman[b] that experimentally demonstrated aperiodic material structures with a new type of long-range order, called *quasicrystals*. In 1984, when studying the electron diffraction spectra from certain metallic alloys (Al6Mn), Shechtman et al. [38], discovered sharp diffraction peaks arranged with icosahedral point group symmetry, which

---

[a]**Tilings** or *tessellations* are a collection of plane figures (i.e., *tiles*) that fill the plane without leaving any empty space. Early attempts to tile planar regions of *finite size* using a combinations of pentagonal and decagonal tiles were already explored by Johannes Kepler, arguably the founder of the mathematical theory of tilings, in his book *Harmonices Mundi* published in 1619.

[b]Shechtman was awarded the Nobel Prize in Chemistry in 2011 for "the discovery of quasicrystals." However, his discovery was very controversial. He was even asked to leave his research group in the course of defending the validity of his findings. Eventually, Shechtman's work forced scientists to redefine their very conception of matter.

is incompatible with lattice periodicity [27, 38]. Moreover, the sharpness of the experimental diffraction peaks, which is a measure of the coherence of spatial interference patterns, turned out to be comparable with the one of ordinary periodic crystals. From a theoretical standpoint, it was subsequently discovered that 3D icosahedral structures can be obtained by projecting *periodic crystals* from a 6D *superspace* according to the mathematical rules of *quasicrystallography* [25, 26]. Stimulated by these findings, Dov Levine and Paul Steinhardt promptly formulated the notion of aperiodic crystals or quasicrystals in a seminal paper titled [39] "Quasicrystals: A New Class of Ordered Structures."

In response to these breakthrough discoveries, the International Union of Crystallography (IUCr) reformulated the concept of crystal structure as "any solid having an essentially discrete diffraction diagram," irrespective of spatial periodicity. This novel definition shifts the key feature of the crystal structure from direct to reciprocal Fourier space [13] and regards the presence of a pure-point diffraction spectrum as the essential attribute of crystalline order, either periodic or quasiperiodic.

3.3 Classification of Aperiodic Structures

Traditionally, optical media have been classified as either periodic or aperiodic, without the need of further distinctions. However, the word "aperiodic" encompasses a very broad range of different concepts that describe complex structures characterized by varying degrees of order and spatial correlations, ranging from quasiperiodic crystals to more disordered amorphous materials with diffuse diffraction spectra.

Moreover, structures featuring mixed spectra, containing discrete peaks embedded within a diffuse background, have also been investigated in science and technology [13]. Therefore, the *dichotomy between periodic and amorphous structures is completely inadequate and needs to be surpassed* in favor of a more accurate classification. Aperiodic structures have been recently classified according to the nature of their (spatial) Fourier and energy

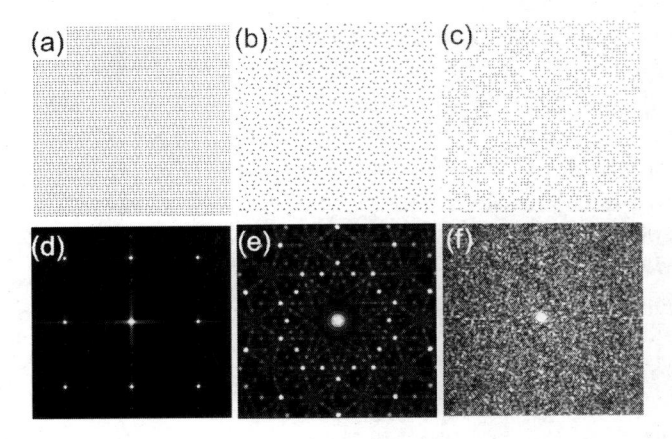

Figure 3.1 (a) Periodic square array, (b) Penrose array, (c) randomized periodic lattice, (d) periodic array reciprocal space, (e) Penrose array reciprocal space, and (f) randomized array reciprocal space.

spectra, which correspond to mathematical *spectral measures.*[a] In optics, these spectral measures are often identified with the characteristic diffraction patterns and/or the optical mode spectra (e.g., local density of states [LDOS]). According to Lebesgue's decomposition theorem [40], any set theoretic measure can be uniquely decomposed into three primitive spectral components, or a mixture of them, namely *pure-point* (μ_P), *singular continuous*[b] (μ_{SC}), and *absolutely continuous* spectral components (μ_{AC}), such that $\mu = \mu_P \cup \mu_{SC} \cup \mu_{AC}$. On the basis of this approach, Maciá Barber recently proposed [12] a classification approach for aperiodic structures that subdivides them according to a matrix with nine entries, corresponding to the possible combinations of the three fundamental spectral measures associated with spatial Fourier and energy spectra (see Fig. 3.2). Random media are characterized by uncorrelated structural fluctuations, resulting in continuous spatial

---

[a]A measure on a set is a function that assigns a nonnegative real number to each suitable subset of that set, intuitively interpreted as its size. The measure function must satisfy certain axioms such as countable additivity and null empty set (i.e., being zero for an empty set). As a result, a measure is a generalization of the usual concepts of length, area, and volume.

[b]Singular-continuous structures support Fourier spectra and can be covered by an ensemble of open intervals with an arbitrarily small total length.

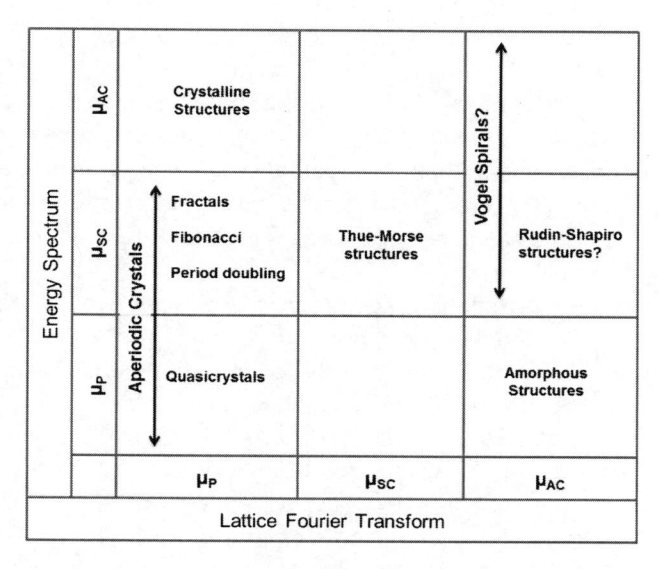

Figure 3.2 Classification matrix of aperiodic systems based on their spectral properties. Structures are distinguished according to the spectral measures of their Fourier spatial spectra and Hamiltonian energy spectra. Adapted from Ref. [12] with kind permission of Enrique Macia Barber.

Fourier spectra. However, random or amorphous systems have singular energy spectra (i.e., pure-point), since disorder-induced localized states appear at discrete resonant frequencies whenever the electronic/optical diffusive transport breaks down (i.e., in the limit of strong multiple scattering). On the other hand, periodic structures support well-defined and sharp diffraction Bragg peaks associated to long-range periodic order, but their energy spectra form continuous energy bands. We represent these two extreme cases by the lower-right and top-left entries in the classification diagram shown in Fig. 3.2. After the discovery of quasicrystals, it was clearly realized that the rich domain of deterministic aperiodic structures could bridge the remaining gaps in the spectral theory of aperiodic systems beyond random and periodic structures.

In fact, the energy spectrum of the self-similar (i.e., fractal) systems considered to date appears to be singular continuous and supported on a Cantor set of zero Lebesgue measures. As a result, this spectrum exhibits an infinity hierarchy of gaps with vanishing bandwidth in the thermodynamic limit (i.e., for infinitely large systems). This fascinating property has been rigorously demonstrated for aperiodic systems based on the Fibonacci, Thue–Morse (which additionally features a singular continuous spatial Fourier spectrum), and period-doubling sequences. It is interesting to note that at present no rigorous results exist on the energy spectrum of the elementary excitations of aperiodic structures with *aperiodic spiral order*.

Spiral arrays are long-range-ordered but do not possess translational and rotational symmetries. In particular, their spatial Fourier transform does not exhibit well-defined diffraction peaks but rather diffuse rings similar to amorphous materials with absolutely continuous spatial spectra. Our recent theoretical and experimental works on the optical resonances of aperiodic Vogel spirals, later reviewed in this chapter, show that such systems can support energy spectra with tunable characteristics and allow us to tentatively classify them, as represented in Fig. 3.2.

A key question in the theory of aperiodic systems regards the relationship between their atomic topological order, determined by a given aperiodic density function, and the physical properties stemming from their structure [41]. Of particular interest is the gap-labeling theorem, which provides a relationship between reciprocal space (Fourier) spectra and Hamiltonian energy spectra. This theorem, which originates within the tight-binding theory of aperiodic electronic systems, relates the position of a number of gaps in the energy spectra of elementary excitations to the singularities of the Fourier transform of the structure [42, 43].

The nature of the spectral measure associated to the spatial Fourier spectrum of a given structure, which is related to the main features of the diffraction pattern, can be quantitatively described by considering the expression

$$I_N(q) = |F_N(q)|^2, \tag{3.1}$$

where $I_N(q)$ is the intensity of the diffraction peaks, N measures the system size, q denotes the scattering wavevector, and F_N is the Fourier transform of the appropriate *density function* that describes the geometry of the structure (e.g., atomic or electronic density, permittivity or refractive index modulation, etc.). The diffraction spectra in the thermodynamic limit can be conveniently described by the *integrated intensity (counting) function* (IIF) defined as [13]

$$H(q) = \lim_{N \to \infty} \int_0^q \frac{I_N(q')}{N} dq'. \tag{3.2}$$

This function represents the normalized distribution of the diffracted intensity peaks up to a given point (q) in the reciprocal space. A completely analogous function can be introduced to study the energy spectra by replacing the normalized intensity distribution with the appropriate spectral measure associated with the density of electronic, vibrational, or optical states.

On the basis of this approach, we can directly visualize the main features of different Lebesgue measures. In what follows we will illustrate this fact in relation to diffraction measures (i.e., spatial Fourier spectra). In particular, in the case of both periodic and quasiperiodic crystals, there will be reciprocal space intervals where the diffraction intensity vanishes along a given q axis, so the IIF remains constant. These intervals are separated by diffraction Bragg peaks where the IIF has finite jumps, according to the equation

$$H(q) = \int_0^q \sum_n c_n \delta(q' - q_n) dq', \tag{3.3}$$

where the sum runs over a countable set of Bragg peaks. A measure described by Eq. 3.3 is said to be *pure-point*. As an example, the spatial Fourier spectrum of Fibonacci quasiperiodic structures has a pure-point measure displaying a countable set of diffraction peaks.

On the other hand, the spatial Fourier spectrum of Thue–Morse aperiodic structures is no longer composed of a countable set of Bragg points separated by well-defined intervals. In fact, it does not contain δ-peaks and it has a structure similar to a Cantor set. More precisely, in the thermodynamic limit Thue–Morse structures are characterized by a *singular-continuous measure*, meaning that the support of their Fourier spectrum can be covered by an ensemble of open intervals with an arbitrary small total length [13].

The third type of primitive spectral measure component is exemplified by the Rudin–Shapiro structures, which possess absolutely continuous Fourier spectra. In the case of a diffuse Fourier spectrum the contribution to the IIF of any interval on a given q axis is roughly proportional to its length, making the IIF plot to appear linear. By

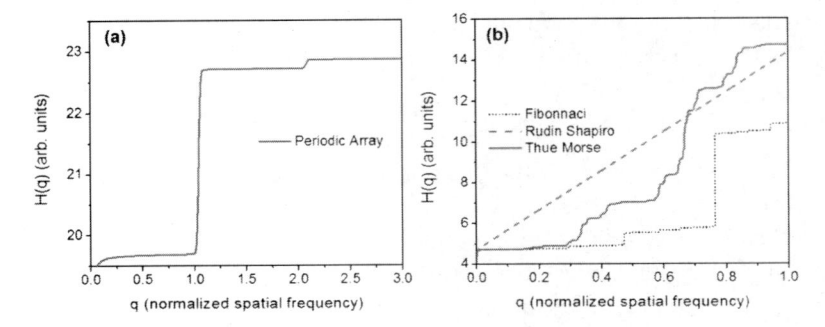

Figure 3.3 (a) H(q) plot for periodic square array, plotting the first three Brillouin zones. (b) $H(q)$ plots for Fibonacci, Rudin–Shapiro, and Thue–Morse arrays, plotting spatial frequencies up to $1/\Delta$, where Δ is the average minimum center-to-center separation.

making these intervals arbitrarily small, it can be proved that any single point in the diffraction spectrum has zero weight and the IIF is *both continuous and differentiable*. Interestingly, this property is also shared with disordered random systems. In Fig. 3.3a we show the calculated IIF curves for a square periodic array limited to contributions from the first three Brillouin zones. In Fig. 3.3b we plot the IIF restricted to the first pseudo-Brillouin zones for Fibonacci, Rudin–Shapiro, and Thue–Morse arrays. The results in Fig. 3.3 clearly demonstrate the distinctive spectral features of the structures.

3.4 Rotational Symmetry in Aperiodic Structures

In 3D space, crystal structures are mathematically described by their 32 point-group symmetries,[a] which are combinations of pure rotation, mirror, and roto-inversion operations fully compatible with the translational symmetry of the 14 Bravais lattices. The addition

---

[a]A point group is a group of geometric symmetries (i.e., isometries) leaving a point fixed. Point groups can exist in Euclidean space of any dimension. The discrete point groups in two dimensions are used to describe the symmetries of an ornament. There are infinitely many discrete point groups in each number of dimensions. However, the crystallographic restriction theorem demonstrates that only a finite number of them is compatible with translational invariance symmetry.

of translation operations defines the crystallographic space groups, which have been completely enumerated in 230 different types by Fedorov and Schoenflies in 1890. One of the deepest results of classical crystallography states that the combination of translations with rotations restricts the total number of available rotational symmetries to the ones compatible with the periodicity of the lattice [27]. This important result is known as the *crystallographic restriction*. We say that a structure possesses an n-fold rotational symmetry if it is left unchanged when rotated by an angle $2\pi/n$, and the integer n is called the *order* of the rotational symmetry (or the order of its symmetry axis). It can be shown that only rotational symmetries of order $n = 2, 3, 4, 6$ fulfill the translational symmetry requirements of 2D and 3D periodic lattices in Euclidean space [25–27], therefore excluding $n = 5$ and $n > 6$. As a result, the pentagonal symmetry, very often encountered in biological systems as in the *pentamerism* of viruses, microorganisms such as *radiolarians*, plants, and a number of marine animals (i.e., sea stars, urchins, crinoids, etc.) has been long neglected in inorganic materials until noncrystallographic symmetries were discovered in quasicrystals.

Aperiodic tilings displaying an *arbitrary degree of rotational symmetry* can be deterministically constructed using a purely algebraic approach [44]. In Fig. 3.4 we display four remarkable aperiodic point patterns featuring increasing rotational symmetry, along with the corresponding spatial Fourier spectra. We notice that aperiodic structures feature a nonperiodic reciprocal space, which can accommodate a larger number of rotational symmetries (as well as more abstract types of group symmetries). However, the diffraction **k**-vectors of aperiodic Fourier space lose their *global meaning* and should be regarded merely as *locally defined spatial frequency components*.

In Fig. 3.4a we show a deterministic point pattern obtained using the Danzer inflation rule [45], which has 14-fold rotational symmetry. The corresponding reciprocal space is shown in Fig. 3.4h. Deterministic point patterns with increasing degree of rotational symmetry up to infinite order (i.e., continuous circular symmetry) have also been demonstrated [46] using a simple iterative procedure that decomposes a triangle into congruent copies.

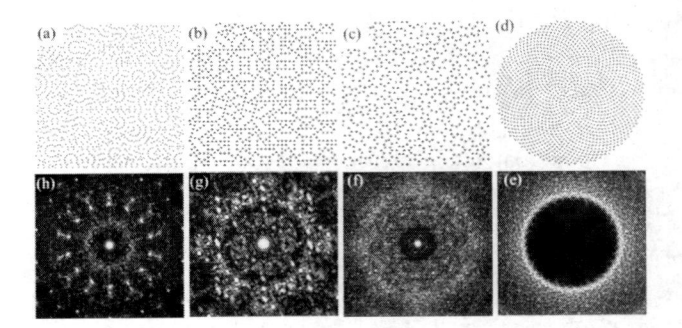

Figure 3.4 (a) Danzer array, generation 4; (b) pinwheel array (5th iteration order); (c) Delaunay triangulated pinwheel centroid (DTPC) array; (d) golden angle Vogel spiral; (e) golden angle Vogel spiral reciprocal space; (f) DTPC reciprocal space; (g) pinwheel reciprocal space; and (h) Danzer reciprocal space.

The resulting structure, called pinwheel tiling, has triangular elements (i.e., tiles) that appear in infinitely many orientations, and in the limit of infinite size, its diffraction pattern displays continuous (infinity-fold) rotational symmetry. Radin has shown that there are no discrete components in the pinwheel diffraction spectrum [46]. However, it is currently unknown if the spectrum is continuous or singular continuous. A point pattern obtained from the pinwheel tiling and the corresponding Fourier spectrum are displayed in Fig. 3.4b,g. Our group recently investigated [47] pinwheel arrays of resonant metallic nanoparticles and reported isotropic structural coloration of metal films using a *homogenized pinwheel pattern*. This pattern, which is called Delaunay triangulated pinwheel centroid (DTPC), is obtained from a regular pinwheel lattice by a homogenization procedure that performs a Delaunay triangulation of the array and positions additional nanoparticles in the center of mass (i.e., baricenter) of the triangular elements (Fig. 3.4c). The DTPC, which is an example of a deterministic isotropic and homogeneous particle array, shares the same rotational symmetry of the regular pinwheel array but features a more uniform spatial distribution of nanoparticles with strongly reduced clustering, as evident in Fig. 3.4c. The calculated reciprocal space of the DTPC pattern is displayed in Fig. 3.4f. We notice the higher degree of spatial uniformity of the DTPC when compared to the regular pinwheel (Fig. 3.4b) pattern. Local structural correlations, which are due to the finite size of the regular pinwheel array, results in well-defined scattering peaks in the pinwheel reciprocal space, shown in Fig. 3.4g. These correlations are absent in the reciprocal space of the DTPC array (Fig. 3.4f) due to its higher degree of spatial uniformity that better approximates uncorrelated disorder. Using DTPC plasmonic arrays, it was possible to obtain bright-green coloration of Au films with greatly reduced angular sensitivity and enhanced color uniformity compared to both periodic and random arrays [47]. Finally, we can appreciate from Fig. 3.4 that the Fourier spectra become more diffuse when increasing the degree of rotational symmetry toward isotropic structures such as the golden angle (GA) Vogel spiral, shown in panels (d and c). This general trend is manifested by amorphous/liquid random media and it is displayed by the broad family of deterministic aperiodic structures known as Vogel spirals. The structural and optical properties of Vogel spiral point patterns will be discussed in great detail in the rest of this chapter.

3.4.1 *Aperiodic Spiral Order: From Phyllotaxis to Nanophotonics*

Spiral lattices are interesting examples of long-range ordered systems where both translational and orientational symmetries are missing. Spiral curves can be generated by simple mathematical rules, which can adopt many forms in polar coordinates, such as $r = a\theta$ (Archimedean), $r = a\sqrt{\theta}$ (Fermat's or parabolic), $r = ae^{\theta}$ (logarithmic), etc. A spiral array is generated from a spiral curve simply by restricting r and θ according to an integer quantization condition $\theta_n = \phi n$ and $r_n = f(n)$, where $\phi = 2\pi/\gamma$ is the divergence angle that measures the angular separation between consecutive radius vectors and $f(n)$ is a prescribed function (e.g., linear, quadratic, etc.). When γ is an irrational number, the divergence angle yields irrational fractions of 2π and the resulting spiral array entirely lacks rotational symmetry.

The spiral geometry is manifested by a large number of physical systems (Fig. 3.5) and it plays an important role in

Figure 3.5 (a) Seed head of sunflower, *Helianthus annuus*, beautifully featuring the golden angle (GA) spiral geometry (b) phyllotaxis of spiral aloe, *Aloe polyphylla*. (c) Nautilus shell's logarithmic growth spiral. (d) Pinwheel spiral galaxy (also known as NGC 5457). Source: Wikipedia.

the morphological studies of plants and animals [48]. The most interesting examples of aperiodic spiral arrays are given by the so-called Vogel spirals, which have been deeply investigated by mathematicians, botanists, and theoretical biologists [49] in relation to the fascinating geometrical problems of phyllotaxis [48, 50, 51]. Phyllotaxis (from the Greek *phullon*, leaf, and *taxis*, arrangement) is concerned with understanding the spatial patterns of leaves, bracts, and florets on plant stems (e.g., the spiral arrangement of florets in the capituli of sunflowers and daisies). The field of phyllotaxis goes back to Leonardo Da Vinci (1452–1519), who described the spiral arrangements of leaves, and Kepler (1571–1639), who observed the frequent occurrence of the number 5 in plants [52]. The history of mathematical phyllotaxis started in the first half of the nineteenth century when the brothers L. and A. Bravais presented the first quantitative treatment of the phenomenon and recognized the relevance of the theory of continued fractions in this area (i.e., the cylindrical representation of phyllotaxis).

The science of phyllotaxis, born as a branch of botany, has subsequently evolved into a multidisciplinary research field that combines mathematical crystallography, L-systems[a] theory and computer graphics with anatomical, cellular, physiological, and paleontological observations to explain the outstanding patterns encountered in plant morphogenesis and, more generally, in mathematical biology (e.g., patterns found in the structures of polymers, viruses, jellyfishes, proteins, etc.).

Vogel spiral point patterns are defined in polar coordinates (r, θ) by the following equations [13, 53, 54]:

$$r_n = a_0 \sqrt{n}$$
$$\theta_n = n\alpha,$$
(3.4)

where $n = 0, 1, 2, \ldots$ is an integer index, a_0 is a constant scaling factor, and α is an irrational number known as the divergence angle. This angle specifies the constant aperture between successive point particles in the array. Irrational numbers (ξ) can be used to generate irrational divergence angles $(\alpha^\circ$, in degrees) by the relationship $\alpha^\circ = 360^\circ - [\xi - floor(\xi)] \cdot 360^\circ$. The value $\alpha \approx 137.508^\circ$ approximates the irrational GA, which generates the so-called Fibonacci GA spiral, shown in Fig. 3.4d. Rational approximations to the GA can be obtained by the formula $\alpha = 360 \cdot (1 + p/q) - 1$ where p and $q < p$ are consecutive Fibonacci numbers.

The spatial structure of the GA spiral can be decomposed into clockwise (CW) and counterclockwise (CCW) families of out-spiraling particles lines, known as *parastichies*, which stretch out from the center of the structure. The number of spiral arms in each family of parastichies is given by consecutive Fibonacci numbers [53]. A pair of spiral families (i.e., *parastichy pair*) formed by m-spirals in one direction and n-spirals in the opposite direction is denoted by (m, n).[b]

---

[a]*L-systems*, or Lindenmayer systems, were introduced and developed in 1968 by the Hungarian theoretical biologist and botanist Aristid Lindenmayer (1925–1989). L-systems are used to model the growth processes of plant development but also the morphology of a variety of organisms. In addition, L-systems can be used to generate self-similar fractals and certain classes of aperiodic tilings such as the Penrose lattice.

[b]A large sunflower's head (i.e., the *capitulum*) features two sets of parastichies (i.e., spirals) running in opposite directions with Fibonacci numbers (34, 55).

We notice that since the GA is an irrational number, the GA spiral lacks both translational and rotational symmetry. Accordingly, its spatial Fourier spectrum does not exhibit well-defined Bragg peaks, as for standard photonic crystals and quasicrystals, but rather it features a broad and diffuse circular ring whose spectral position is determined by the particles' geometry.

More generally, different types of aperiodic Vogel spirals can be generated by choosing other irrational values of the divergence angle, giving rise to vastly different spiral geometries all characterized by isotropic Fourier spectra.

A very general mathematical theorem, known as the fundamental theorem of phyllotaxis, relates the numbers of visible parastichy pairs (m, n) with the divergence angle of the spiral [51, 52]. We show in Fig. 3.6 three representative examples of aperiodic spirals generated by irrational divergence angles referred to as τ-, μ-, and π-spiral arrays, obtained using the approximated values listed in Table 3.1. We notice that Vogel spirals with remarkably different structural properties can be obtained by choosing only slightly different values for the divergence angle, thus providing an opportunity to tailor and explore distinctively different degrees of aperiodic structural order.

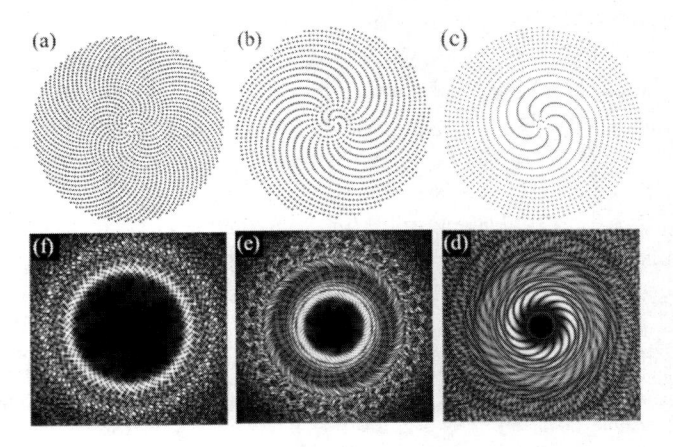

Figure 3.6 (a) τ-spiral array, (b) μ-spiral array, (c) π-spiral array, (d) π-spiral reciprocal space, (e) μ-spiral reciprocal space, and (f) τ-spiral reciprocal space.

Table 3.1 Divergence angles utilized for the irrational and perturbed GA spirals

Label	ξ	Divergence Angle (°)	Label	ξ	Divergence Angle (°)
τ	$(2+\sqrt{8})/2$	210.883118	GA	$(1+\sqrt{5})/2$	137.507764
μ	$(5+\sqrt{29})/2$	290.670335			
π	3.14159...	309.026645			
α_1		137.30000			
α_2		137.369255	β_1		137.523137
α_3		137.403882	β_2		137.553882
α_4		137.473137	β_3		137.569255
GA	$(1+\sqrt{5})/2$	137.507764	β_4		137.600000

Aperiodic Vogel spiral arrays of nanoparticles are rapidly emerging as a powerful nanophotonics platform with distinctive optical properties of interest to a number of engineering applications [36, 55, 56a]. This fascinating category of deterministic aperiodic media features circularly symmetric scattering rings in Fourier space entirely controlled by simple generation rules that induce a very rich structural complexity.

Trevino et al. [37] have recently described the structure of Vogel spiral arrays of nanoparticles using multifractal geometry and discovered that such structures feature a degree of local order in between short-range correlated amorphous/liquid systems and uncorrelated random systems. Moreover, it has been recently demonstrated that Vogel spiral arrays of metallic nanoparticles feature distinctive structural resonances and produce polarization-insensitive, planar light diffraction across a broad spectral range, referred to as circular light scattering [34].

This interesting phenomenon originates from the circular symmetry of the reciprocal space of Vogel spirals, and it can be simply understood within standard Fourier optics (i.e., neglecting near-field interactions among neighboring particles and the vector character of light). In fact, the condition for light waves normally incident on an array of particles to be diffracted into the plane of the array requires the longitudinal wavevector component of light to identically vanish, that is, $k_z = 0$. This requirement is equivalent to the well-known Rayleigh condition that determines the transition between the propagation and cutoff of the first diffractive order in periodic gratings so that diffracted waves travel along the grating surface [56b]. The Rayleigh cutoff condition depends on the wavelength λ and on the transverse spatial frequencies v_x and v_y of the diffracting element, according to

$$k_z = 2\pi \sqrt{(1/\lambda)^2 - v_x^2 - v_y^2} = 0 \qquad (3.5)$$

Equation 3.5 is satisfied on a circle of radius $1/\lambda$ in the reciprocal space of the array. As a result, diffractive elements possessing circularly symmetric Fourier space naturally satisfy the Rayleigh condition irrespective of the incident polarization and diffract normal incident radiation into evanescent surface modes.

We say that the resonant condition expressed by Eq. 3.5 gives rise to *planar omnidirectional diffraction*. It is important to notice that differently from periodic crystals and quasicrystals with rotational symmetries of finite order, Vogel spirals satisfy the resonant condition for planar diffraction over a broader range of wavelengths because they display thick isotropic rings in reciprocal space (see Fig. 3.7) uniquely determined by the generation parameters of the structures. The planar diffraction property of Vogel spirals is ideally suited to enhance light–matter interactions on planar substrates [34] and recently led to the demonstration of thin-film solar cell absorption enhancement [36], light emission [55, 57], and second harmonic generation enhancement [58] using metal–dielectric arrays. Another fascinating feature of Vogel spiral diffracting elements is their ability to support distinctive scattering resonances carrying well-defined numerical sequences of orbital angular momentum (OAM) of light, potentially leading to novel applications in singular optics and optical cryptography [59–61].

3.4.2 *Structural Properties of Vogel Spirals*

Previous studies have focused on the three most investigated types of aperiodic spirals, including the GA spiral and two other Vogel spirals obtained by the following choice of divergence angles: 137.3° (i.e., α_1-spiral) and 137.6° (i.e., β_4-spiral) [34, 50, 53]. The α_1 and β_4-spirals are called nearly golden spirals because their divergence angles are numerically very close to the GA value, but their families

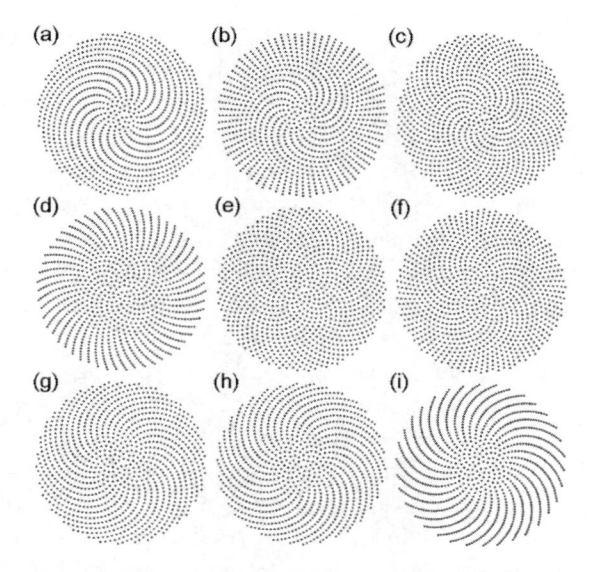

Figure 3.7 Vogel spiral array consisting of 1000 particles, created with a divergence angle of (a) 137.3° (α_1), (b) 137.3692546° ((α_2)), (c) 137.4038819° ((α_3)), (d) 137.4731367° ((α_4)), (e) 137.5077641° (GA), (f) 137.5231367° (β_1), (g) 137.553882° (β_2), (h) 137.5692547° (β_3), and (i) 137.6° (β_4). Reproduced from Ref. [37] with permission from OSA.

of diverging arms, known as parastichies, are considerably fewer. In a recent paper Trevino et al. [37] extended the analysis of aperiodic Vogel spirals to structures generated with divergence angles that are equispaced between the α_1-spiral and the GA spiral and between the GA and β_4, as summarized in Table 3.1.

These spiral structures, shown in Fig. 3.7, can be considered as one-parameter (i.e., the divergence angle α) *structural perturbation of the GA spiral* and possess fascinating geometrical features, which are responsible for unique mode localization properties and optical spectra [37].

When the divergence angle is varied either above (supra-GA or β-series) or below the GA (sub-GA or α-series), the center region of the spiral where both sets of parastichies (CW and CCW) exist shrinks to a point. The outer regions are left with parastichies that rotate only CW for divergence angles greater than the GA and CCW for those below, thus providing deterministic aperiodic structures *that display a distinctively chiral geometry.* For the spirals with larger deviation from the GA (α_1 and β_4 in Fig. 3.7), gaps appear in the center head of the spirals and the resulting point patterns mostly consist of either CW- or CCW-spiraling arms. We also notice that stronger structural perturbations (i.e., further increase in the diverge angle) lead to less interesting spiral structures containing only radially diverging parastichies (not investigated here).

To better understand the consequences of the divergence angle perturbation on the optical properties of Vogel spiral arrays, we have investigated their spatial Fourier spectra. Figure 3.8 displays the 2D spatial Fourier spectra obtained by calculating the amplitude of the discrete Fourier transform (DFT) of the spiral arrays shown in Fig. 3.7.

We can appreciate that all the spectra in Fig. 3.8 lack Bragg peaks and display diffuse circular rings (Fig. 3.8a–i). The many spatial frequency components in Vogel spirals give rise to a diffuse background, as for amorphous and random systems. Interestingly, despite the lack of rotational symmetry of Vogel spirals, their Fourier spectra are highly isotropic (approaching circular symmetry) as a consequence of a high degree of statistical isotropy [34, 60].

As previously reported [34, 60, 62], the GA spiral features a well-defined and broad scattering ring in the center of the reciprocal space (Fig. 3.8e), which corresponds to the dominant spatial frequency of the structure [60].

Perturbing the GA spiral by varying the divergence angle from the GA creates more *disordered* Vogel spirals and results in the formation of multiple scattering rings, associated with additional characteristic length scales, embedded in a diffuse background of fluctuating spots with weaker intensity. In the *perturbed Vogel spirals* (i.e., Fig. 3.8a,b,d,g–i) different patterns of spatial organization at finer scales are clearly discernable in the diffuse background. The onset of these substructures in Fourier space reflects the gradual removal of statistical isotropy of the GA spiral, which transitions to less homogeneous substructures with variable degrees of local order.

To better characterize the degree of local order in Vogel spirals we resort to analytical tools that are more suitable for the detection of *local spatial variations*. Our group has recently studied the local

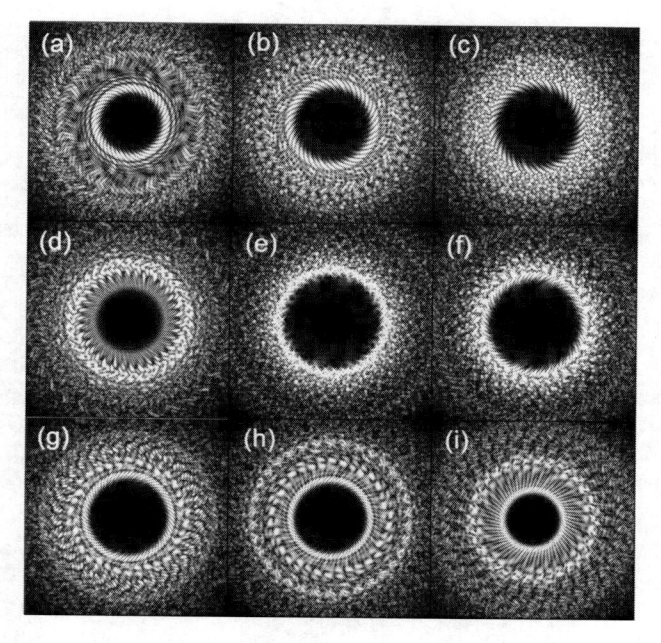

Figure 3.8 Calculated spatial Fourier spectrum of the spiral structures show in Fig. 3.7. The reciprocal space structure of an (a) α_1-spiral, (b) α_2-spiral, (c) α_3-spiral, (d) α_4-spiral, (e) GA spiral, (f) β_1-spiral (g) β_2-spiral, (h) β_3-spiral, and (i) β_4-spiral are plotted. Reproduced from Ref. [37] with permission from OSA.

geometrical structure of Vogel spirals by the powerful methods of spatial correlation functions [37]. A comprehensive discussion of all these aspects can be found in Ref. [37].

The pair correlation function, $g(r)$, also known as the radial density distribution function, is employed to evaluate the probability of finding two particles separated by a distance r, thus measuring the local (correlation) order in the structure. Figure 3.9a displays the calculated $g(r)$ for spiral arrays with irrational divergence angles τ, μ, and π (see Table 3.1), while in Fig. 3.9b we show the results of the analysis for arrays generated with divergence angles between α_1 and the GA (α-series). To better capture the geometrical features associated with the structure (i.e., array pattern) of Vogel spirals, the $g(r)$ was calculated directly from the array point patterns (i.e., no element pattern associated with finite-size particles) using the

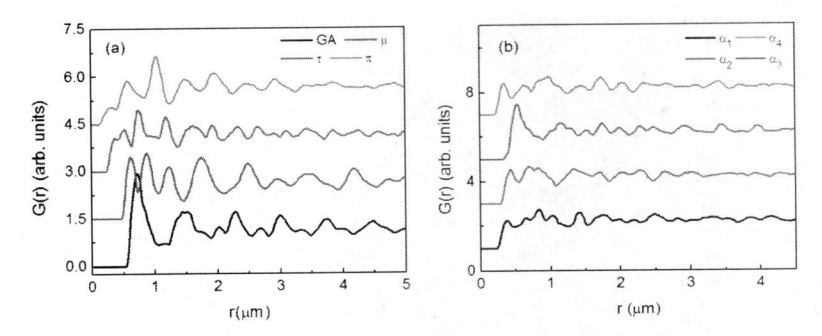

Figure 3.9 (a) Pair correlation function $g(r)$ for irrational angle spiral arrays and (b) arrays with divergence angles between α_1 and the golden angle.

library *spatstat* [63] within the R statistical analysis package. The pair correlation function is calculated as

$$g(r) = \frac{K'(r)}{2\pi r},$$ (3.6)

where r is the radius of the observation window and $K'(r)$ is the first derivative of the reduced second moment function (Ripley's K function) [64]. The results of the pair correlation analysis shown in Fig. 3.9 reveal a fascinating aspect of the geometry of Vogel spirals, namely their structural similarity to monoatomic gases and liquids. We can clearly appreciate from Fig. 3.9 that the GA spiral exhibits several oscillating peaks, indicating that for certain radial separations, corresponding to local coordination shells, it is more likely to find particles in the array. A similar oscillating behavior for $g(r)$ can be observed when studying the structure of liquids by X-ray scattering [65]. We also notice that the $g(r)$ of the most perturbed (i.e., more disordered) Vogel spiral (Fig. 3.9b, α_1-spiral) features strongly damped oscillations against a constant background, similarly as the $g(r)$ measured for a gas of random particles. Between these two extremes (α_2 to α_4) a varying degree of local order can be observed for the other spirals in the series. These results demonstrate that the degree of local order in Vogel spiral structures can be deterministically controlled between the correlation properties of photonic amorphous structures [66, 67] and uncorrelated random systems by continuously varying the divergence angle α, which acts as an *order parameter*.

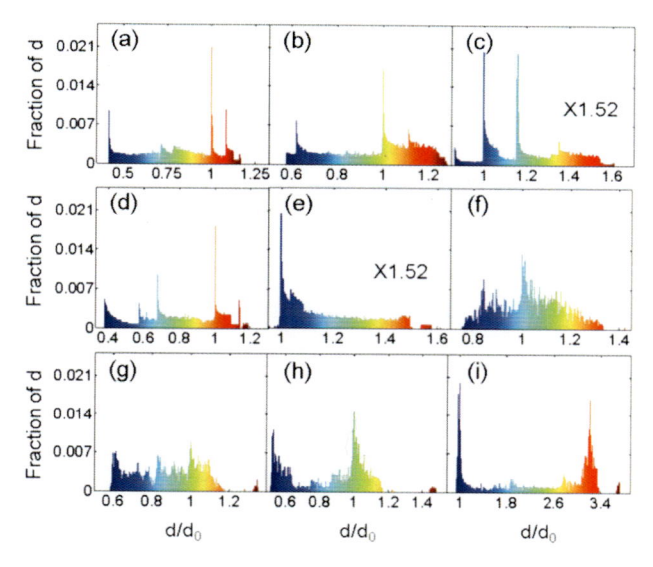

Figure 3.10 Statistical distribution of spiral structures shown in Fig. 3.7. Values represent the distance between neighboring particles, d, normalized to the most probable value d_0, obtained by Delaunay triangulation (increasing numerical values from blue to red color). The y axis displays the fraction of d in the total distribution. Reproduced from Ref. [37] with permission from OSA.

We have also investigated the spatial distribution of the distance d between first neighboring particles by performing a Delaunay triangulation of the spiral array [60, 62]. This technique provides information on the statistical distribution of the first neighbor distance d and provides a measure of the spatial uniformity of point patterns [68]. In Fig. 3.10 we show the calculated statistical distribution, obtained by the Delaunay triangulation, of the parameter d normalized by d_0, which corresponds to the most probable value (where the distribution is peaked). In all the investigated structures (shown in Fig. 3.7), the most probable value d_0 is generally found to be close to the average interparticle separation. However, the distributions of neighboring particles in Fig. 3.10 are distinctively non-Gaussian in nature and display slowly decaying tails, similar to the heavy tails often encountered in mathematical finance. These characteristic fluctuations are very pronounced for the two series of

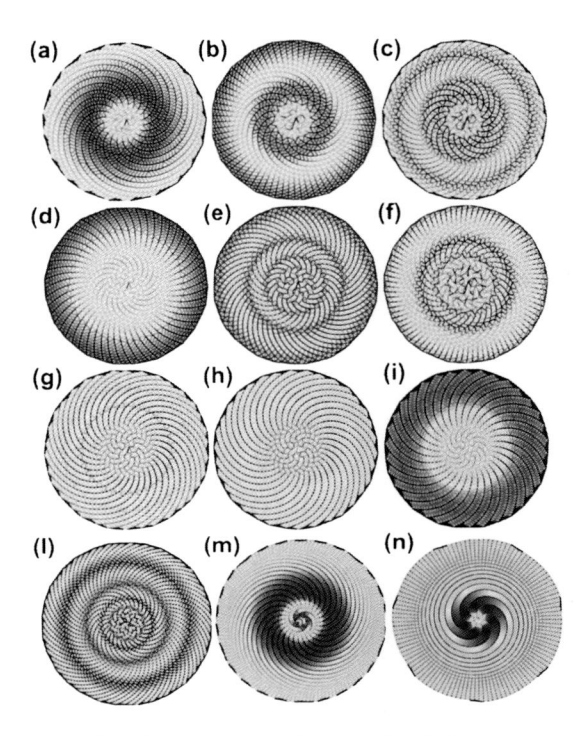

Figure 3.11 (a–i) Delaunay triangulation of spiral structures shown in Fig. 3.7. (l–n) Delaunay triangulation of irrational angle spiral structures τ, μ, and π. The line segments that connect neighboring circles are color-coded by their lengths d. The colors are consistent with those in Fig. 3.10.

perturbed GA spirals, consistent with their reduced degree of spatial homogeneity.

All the distributions in Fig. 3.10 are broad with varying numbers of sharp peaks corresponding to different correlation lengths, consistent with the presence of the fine substructures already captured in Fourier space (Fig. 3.8).

Next, we performed *spatial Delaunay triangulation analysis* to visualize the spatial locations on the spirals where the different correlation lengths (i.e., distribution spikes) appear more frequently. In Fig. 3.11 we directly visualize the spatial map of the first neighbors' connectivity of the Vogel spirals. Each line segment in Fig. 3.11 connects two neighboring particles on the spirals, and the connectivity length d is color-coded consistently with the scale of Fig. 3.10 (i.e., increasing numerical values from blue to red color). The nonuniform color distributions shown in Fig. 3.11 graphically represent the distinctive spatial order of Vogel spirals. In particular, we clearly notice that circular symmetry is found in the distribution of particles for all the spirals, including the strongly inhomogeneous α- and β-series as well as the irrational angle spirals τ, μ, and π. As recently demonstrated by Trevino et al. [37], regions of markedly different values of d define *radial heterostructures* that can efficiently trap radiation in regions with different lattice constants, similar to the case of the concentric rings of omniguide Bragg fibers. The index contrast between radially adjacent rings traps radiation by Bragg scattering along different circular loops. The circular regions evidenced in the spatial map of local particle coordination in Fig. 3.11 well correspond to the scattering rings observed in the Fourier spectra (Fig. 3.8). This analysis unveils very general structural features of aperiodic Vogel spirals generated by irrational divergence angles and can guide the design of photonic–plasmonic structures that support a large density of distinctively localized optical resonances with a large degree of azimuthal symmetry. In particular, the radially localized azimuthal modes of perturbed and irrational-angle aperiodic spirals are extremely attractive for the engineering of novel light sources, laser devices, and optical sensors that combine a broad spectrum of localized eigenmodes with open dielectric pillar structures for increased refractive index sensitivity to environmental perturbations.

3.5 Optical Resonances of Vogel Spiral Arrays

PBG structures have received a lot of attention in recent years [1]. The ability to engineer spectral gaps in the electromagnetic wave spectrum and create highly localized modes opens the door to numerous exciting applications, including high-Q cavities [69], PBG novel optical waveguides [70] and enhanced light-emitting diodes (LEDs) [71] and lasing structures [72]. Many of these applications rely on photonic crystals that possess a complete PBG, which is readily achieved in quasiperiodic lattices with a higher degree of rotational symmetry [73]. However, while photonic quasicrystals have been mostly investigated for the engineering of isotropic PBGs, the more general study of 2D structures with deterministic aperiodic order offers additional opportunities to manipulate light transport by engineering a broader spectrum of optical modes with distinctive localization properties [16].

In this section, we review our systematic study of the structural properties, photonic gaps, and band-edge modes of 2D Vogel spiral arrays of dielectric cylinders in air. Specifically, a number of Vogel spiral arrays generated by a gradual structural perturbation of the GA spiral will be studied.

The divergence angle is varied in a equispaced fashion between $\alpha_1(137.3°)$ and the GA and also between the GA and $\beta_4(137.6°)$, as summarized in Table 3.1. The optical properties of Vogel spirals are now investigated by numerically calculating their LDOS across the large wavelength interval from 0.4 μm to 2 μm. This choice is mostly motivated by the engineering polarization insensitive and localized band-edge modes for applications to broadband solar energy conversion. Calculations were performed for all arrays shown in Fig. 3.7, consisting of $N = 1000$ dielectric cylinders, 200 nm in diameter, with a permittivity $\epsilon = 10.5$ embedded in air, for which transverse magnetic (TM) PBGs are favored.

All arrays are generated using a scaling factor, α, equal to 3×10^{-7}. The LDOS is calculated at the center of the spiral structure using the well-known relation $g(r, \omega) = (2\omega/\pi c^2)\text{Im}[G(r, r', \omega)]$, where $[G(r, r', \omega)$ is Green's function for the propagation of the E_z component from point r to r'. The numerical calculations are implemented using the finite-element method within COMSOL Multiphysics (version 3.5a). A perfectly matched layer (PML) is utilized to absorb all radiation leaking toward the computational window. In Fig. 3.12a,b we display the calculated LDOS for the spiral arrays in the α- and β-series as a function of frequency (ω) normalized by the GA spiral bandgap center frequency ($\omega_0 = 13.2 \cdot 10^{14}$ Hz), respectively. For comparison, the LDOS of the GA spiral is also reported in both panels.

The results in Fig. 3.12 demonstrate the existence of a central large LDOS bandgap for all the investigated structures, which originates from the Mie resonances of the individual cylinders,

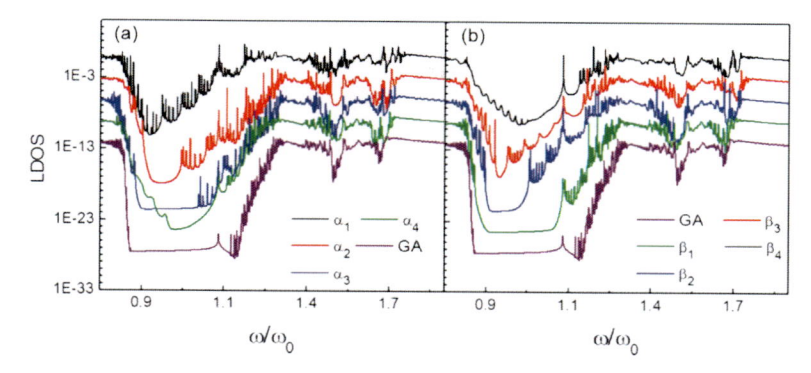

Figure 3.12 LDOS calculated at the center of the each spiral array as a function of normalized frequency for spiral arrays with divergence angles between (a) α_1 and the golden angle and (b) between the golden angle and β_4. Reproduced from Ref. [37] with permission from OSA.

as previously demonstrated by Pollard et al. for the GA spiral [62]. However, we also observed that the edges of these bandgaps split into a large number of secondary gap regions of smaller amplitudes (i.e., subgaps) separated by narrow resonant states that reach, in different proportions, into the central gap region as the inhomogeneity of the structures is increased from the GA spiral. The width, shape, and fine resonant structure of these band-edge features are determined by the unique array geometries. A large peak located inside the gap at $\omega/\omega_0 = 1.122 (1.273 \ \mu m)$ represents a defect mode localized at the center of the spiral array where a small air region free of dielectric cylinders acts as a structural defect. Several peaks corresponding to localized modes appear both along the band edges and within the gap. These dense series of photonic band-edge modes have been observed for all types of Vogel spirals and correspond to spatially localized modes due to the inhomogeneous distribution of neighboring particles. A more detailed study of the formation of mechanism of these modes in a GA spiral is presented in Ref. [60]. The results are extended to all the investigated Vogel spirals based on the knowledge of their first neighbor connectivity structure, shown in Fig. 3.11. In particular, we note that localized band-edge modes are supported when ring-shaped regions of similar interparticle separation d in

Fig. 3.11 are sandwiched between two other regions of distinctively different values of d, thus creating a photonic heterostructure that can efficiently localize optical modes. In this picture, the outer regions of the spirals act as effective barriers that confine different classes of modes within the middle spirals regions. According to this mode localization mechanism, the reduced number of band-edge modes calculated for spirals α_4 and β_4 is attributed to the monotonic decrease (i.e., gradual fading) of interparticle separations when moving away from the central regions of the spirals, consistent with the corresponding Delaunay triangulation maps in Fig. 3.11. In particular, since these strongly perturbed spiral structures do not display clearly contrasted (i.e., sandwiched) areas of differing interparticle separations, their band-edge LDOS is strongly reduced and circularly symmetric band-edge modes cannot be formed. In the next section it will be demonstrated that the distinctive geometrical structure of Vogel spirals and their photonic LDOS spectra display multifractal scaling.

3.5.1 *Multifractal Scaling of Vogel Spirals*

In this section, we review our work on the application of multifractal analysis as a way to characterize the inhomogeneous structures of Vogel spirals arrays as well as their LDOS spectra [37]. Geometrical objects display fractal behavior if they display scale-invariance symmetry, or self-similarity, that is, a part of the object resembles the whole object [74]. Fractal objects are described by noninteger fractal dimensions and display power-law scaling in their structural (i.e., density–density correlation, structure factor) and dynamical (i.e., density of modes) properties [65, 75].

The relevance of fractals to physical sciences and other disciplines (i.e., economics) was originally pointed out by the pioneering work of Mandelbrot [76]. However, the complex geometry of physical structures and multiscale physical phenomena (i.e., turbulence) cannot be entirely captured by homogeneous fractals with a single fractal dimension (i.e., monofractals). In general, a spectrum of local scaling exponents associated with different spatial regions needs to be considered. For this purpose, the concept of multifractals, or inhomogeneous fractals, has been more recently introduced [77,

78] and a rigorous multifractal formalism has been developed to quantitatively describe local fractal scaling [79].

In general, when dealing with multifractal objects on which a local measure μ is defined (i.e., a mass density, a velocity, an electrical signal, or some other scalar physical parameter defined on the fractal object), the (local) *singularity strength* $\alpha(x)$ of the multifractal measure μ obeys the local scaling law:

$$\mu(B_x(\epsilon)) \approx \epsilon^{\alpha(x)}, \tag{3.7}$$

where $B_x(\epsilon)$ is a ball (i.e., interval) centered at x and of size ϵ. The smaller the exponent $\alpha(x)$, the more singular the measure around x (i.e., local singularity). The *multifractal spectrum* $f(\alpha)$, also known as the *singularity spectrum*, characterizes the statistical distribution of the singularity exponent $\alpha(x)$ of a multifractal measure. If we cover the support of the measure μ with balls of size ϵ, the number of balls $N_\alpha(\epsilon)$ that, for a given α, scales like ϵ^α behaves as

$$N_\alpha(\epsilon) \approx \epsilon^{-f(\alpha)}. \tag{3.8}$$

In the limit of vanishingly small ϵ, $f(\alpha)$ coincides with the fractal dimension of the set of all points x with scaling index α [75]. The spectrum $f(\alpha)$ was originally introduced by Frisch and Parisi [79] to investigate the energy dissipation of turbulent fluids. From a physical point of view, the multifractal spectrum is a quantitative measure of structural inhomogeneity. As shown by Arneodo et al. [80], the multifractal spectrum is well suited for characterizing complex spatial signals because it can efficiently resolve their local fluctuations. Examples of multifractal structures/phenomena are commonly encountered in dynamical systems theory (e.g., strange attractors of nonlinear maps), physics (e.g., diffusion-limited aggregates, turbulence), engineering (e.g., random resistive networks, image analysis), geophysics (e.g., rock shapes, creeks), and even finance (e.g., stock market fluctuations). In the case of singular measures with a recursive multiplicative structure (i.e., the devil's staircase), the multifractal spectrum can be calculated analytically [81]. However, in general, multifractal spectra are computed numerically.

In our work, the multifractal spectra of Vogel spirals and of their LDOS spectra were calculated for the first time. The multifractal singularity spectrum of each spiral structure was calculated from the corresponding 600 dpi bitmap image, using the direct Chhabra–Jensen algorithm [82] implemented in the routine FracLac (version 2.5) [83] developed for the National Institutes of Health (NIH)-distributed Image-J software package [84]. All relevant implementation details can be found in Ref. [37]. Multifractal measures involve singularities of different strengths and their $f(\alpha)$ spectrum generally displays a single humped shape (i.e., downward concavity), which extends over a compact interval $[\alpha_{min}, \alpha_{max}]$, where α_{min} (respectively, α_{max}) corresponds to the strongest (respectively, the weakest) singularities. The maximum value of $f(\alpha)$ corresponds to the (average) box-counting dimension of the multifractal object, while the difference $\Delta\alpha = \alpha_{max} - \alpha_{min}$ can be used as a parameter reflecting the fluctuations in the length scales of the intensity measure [85]. The calculated multifractal spectra are shown in Fig. 3.13a,b for the spiral arrays in the α- and β-series, respectively. All spirals exhibit clear multifractal behavior with singularity spectra of characteristic downward concavity, demonstrating the multifractal nature of the geometrical structure of Vogel spirals. We notice in Fig. 3.13a,b that the GA spiral features the largest fractal dimensionality ($D_f = 1.873$), which is consistent with its more regular structure. We also notice that the $\Delta\alpha$ for the GA spiral is the largest, consistently with the diffuse nature (absolutely continuous) of its Fourier spectrum. On the other hand, the less homogeneous α_1-spiral structure features the lowest fractal dimensionality ($D_f = 1.706$), consistent with a larger degree of structural disorder. All other spirals in the α-series were found to vary in between these two extremes. These results demonstrate that multifractal analysis is suitable to detect the small local structural differences among Vogel spirals obtained by very small variations in the divergence angle α.

In Ref. [37] we have also demonstrated the multifractal geometry of the LDOS spectra of Vogel spirals. The connection between the multifractality of geometrical structures and the corresponding energy or LDOS spectra is far from trivial in general. In fact, multifractal energy spectra have been discovered for deterministic quasiperiodic and aperiodic systems that do not display any fractality in their spatial arrangement, despite the fact that they

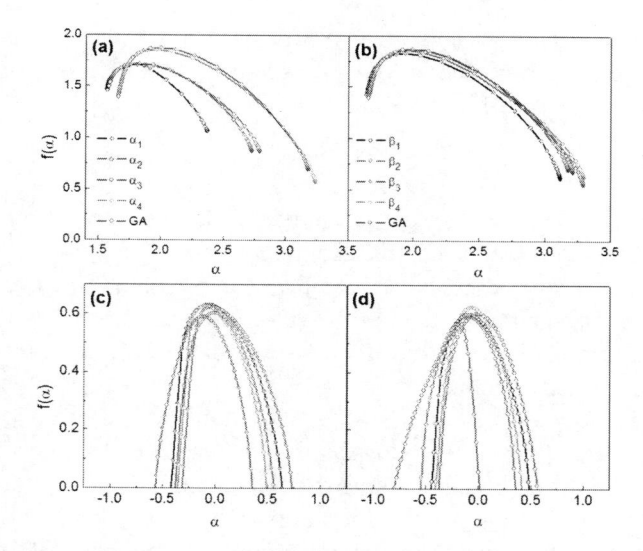

Figure 3.13 Multifractal singularity spectra f_α of direct space spiral arrays ($N = 1000$) with divergence angles between (a) α_1 and the golden angle and (b) between the golden angle and β_4. Multifractal spectra for spiral LDOS with divergence angles between (c) α_1 and the golden angle and (d) between the golden angle and β_4. Reproduced from Ref. [37] with permission from OSA.

are generated by fractal recursion rules. Typical examples are Fibonacci optical quasicrystals and Thue–Morse structures [85]. Optical structures with multifractal eigenmode density (or energy spectra) often display a rich and fascinating behavior, leading to the formation of a hierarchy of satellite pseudogaps, called fractal gaps, and of critically localized eigenmodes when the size of the system is increased [86]. Moreover, dynamical excitations in fractals, or fracton modes, have been found to originate from multiple scattering in aperiodic environments with multiscale local correlations, which are described by multifractal geometry [54].

To demonstrate the multifractal character of the LDOS spectra of Vogel spirals, we performed wavelet-based multifractal analysis [87], as detailed in Ref. [37]. This approach is particularly suited to analyze signals with nonisolated singularities, such as the LDOS spectra shown in Fig. 3.12. The wavelet transform (WT) of a function f is a decomposition into elementary *space–scale contributions*, associated with so-called wavelets that are constructed from one single function Ψ by means of translations and dilation operations. The WT of the function f is defined as

$$W_\Psi[f](b, a) = \frac{1}{a} \int_{-\infty}^{+\infty} \bar{\Psi}\left(\frac{x-b}{a}\right) f(x)dx, \qquad (3.9)$$

where α is the real scale parameter, β is the real translation parameter, and $\bar{\Psi}$ is the complex conjugate of Ψ. Usually, the wavelet Ψ is only required to be a zero-average function. However, for the type of singularity tracking required for multifractal analysis, it is additionally required for the wavelet to have a certain number of vanishing moments [87]. Frequently used real-valued analyzing wavelets satisfying this last condition are given by the integer derivatives of the Gaussian function, and the first derivative Gaussian wavelet is used in our multifractal analysis of the LDOS. In the wavelet-based approach, the multifractal spectrum is obtained by the so-called wavelet transform modulus maxima (WTMM) method [87] using the global partition function $Z(q, a)$ originally introduced by Arneodo et al. [80, 81]. For each q, from the scaling behavior of the partition function at fine scales one can obtain the scaling exponent $\tau(q)$:

$$Z(q, a) \approx a^{\tau(a)} \qquad (3.10)$$

The singularity (multifractal) spectrum $f(\alpha)$ is derived from $\tau(q)$ by a Legendre transform [87, 88]. To analyze the LDOS of photonic Vogel spirals, the aforementioned WTMM method within the free library of Matlab wavelet routines WaveLab850 [89] has been implemented. The code has been carefully tested against a number of analytical multifractals (e.g., devil's staircase) and found to generate results in excellent agreement with known analytical spectra. The calculated LDOS singularity spectra are shown in Fig. 3.13c,d for the α- and β-spiral series, respectively. For comparison, the LDOS of the GA spiral is also reported in both panels. The data shown in Fig. 3.13c,d demonstrate the multifractal nature of the LDOS spectra of Vogel spirals with singularity spectra of characteristic downward concavity. The average fractal dimensions of the LDOS were found in the range of $D_f \approx 0.6 - 0.74$, with the two extremes belonging to the α-series (i.e., α_1 and α_2, respectively). The strength of the LDOS singularity is measured by the value of

$\alpha_0 = \alpha_{\max}$, which is the singularity exponent corresponding to the peak of the $f(\alpha)$ spectrum. In Fig. 3.13c, we notice that the singular character of the LDOS spectra steadily increases from spiral α_1 to the GA spiral across the α-series. On the other hand, a more complex behavior is observed across the β-series, where the strength of the LDOS singularity increases from β_2 to β_4-spirals.

3.5.2 *Optical Mode Analysis of Vogel Spirals*

In this section we summarize our main results from the analysis of optical modes in Vogel spirals. In particular, we focus on the high-frequency multifractal band edge, where the field patterns of the modes show the highest intensity in the air regions between the dielectric cylinders and thus are best suitable for sensing and lasing applications where gain materials can easily be embedded between rods [90]. The spatial profile of the modal fields and their complex eigenfrequencies $\omega = \omega_r + i\omega_i$ were calculated using eigenmode analysis within COMSOL. Complex mode frequencies naturally arise from radiation leakage through the open boundary of the arrays. The imaginary components of the complex mode frequencies give the leakage rates of the mode, from which the quality factor can be defined as $Q = \omega_r / 2\omega_i$. The calculated quality factors of the modes are plotted in Fig. 3.14a for the α_1-spiral and in Fig. 3.14b for the GA spiral and the β_4-spiral as a function of normalized frequency. The analysis is limited to only these three structures since they

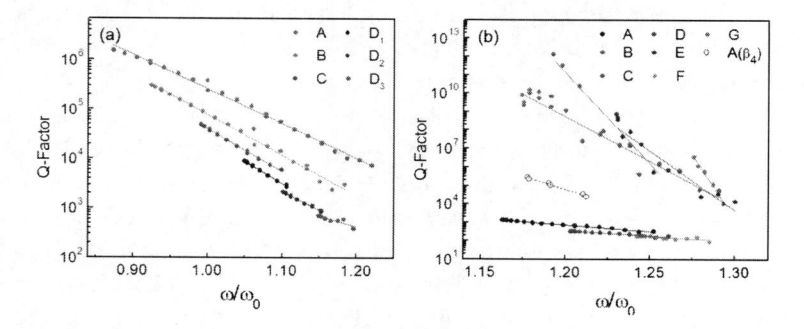

Figure 3.14 Quality factors of the air band edge modes for (a) α_1-spiral and (b) GA spiral and β_4-spiral versus normalized frequency. Reproduced from Ref. [37] with permission from OSA.

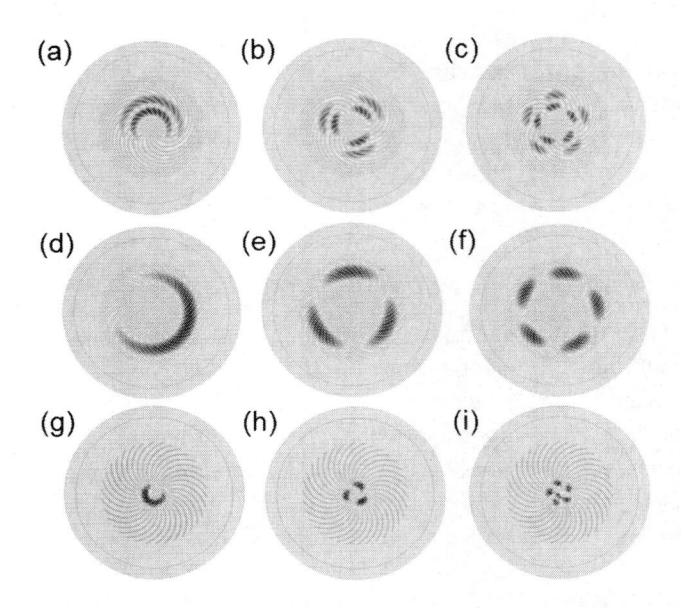

Figure 3.15 Spatial distributions of electric field E_z for the first three band-edge modes of (a–c) class B in an α_1-spiral, (d–f) class A in a g-spiral, and (g–i) class A in a β_4-spiral. Spectrally located at ω/ω_0 (a) 0.9248, (b) 0.9290, (c) 0.9376, (d) 1.1629, (e) 1.1638, (f) 1.1657, (g) 1.1781, (h) 1.1900, and (i) 1.2152. Reproduced from Ref. [37] with permission from OSA.

cover the full perturbation spectrum and are representative of the general behavior of the localized band-edge modes in Vogel spirals. By examining the spatial electric field patterns of the modes across the air band edge of Vogel spirals we discovered that it is possible to group them into several different classes [60].

Moreover, the Q-factors of modes in the same class depend linearly on frequency, as shown in Fig. 3.15. In particular, their quality factors are found to linearly decrease as the modes in each class move further away from the central PBG region. The frequency range spanned by each class of modes depends on the class and the spiral type. As an example, in Fig. 3.15 we show the calculated electric field distributions (E_z component) for the first three band-edge modes in class B of the α_1-spiral and GA spiral, as well as the first three band-edge modes in class A of the β_4-spiral. Each spiral band-edge mode is accompanied by a degenerate mode at the same frequency but with a complementary spatial pattern, rotated approximately by $180°$ (not shown here). We notice in Fig. 3.15 that modes belonging to each class are (radially) confined within rings of different radii and display more azimuthal oscillations as the frequency moves away from the center of the PBG. A detailed analysis of the mechanism of mode confinement and mode splitting into different classes for the GA spiral can be found in Ref. [60]. In particular, we demonstrate that for the GA spiral the field patterns of band-edge modes originate via Bragg scattering occurring perpendicular to the curved parastichies formed by the dielectric pillars. We have also established that a similar behavior occurs for all Vogel spirals with an irrational divergence angle, examined here, each characterized by a unique configuration of parastichies that reflect into characteristic spatial patterns of the modes. As an example, we show in Fig. 3.16 representative localized eigenmodes at the band edge of μ-, τ-, and π-spirals as well.

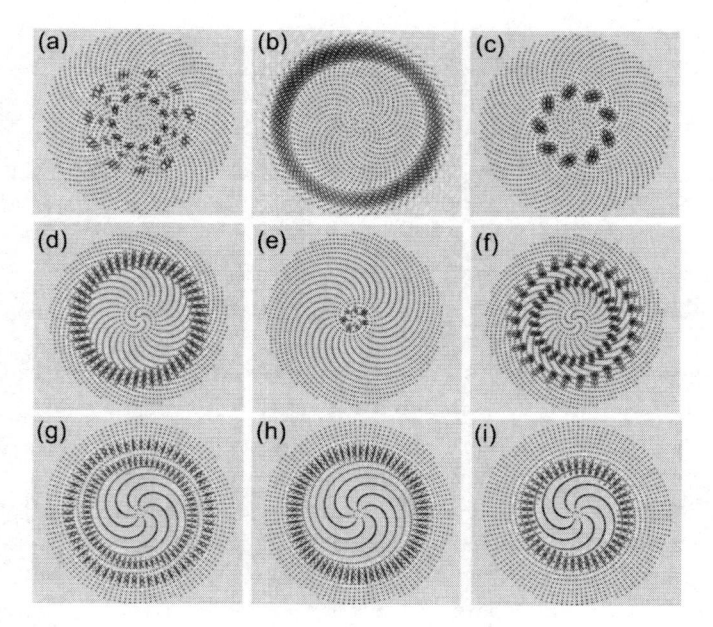

Figure 3.16 Spatial distributions of electric field E_z for the first three band-edge modes of (a–c) in a τ-spiral, (d–f) class A in a τ-spiral, and (g–i) class A in a π-spiral.

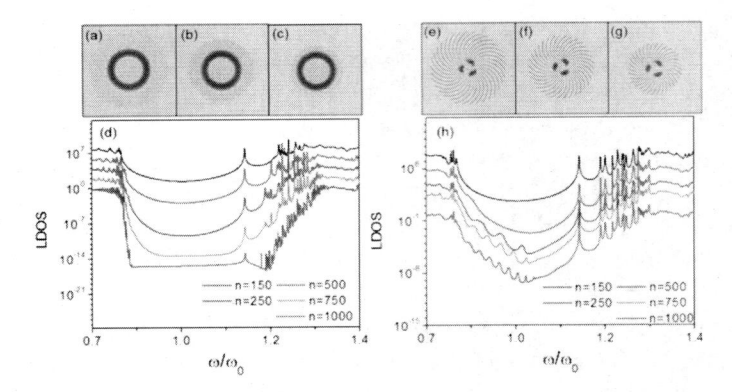

Figure 3.17 (a–c) Spatial distributions of electric field E_z class B band-edge modes in a GA spiral with increasing number of particles. (d) LDOS calculated at the center of GA spirals with increasing particle numbers. (e–g) Spatial distributions of electric field E_z class A band-edge modes in a β_4-spiral with increasing number of particles and (h) calculated LDOS spectra. Adapted from Ref. [37] with permission from OSA.

Finally we discuss the size scaling of the LDOS and of the band-edge modes for the three most representative spiral structures (GA, α_1, and β_4). The LDOS is again computed at the center of the array, utilizing the same methodology described previously. In Fig. 3.17a–h the calculated LDOS for the three spiral types with a progressively decreasing number of cylinders from 1000 to 150 is shown. Also included in Fig. 3.17 are representative air band-edge modes calculated for a spiral with decreasing size (from panels a to c). Examining the size-scaling behavior of the LDOS of the spiral arrays shown in Fig. 3.17a–h certain general characteristics can be noticed. First of all, for each spiral the frequency positions and overall widths of the principal TM gaps remain unaffected when scaling the number of particles, but the gaps become deeper as the number of particles is increased. This is consistent with the known fact that the main gaps supported by arrays of dielectric cylinders are dominated by the single-cylinder Mie resonances for TM polarization. Moreover, the frequency position of the most localized resonant mode inside the gap remains almost constant when varying the particles number, while its intensity decreases with the bandgap depth. This implies that this mode is created by the

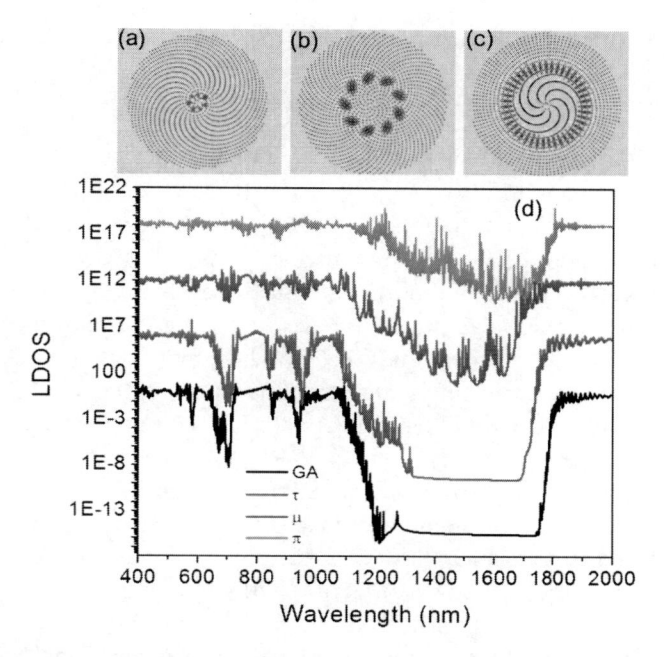

Figure 3.18 (a–c) Spatial distributions of electric field E_z for localized band-edge modes in τ-, μ-, and π-spirals, respectively. (d) Calculated LSOD spectra (1000 particles).

small number of cylinders at the center, which is defined in the first few generated particles. However, the most striking feature of the LDOS scaling, evident in Fig. 3.17, is the generation of a multitude of secondary subgaps of smaller intensity as the number of cylinders is increased. As the number of dielectric cylinders is increased, regions with different spatial distributions of cylinders are created in the spirals, resulting in many more spatial frequency components. As previously shown, these are key components in creating new classes of modes, leading to the distinctive behavior of fractal band-edge modes. This phenomenon has also been recently observed in τ-, μ-, and π-spirals (Fig. 3.18), reflecting a general feature of Vogel arrays with irrational divergence.

The results reviewed in this section demonstrate the localized nature of the air band-edge modes that densely populate the multifractal LDOS spectra of Vogel spirals.

3.6 Device Applications of Vogel Spirals

3.6.1 *Absorption Enhancement in Thin-Film Silicon*

The solar cell market is predominately based on crystalline silicon (c-Si) wafers with absorbing layer thickness in between 100 μm and 300 μm to guarantee complete light absorption and effective carrier collection. However, the increasing material costs related to the fabrication of highly pure c-Si and the ever-growing efficiency demands of the solar industry have motivated novel approaches to increase light absorption in cells with reduced active thickness. In particular, promising approaches consist of the engineering of thin-film noncrystalline (amorphous or polycrystalline Si) and nanocrystalline Si (Si-ncs) structures [91, 92].

Amorphous and nanocrystalline materials can be fabricated with strongly reduced thermal budgets, costs, and much larger volumes compared to traditional Si wafers. However, their shorter diffusion lengths, limited by defects and grain boundaries, restrict the active cell thicknesses to approximately a few hundreds of nanometers, severely decreasing the probability of photon absorption. This has recently spurred the search for advanced photon-recycling and light-trapping schemes capable of increasing the optical paths of photons, and therefore the absorption probability, in ultrathin-film Si solar cells (<200 nm thick) [93–97]. Recent studies have shown that metal nanostructures can lead to effective light trapping into thin-film solar cells, improving the overall efficiency due to the enhancement of optical cross sections associated with the excitation of localized surface plasmon (LSP) modes [98–101].

A commonly utilized geometry consists of the fabrication of metal–dielectric nanoparticles on the front/bottom surface of the absorbing cell structure. When the nanoparticles shapes and filling fraction are correctly designed, incident light is preferentially scattered into the thin-film absorbing Si layer over an increased angular range, effectively enhancing the material absorption into thin films [102–105]. Plasmon-enhanced light absorption in thin-film Si solar cells has been demonstrated using periodic arrays of gold (Au) or silver (Ag) nanoparticles, which give rise to the best enhancement in the spectral regions where evanescent diffraction grating orders spectrally overlap the broader LSP resonances characteristic of metallic nanoparticles. However, polarization sensitivity and the narrow frequency range for effective photonic–plasmonic coupling in periodic grating structures inherently limit these approaches. Recently, quasiperiodic nanohole arrays in metallic thin membranes have also been investigated and have led to significant enhancement of light absorption in thin-film organic solar cell structures [106].

To broaden the spectral region of enhancement, it is crucial to engineer aperiodic nanoparticle arrays with a *higher density of spatial frequencies and a high degree of rotational symmetry* without resorting to uncontrollable random systems. Recent studies have proposed to utilize plasmonic arrays with aperiodic quasicrystal structures, such as Penrose lattices, which exhibit noncrystallo-graphic rotational symmetries [97, 106]. Such arrays give rise to enhanced scattering along multiple directions and over a broader wavelength range, producing coherent wave interactions among an increased number of neighboring particles. Densely packed arrays of nanoparticles/nanoholes with *continuous circular symmetry* would provide an ideal solution for thin-film solar cells with polarization isotropy and efficient planar light diffraction within targeted spectral and angular ranges.

Trevino et al. [36] have recently proposed to utilize plasmonic aperiodic Vogel spiral arrays as a viable strategy to achieve wide-angle light scattering for broadband and polarization-insensitive absorption enhancement in thin-film Si solar cells. This property is ideal to manipulate light trapping in thin-film Si solar cells. Following this approach, our group recently designed and fabricated GA arrays of Au nanoparticles atop an ultrathin-film (i.e., 50 nm thick absorbing amorphous Si layer) silicon-on-insulator (SOI) Schottky photodetector structure, as sketched in Fig. 3.19, and demonstrated experimentally larger photocurrent enhancement in the 600–950 nm spectral range compared to optimized nanoparticle gratings [36]. The relation between the spatial Fourier spectrum of GA arrays of Au nanoparticles and the large angular distribution of scattered radiation in the forward-scattering hemisphere has been rigorously investigated by calculating the angular radiation diagrams using the coupled dipole approximation (CDA) methods for particles with ellipsoidal shape [107].

This CDA approach is particularly suited to efficiently treat large-scale plasmonic systems of small and well-separated nanoparticles, and it has been previously validated against semianalytical multiple scattering methods [108]. In our work, all nanoparticles were modeled by oblate spheroids with a 100 nm diameter and a 30 nm height. Moreover, the arrays are embedded in Si and normally excited by a linearly polarized plane wave. The parameter of interest for the understanding of the angular scattering properties is the differential scattering cross section, which describes the angular distribution of electromagnetic power density scattered at a given wavelength within a unit solid angle centered around an angular direction (θ, ϕ) per unit incident irradiance [47]. In the case of arrays composed of dispersive metal nanoparticles, the power scattered from a particular structure is generally a function of both the geometrical parameters of the array and the wavelength of the incident radiation. Full information on angular scattering is thus captured by calculating the averaged differential scattering cross section, where the average is performed on the azimuthal angle ϕ and the scattered intensity is normalized to the maximum value (i.e., forward-scattering peak). By plotting the azimuthally averaged differential scattering cross section versus the inclination angle, we obtain the radiation diagrams of the arrays. In Fig. 3.19c,d we show (plotted in decibel scale) the calculated radiation diagrams for a periodic grating structure and the optimized GA arrays at three different wavelengths $\lambda_B = 480$ nm, $\lambda_G = 550$ nm, and $\lambda_R = 610$ nm (i.e., corresponding to blue, green, and red colors), respectively. Differently from the well-known case of periodic structures, where the scattered radiation is preferentially redistributed along the directions of coherent Bragg scattering (Fig. 3.19c), the radiation diagram of GA arrays is significantly broadened at large angles (i.e., $>30°$) for all the investigated wavelengths, demonstrating broadband wide-angle scattering behavior (Fig. 3.19d). By combining experimental absorption enhancement and photocurrent measurements with CDA and full-vector 3D finite-difference time domain (FDTD) simulations, we recently demonstrated [36] that broadband wide-angle scattering in GA spiral arrays redirects a larger fraction of the incident radiation into the absorbing Si substrate. Moreover, this

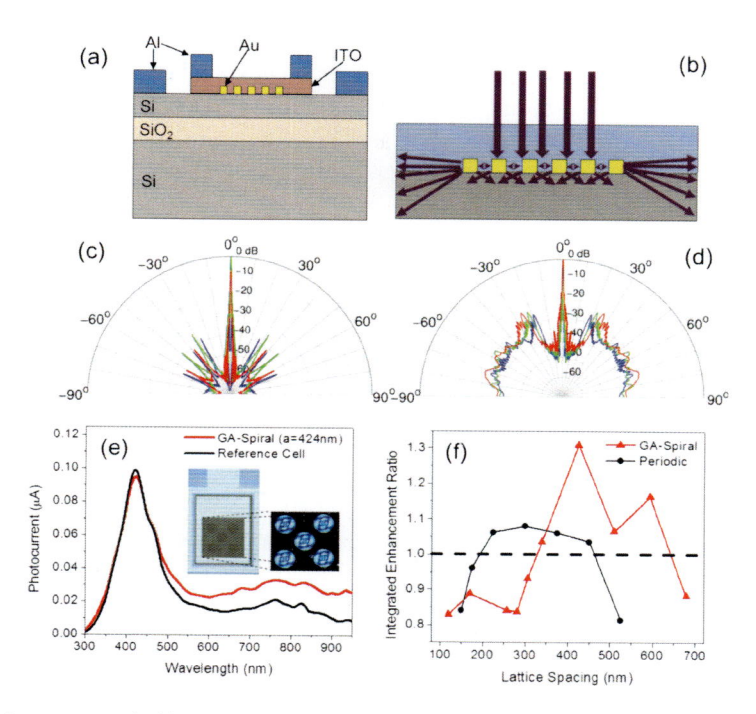

Figure 3.19 (a, b) Schematics of the device cross section of the SOI Schottky photodetector with plasmonic arrays integrated onto the absorbing surface. (c, d) Calculated radiation diagrams as a function of the inclination angle for a GA spiral (c) and a periodic array (d) for three different wavelengths, namely 480 nm (blue), 520 nm (green), and 650 nm (red). (e) Spiral array photocurrent with average center-to-center spacing of 425 nm (red) and empty neighbor reference cells (black). (f) Integrated photocurrent enhancement ratio for GA spiral (red) and periodic (black dashed) arrays of different center-to-center particle spacing. Adapted from Ref. [36] with permission from OSA.

effect increases the optical path of photons in the photodetector, as well as enhancing the coupling to LSPs in the array plane. In Fig. 3.19e we show the measured photocurrent spectra for the best-performing GA spiral photodetector device. The reference line shows the photocurrent spectra measured on the unpatterned devices in the nearest reference cells. A maximum photocurrent enhancement of approximately a factor of 3 is measured for the GA spiral at 950 nm. Such increase in photocurrent results from the interplay of plasmonic–photonic coupling in GA spirals, which contribute to the overall enhancement as follows: by providing more efficient coupling of incident radiation into the thin Si layer due to photonic wide-angle scattering and by enhancing the intensity of the near fields around the nanoparticles at the Si interface owing to better LSP coupling in the plane of the array compared to periodic structures.

A detailed comparison with optimized periodic grating structures is discussed in Ref. [36]. In Fig. 3.19f we show the integrated photocurrent enhancement ratio, calculated by the ratio of the integrated photocurrent spectrum of the device with and without plasmonic arrays. The ratios falling below the dotted line indicate devices with overall reduced performance when compared to their neighboring empty reference cells. We see in Fig. 3.19f that both GA spirals and periodic arrays exhibit an optimization trend with respect to the interparticle spacing, yielding maximum integrated enhancements of 8% and 31% over the reference cells, respectively. We observe finally that the experimentally measured 31% integrated enhancement of GA spiral arrays has been demonstrated covering only 25% of the active photodetector area by GA arrays (Fig. 3.19e, inset). The significant photocurrent enhancement values demonstrated for devices coupled to GA spirals highlight the potential for even greater performances in the case of complete coverage of the device area. Moreover, rapid and scalable room-temperature nanoimprint techniques have been successfully utilized to achieve cost-effective replication of spiral nanoparticle arrays [109].

3.6.2 *Radiation Engineering with Erbium-Doped Vogel Arrays*

In this section we summarize our main results on the engineering of active Vogel spiral nanostructures for enhanced omnidirectional light extraction and the coupling of 1.55 μm radiation into distinctive optical resonances carrying of OAM of light in Si-based materials. In particular, we review the light extraction and emission properties of active arrays by focusing on Er-doped Si-rich nitride (Er:SRN) nanopillar structures, which can be reliably fabricated with a high degree of accuracy using well-established Si-compatible processing. To design active pillar arrays for light extraction, the pillar height and the geometrical parameters of the arrays have been independently designed. FDTD simulations were utilized to optimize the height of the nanopillars, while extraction calculations based on Bragg scattering theory are employed to optimize the pillar spacing for a given array geometry [55]. A schematic of the FDTD-simulated structures is shown in the inset of Fig. 3.20a, consisting of a thin waveguiding layer of SRN ($n = 2.1$) on an SiO_2 substrate ($n = 1.4$) with a periodic grating of SRN pillars placed atop in order to extract the guided radiation.

A mode source is injected into the SRN waveguiding layer and the radiation extracted from this mode, both into air and into the substrate, is quantified by calculating the optical power transmitted through two power monitors positioned 10 μm above and below the device. The extracted power at 1.55 μm is plotted in Fig. 3.20a as a function of the pillar height. The results show a maximum in the power diffracted into air for a pillar height of 350 nm, exceeding the amount of power diffracted into the substrate by a factor of 2. However, the large optical size (i.e., geometrical size with respect to the wavelength) of the investigated aperiodic arrays prevents direct numerical simulations. Therefore, we have identified the optimal pillar spacing for any given array geometry by resorting to analytical Bragg scattering theory [110], which is a powerful and efficient method utilized for the design of LED extraction when strong photonic coupling effects can be neglected (i.e., weak coupling regime or single scattering). Calculation of extracted light using the Bragg scattering method enables the exploration of a large parameter space at relatively low computational cost and provides an opportunity to optimize the lattice spacing and filling fraction of different aperiodic arrays.

In the case of weak scattering, the extraction of light is described by Bragg's law of diffraction: $\beta_d = \beta_i + G$, where β_i is the input waveguide **k**-vector, β_d is the extracted (i.e., final) **k**-vector, and **G** is a reciprocal space vector obtained by computing the Fourier transform of the corresponding array. On the basis of Bragg's law of diffraction, we notice that if the input waveguide **k**-vector with amplitude $|\beta_i > |k_0|$ ($k_0 = 2\pi/\lambda$ is the free-space wavevector at

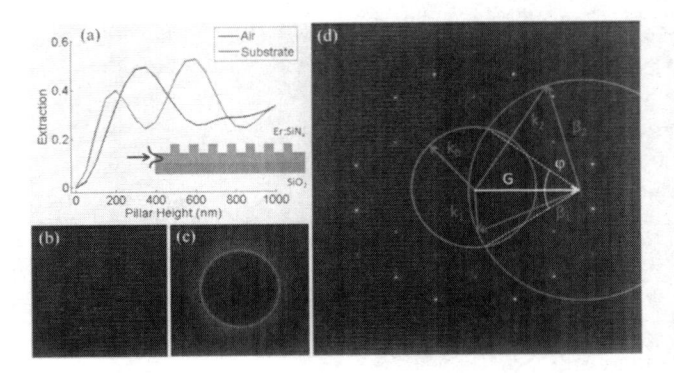

Figure 3.20 (a) Extraction of a guided mode by SRN pillars to air and the substrate, with an inset of the device showing a mode injected into the SRN layer. Fourier transforms of an (b) A7 Archimedean lattice and a (c) GA spiral. (d) Representation of vector arrangement relevant to the Bragg scattering calculations. Reprinted with permission from Ref. [55]. Copyright 2012, AIP Publishing LLC.

wavelength λ) interacts with a suitable reciprocal space vector **G**, giving rise to a final **k**-vector β_d with amplitude $|\beta_d| < |k_0|$, then the input waveguide mode is folded back into the light cone and can be extracted from the device. Therefore, knowing the input waveguide **k**-vector β_i and the reciprocal vectors $\bar{G}$ of the structure, we can design the overall extraction efficiency of a given array exclusively on the basis of geometrical arguments. Figure 3.20d illustrates an example of such Bragg scattering calculations, where the points along the green circle correspond to possible **k**-vectors, resulting from the interaction $\beta_i + G$. As an example, in Fig. 3.20d we show two vectors, named $\bar{k}_1 = \bar{\beta}_1 + \bar{G}$ and $\bar{k}_2 = \bar{\beta}_2 + \bar{G}$, of which $\bar{k}_1$ is extracted to the light cone, while $\bar{k}_2$ is not. On the basis of this geometrical construction, we can define the extraction efficiency for the waveguide mode β_1 as

$$\kappa(\beta_i, G) \propto \begin{cases} |\epsilon_G|^2 & \text{if } |\beta_i - G| < k_0 \\ 0 & \text{else} \end{cases}, \tag{3.11}$$

where ϵ_G is the complex amplitude of the array Fourier transform at $\bar{G}$. The arc length $\ell = \phi\beta_i$, shown in Fig. 3.20d in white, determines the in-plane directions for β_i that can be extracted by a given $\bar{G}$. Maximum efficiency is achieved for $|\bar{G}| = |\beta_i|$, which determines the optimal vector amplitude for extracting a randomly oriented input. The total extraction efficiency κ_{tot} of a given array geometry is therefore calculated by the integration over all the reciprocal space vectors compatible with the arc length ℓ [55]:

$$\kappa_{\text{tot}}(\beta_i) \propto \beta_i \sum_G \phi(\beta_i, G)|\epsilon_G|^2 \qquad (3.12)$$

This analysis makes us clearly appreciate that *the optimal Fourier space for light extraction consists of a circle of radius β_i*. This condition guarantees the highest Bragg extraction efficiency for a randomly oriented set of input wavevectors, and it is naturally met by engineering isotropic Fourier space. A detailed comparison of Bragg light extraction for a number of deterministic aperiodic arrays with varying rotational symmetry in Fourier space can be found in Ref. [55]. These results demonstrate clearly the importance of the arrays' aperiodic isotropic geometry for light extraction. However, we also note that for a given array geometry, the effectiveness of Fourier components in determining the overall extraction depends on the refractive index contrast (i.e., choice of materials). In particular, light extraction with isotropic Fourier space is expected to be even more efficient in the low-refractive-index regime, where a longer interaction length enables better sampling of the extended aperiodic geometry.

To demonstrate the concept of *omnidirectional light extraction* in Si-based devices, aperiodic pillar arrays were fabricated from Er:SRN using radio frequency (RF) magnetron sputtering, electron beam lithography (EBL), and reactive ion etching (RIE) [55]. Scanning electron micrographs (SEMs) of fabricated pinwheel and GA spiral devices are shown in Fig. 3.21a–d. All the fabricated arrays have been truncated using circular masks of identical size in order to ensure that pumping conditions are identical during the photoluminescence (PL) optical measurements.

The optical characterization of the fabricated structures is performed by leaky-mode PL spectroscopy to quantify the extraction enhancement of nanopillar arrays with increasing degrees of rotational symmetry in Fourier space [55]. The emission spectra of Er corresponding to the optimized array structures are displayed in Fig. 3.21e, together with the emission of the reference unpatterned device. All pillar arrays show a large increase in the emission signal due to enhanced extraction. Figure 3.21f demonstrates a significant increase in extraction efficiency for all the arrays with a maximum

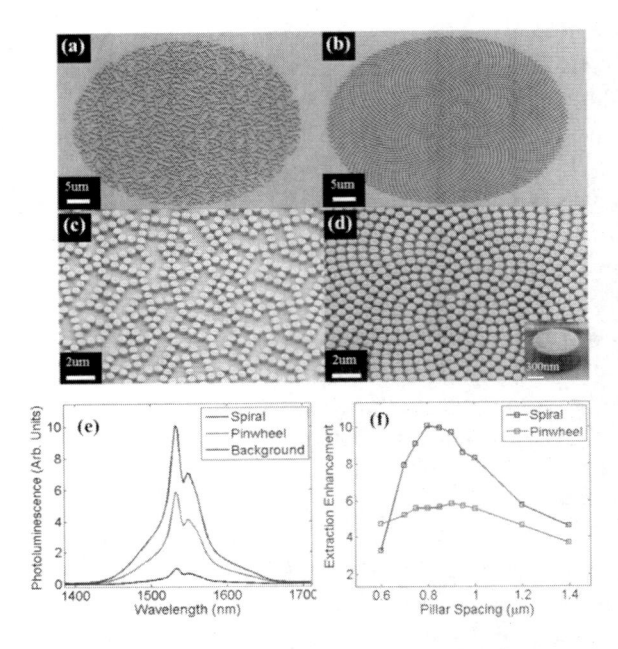

Figure 3.21 (a) Pinwheel nanopillar array and (c) GA spiral nanopillar array fabricated in Er:SRN. (c, d) Close-ups of each array to highlight the geometry. An inset of a single pillar is shown in the lower-right corner of (d) with a height of 350 nm. The arrays are 50 µm in diameter. (f) Measured Er emission extraction coefficient for GA spiral and pinwheel active pillar arrays of varying nanopillar spacing. (e) Measured Er emission spectra from optimized pillar arrays and unpatterned (background) sample. Reprinted with permission from Ref. [55]. Copyright 2012, AIP Publishing LLC.

extraction of approximately a factor of 10 achieved for the optimal GA isotropic geometry.

The transmission spectra of the arrays were also measured at the laser pump wavelength and showed only a very minor variation in pumping efficiency among the different structures. Therefore, the trends in Fig. 3.21 result directly from the optimization of the array geometry for extraction efficiency, in very good agreement with the theoretical trends based on the Bragg scattering analysis. In addition, by imaging Er radiation in direct and reciprocal space, in Ref. [55] we show that active GA spirals support *angularly isotropic emission patterns* with distinctive optical resonances carrying OAM of light. These findings offer unique opportunities for the engineering of novel active nanostructures that leverage isotropic emission patterns and structured light for secure optical communication, sensing, imaging, and enhanced light sources on a Si platform [61].

3.6.3 *Engineering the Orbital Angular Momentum of Light*

It has long been established that electromagnetic fields carry linear momentum and angular spin. However, the possibility for optical fields to additionally carry OAM has only recently been realized [111, 112].

In general, OAM arises through the azimuthal phase dependence of the complex optical field. Optical OAM has recently found uses in rotating optical traps [113], secure optical communication [114], and increasing data transfer rates through OAM multiplexing for fiber-based systems [115]. Moreover, recent advances in the science and technology of OAM have provided the possibility to detect light waves carrying simultaneously multiple OAM values [116]. However, the controlled generation of optical waves that can simultaneously carry large values of OAM still remains very challenging.

Currently, the generation of OAM is achieved by converting Gaussian laser modes to Laguerre–Gaussian (LG) modes with explicit azimuthal phase dependence. This can be accomplished using a system of cylindrical lenses [117] or spatial light modulators (SLMs) [118]. However, SLMs are expensive and their pixel size limits the complexity of patterns that can be created. More recently, the generation of OAM from planar plasmonic devices has also been demonstrated [34, 119]. Unfortunately, though, most of the current generation methods are limited to creating OAM states with only a few azimuthal values.

Our group recently discovered that Vogel spiral arrays of metallic nanoparticles, when illuminated by optical beams, give rise to diffracted light carrying OAM patterns with large and tunable azimuthal numbers [34, 59, 60]. Dal Negro et al. [59] developed an analytical model that captured in closed-form solution the Fourier–Hankel spectral properties of arbitrary Vogel spiral arrays and applied it to the engineering of complex OAM states in the far-field diffraction region. Within the framework of scalar Fourier optics, the Fourier spectrum (i.e., Fraunhofer diffraction pattern) of Vogel spirals is described by the complex sum [59]

$$E_\infty(v_r, v_\theta) = E_0 \sum_{n=1}^{N} e^{j2\pi \sqrt{n} a_0 v_r \cos(v_\theta - n\alpha)}, \tag{3.13}$$

where the variables (v_r, v_θ) are Fourier conjugates of the direct-space cylindrical coordinates (r, θ) used to represent the Vogel spiral array, α is the irrational divergence angle, a_0 is a constant scaling factor, and N is the number of particles in the array [59].

Fourier–Hankel modal decomposition can be used to analyze a superposition of OAM states in the far field and to determine their relative contributions to the overall diffracted beam. Decomposition of $\rho(r, \theta)$ into a basis [60, 120] set with helical phase fronts is accomplished through Fourier–Hankel decomposition (FHD) according to

$$f(m, k_r) = \frac{1}{2\pi} \int_0^\infty \int_0^{2\pi} r \, dr \, d\theta \rho(r, \theta) J_m(k_r r) e^{im\theta}, \tag{3.14}$$

where J_m is the mth order Bessel function. In this decomposition, the mth order function identifies OAM states with azimuthal number m, accommodating both positive and negative integer m values.

By analytically performing FHD analysis, Dal Negro et al. [59] demonstrated that diffracted optical beams by Vogel spirals carry OAM values distributed according to aperiodic numerical sequences determined by an irrational divergence angle α. More precisely, OAM values transmitted in the far-field region are given by the rational approximations of the continued fraction expansion of the irrational divergence angles of Vogel spirals [59]. In particular, wave diffraction by GA arrays generates a Fibonacci sequence of OAM values in the Fraunhofer far-field region. This fascinating property of Vogel spirals can be understood quantitatively by considering the analytical solution of the FHD of the far-field radiation pattern, given by [59]

$$f(m, k_r) = \sum_{n=1}^{N} A(k_r) e^{imn\alpha}, \tag{3.15}$$

where $A(k_r)$ is a k_r-dependent coefficient, which can be ignored since we are concerned with the azimuthal dependence contained in $f(m)$.

We see from the result in Eq. 3.15 that when the product $m\alpha$ is an integer, the N contributing waves will be exactly in phase to produce an OAM peak with azimuthal number m. However, for an irrational angle α, this condition will never be exactly met. Nevertheless, using the theory of continued expansion in rational fractions we can design structures that approximately match the required integer condition for the $m\alpha$ product, *thus producing well-defined OAM peaks by design*. An arbitrary divergence angle α is uniquely associated to an irrational number ξ that admits precisely one infinite continued fraction representation (and vice versa) in the form [32, 33]

$$\xi = [a_0; a_1, a_2, a_3, \ldots] = a_0 + \cfrac{1}{a_1 + \cfrac{1}{a_2 + \cfrac{1}{a_3 + \cfrac{1}{a_4 + \cdots}}}} \tag{3.16}$$

The rational approximations (i.e., fractions) are called the *convergents* of the continued fraction. Once the continued fraction expansion of ξ has been obtained, well-defined recursion rules exist to quickly generate the successive convergents. In fact, each convergent can be expressed explicitly in terms of the continued fraction as the ratio of certain multivariate polynomials called continuants. If two convergents are found, with numerators p_1, p_2, $\ldots$ and denominators q_1, q_2, $\ldots$, then the successive convergents are given by the formula

$$\frac{p_n}{q_n} = \frac{a_n p_{n-1} + p_{n-2}}{a_n q_{n-1} + q_{n-2}}, \tag{3.17}$$

Thus to generate new terms into a rational approximation only the two previous convergents are necessary. The initial or seed values required for the evaluation of the first two terms are $(0,1)$ and $(1,0)$ for (p_{-2}, p_{-1}) and (q_{-2}, q_{-1}), respectively. It is clear from the discussion above that for spirals generated by an arbitrary irrational number ξ, azimuthal peaks of order m (i.e., Bessel order m) will appear in its FHD due to the denominators q_n of the rational approximations (i.e., the convergents) of $\xi \approx p_n/q_n$. In fact, for all integer Bessel orders $m = q_n$ the exponential sum in Eq. 3.15 will give in-phase contributions to the FHD and produce strong azimuthal peaks. Therefore, Vogel spirals "encode" in their OAM spectra well-defined numerical sequences of azimuthal orders determined by the denominators q_n in Eq. 3.15. It is important to

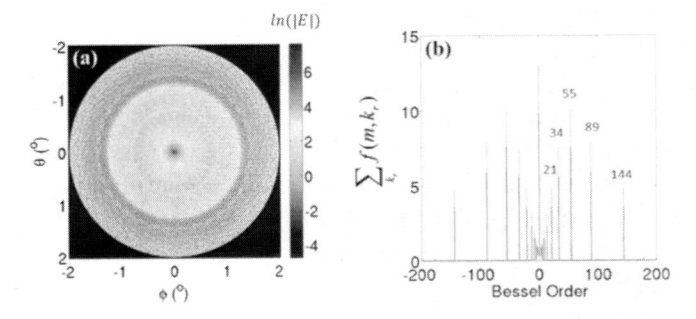

Figure 3.22 (a) Analytically calculated far-field radiation pattern of GA spiral with 2000 particles at a wavelength of 633 nm for a structure with $a_0 = 14.5$ μm. The far-field radiation pattern has been truncated with an angular aperture of 4°. (b) Fourier–Hankel transform of far-field scattered radiation by GA spiral summed over the radial wavenumbers k_r. The numbers in the figures indicate the azimuthal Bessel order of the corresponding FHD peaks. Adapted from Ref. [59] with permission from OSA.

realize that using the approach just introduced it becomes possible to generate light with controllable OAM spectra by plasmon-enhanced diffraction in aperiodic arrays of nanoparticles. Since we are primarily concerned with the azimuthal component $f(m)$ of OAM, we can sum $f(m, k_r)$ over the radial wavenumbers k_r. Figure 3.22 demonstrates the very rich structure of OAM peaks of the scattered radiation enabled by the aperiodic geometry control of Vogel spirals. These peaks exactly occur at azimuthal numbers (labeled in the figures) corresponding to the different denominators of the rational approximations of the irrational divergence angles used to generate the spirals, as previously discussed. In the particular case of the GA spiral arrays shown in Fig. 3.22b we obtain a Fibonacci sequence of OAM values that is coded in the far-field region of the radiation.

Using phase-delayed interferometric measurements, Lawrence et al. [56a] have recently measured experimentally the complex electric field distribution of scattered radiation by Vogel spirals and demonstrated for the first time structured light carrying multiple values of OAM in the far-field region of arrays of metallic nanoparticles, in excellent agreement with the analytical theory

[59]. In addition, by engineering light-emitting GA nanopillar arrays using Er-doped SiN [55], we have recently demonstrated direct excitation of optical modes carrying Fibonacci-valued OAM sequences.

3.6.4 *Diffracted Beam Propagation from Aperiodic Chiral Spirals*

In this section we will review our recent work addressing the interplay between Vogel spiral order and well-defined CCW chirality in α_1 dielectric spirals [121]. Light scattering and diffraction in these complex optical systems gives rise to fascinating self-imaging properties that can be of interest to a number of applications in singular optics. In particular, we will discuss the evolution of a scattered plane wave as it propagates away from the plane of a diffracting array of apertures with α_1 chiral geometry toward the far-field region. This analysis will also provide physical insights into the formation of the structured far fields with multiple OAM states, which are a distinctive feature of diffracting arrays.

To efficiently investigate the paraxial free propagation of the scattered field intensity over a larger range of distances from the spiral (i.e., object) plane, we have resorted to the method of fractional Fourier transformation (FRFT). This approach provides an equivalent formulation of the paraxial wave propagation and Fresnel scalar diffraction theory [122] and considers light propagation as a process of continual fractional transformation of increasing order. The FRFT is a well-known generalization in fractional calculus of the familiar Fourier transform operation [123], and it has successfully been applied to the study of quadratic phase systems, imaging systems, and diffraction problems in general [122, 124, 125].

Given a function $f(u)$, under the same conditions in which the standard Fourier transform exists, we can define the ath-order FRFT $f_a(u)$ with a being a *real number* in several equivalent ways. The most direct definition of the FRFT is given in terms of the linear integral transform

$$f_a(u) = \int_{\infty}^{\infty} K_a(u, u') f(u') du', \qquad (3.18)$$

with a kernel defined as in Refs. [124, 128]

$$K_a(u, u') \equiv A_\alpha exp[i\pi(cot\alpha u^2 - 2csc\alpha uu' + cot\alpha u'^2)]$$
$$A_\alpha \equiv \sqrt{(1 - icot\alpha)} \tag{3.19}$$
$$\alpha \equiv \frac{a\pi}{2}$$

for $a \neq 2n$ and $K_a(u, u') = \delta(u - u')$ when $a = 4n$ and $K_a(u, u') = \delta(u + u')$ when $a = 4n \pm 2$, with n being an integer. The ath-order fractional transform defined above is sometimes called the αth-order transform and it coincides with the standard (i.e., integer) Fourier transform for $a = 1$ or $\alpha = \pi/2$. More information on alternative definitions, generalizations, and the many fascinating properties of FRFTs can be found in Refs. [122, 126]. The FRFT of a function can be roughly thought of as the Fourier transform to the nth power, where n need not be an integer. Moreover, being a particular type of *linear canonical transformation*, the FRFT maps a function to any intermediate domain between time and frequency and can be interpreted as a *rotation in the time–frequency domain*.

The main advantage of the FRFT method compared to numerical simulation relies on its superior computational efficiency, which enables more extended investigations of the qualitative behavior of the intensity propagation over a large range. In what follows, we computed the 2D FRFT according to Refs. [127, 128] and studied the propagation of the diffracted field up to 50 μm above the spiral plane.

In Fig. 3.23 we show a few representative simulations that display the calculated diffracted intensity at different distances from the spiral plane, as specified in the caption. We can clearly appreciate in Fig. 3.23f,g situations in which the diffracted field patterns are clearly rotated with respect to the geometry of the parastichies arms. The emergence of this characteristic rotation phenomenon, which we also reported using full-vector FDTD calculations [121], is very robust, since the FRFT field simulations do not consider material dispersion properties or the detailed particle shapes. In fact, this intriguing effect originates from the coherent interactions of (singly) diffracted wavelets. In particular, we can appreciate from the field evolution shown in Fig. 3.23 that the central area of the spiral couples radiation into directions that are orthogonal to the

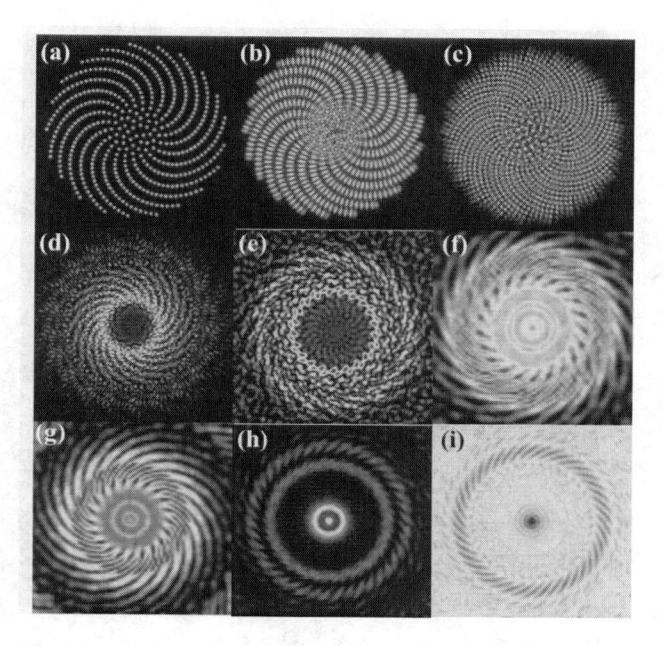

Figure 3.23 Magnitude of electric field propagated to different planes using fractional Fourier transform. Propagation distances for panels (a–h) are 10 nm, 78 nm, 107 nm, 0.457 μm, 0.836 μm, 1.95 μm, 2.68 μm, and 6.5 μm, respectively. Panel (i) shows the analytical far field in log_{10} scale. The structure is excited by a normally incident plane wave at 1550 nm.

surrounding parastichies' arms. These secondary lines of scattered radiation spatially define a complementary set of parastichies' arms that are responsible for the inversion of the intensity pattern at short distances from the object plane. As the intensity propagation unfolds, the diffracted wavelets coherently reinforce each other along, distinctively rotating ring-like structures observed in Fig. 3.23d,e, which gradually transition at larger distances into the circularly symmetric far-field patterns of Vogel spirals (Fig. 3.23i) [34, 37]. This distinct rotation of the diffracted intensity pattern observed in chiral spiral arrays provides clear evidence of net OAM transfer within a few micrometers from the object, which could be exploited to realize optical torques and novel tweezers based on planar plasmonic nanostructures. The near-field coupling and propagation behavior of electromagnetic energy scattered at

1.56 μm by arrays of silicon nitride nanopillars with Vogel spiral geometry has been also investigated experimentally in Ref. [121]. By using scanning near-field optical microscopy in partnership with full-vector FDTD numerical simulations, Intonti et al. [121] demonstrate a characteristic rotation of the scattered field pattern by a Vogel spiral consistent with the predicted net transfer of OAM in the Fresnel zone, within a few micrometers from the plane of the array.

We believe that the unique interplay between aperiodic order and chiral structures such as the investigated α_1-spiral can provide novel opportunities for the manipulation of subwavelength optical fields and disclose richer scenarios for the engineering of focusing and self-imaging phenomena in nanophotonics.

3.6.5 *Experimental Demonstration of OAM Generation and Control*

In this final section we will review our experimental demonstration of broad OAM azimuthal spectra in the radiation scattered by plasmonic Vogel spirals. The controlled generation and manipulation of OAM states with large values of azimuthal numbers have very recently been demonstrated experimentally by Lawrence et al. [56a] using various types of Vogel arrays of metal nanoparticles. The OAM content of a diffracted optical wave at 633 nm was analyzed using phase-stepped interferometric measurements to recover the complex optical field [56a]. Fourier–Hankel modal decomposition of scattered radiation was performed to demonstrate the generation of OAM sequences from Vogel spirals. Nanoparticle arrays with GA, τ, μ, and π aperiodic geometries were fabricated using EBL on quartz substrates. The fabrication process flow is detailed in Ref. [56a]. The SEMs of fabricated arrays are shown in Fig. 3.24a–c. In Fig. 3.25 we compare the measured and calculated far-field diffraction patterns for the fabricated structures. A HeNe laser was utilized in these experiments and weakly focused to a spot size of 50 μm from the rear of the sample and the scattered light from the array was collected by a 50X (NA = 0.75) objective. Two additional lenses were added to image the far field at the plane of the charge-coupled device (CCD) [56a]. In Fig. 3.25a,b we show the calculated Fraunhofer far

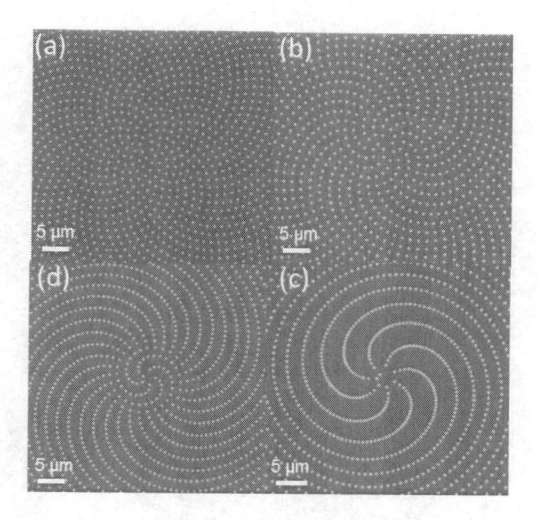

Figure 3.24 SEM micrographs of (a) GA spiral, (b) τ-spiral, (c) π-spiral, and (d) μ-spiral gold nanoparticle arrays on a fused silica substrate.

field and the corresponding azimuthal OAM spectrum of the GA spiral array.

To measure the phase of the far-field scattered radiation, a reference beam was reflected from the piezo-stage-mounted mirrors, expanded, and directed to the CCD. A schematics of the experimental setup is shown in Fig. 3.26a. Multiple interference patterns were collected by increasing the piezo bias voltage, scanning the phase of the reference beam. A phase retrieval algorithm was finally used to recover the phase of the scattered light relative to the one of the reference beam [56a].

As representative examples, we show in Fig. 3.26 the comparison between analytically calculated and experimentally measured phase portraits for the far-field scattered radiation by GA and μ-spirals [56a]. The shaded areas in Fig. 3.26 indicate the limits (angular range) where the measurement of the scattered light is inaccurate due to the limitations of the experimental setup [56a]. Very similar results were obtained for the τ- and π-spirals (not shown here). The insets in the reconstructed phase portraits in Fig. 3.26 highlight the complex distribution of optical vortices at positions that well match the analytical calculations.

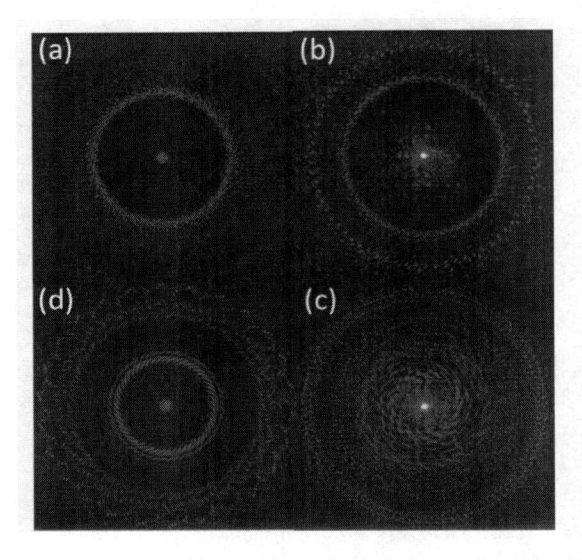

Figure 3.25 Measured far-field diffraction patterns for the (a) GA spiral, (b) τ-spiral, (c) π-spiral, and (d) μ-spiral arrays corresponding to the samples shown in Fig. 3.24.

Modal decomposition was used to analyze a superposition of OAM-carrying modes in the far-field pattern and determine their relative contribution [120]. Decomposition into a basis set with azimuthal dependence is accomplished through FHD according to the Eq. 3.14. We recall that in this decomposition the mth-order function carries OAM with azimuthal number m, accommodating both positive and negative integer values for m. In Figs. 3.27 and 3.28 we summed the FHD of the calculated and experimentally measured far-field radiation from the fabricated spirals. The sum was performed over the radial wavenumbers k_r since we are primarily concerned with the azimuthal component of the field. We should notice in Figs. 3.27 and 3.28 that the azimuthal spectra exhibits a number of peaks at numbers corresponding to the denominators of the rational approximations (i.e., convergents) of the irrational divergence angles used to generate the spirals [59].

For the GA spiral the encoding of the Fibonacci sequence up to azimuthal number 144 was experimentally demonstrated in the measured OAM spectrum. The τ-spiral has a similar number of

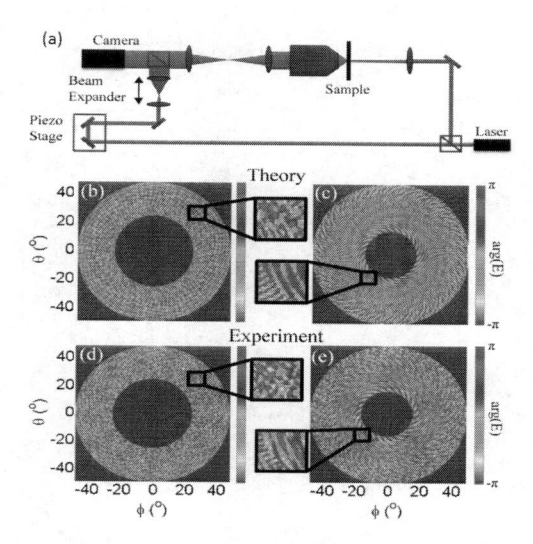

Figure 3.26 (a) Optical setup used for the complex amplitude retrieval. (b, c) Analytically calculated phase portrait of far-field radiation from GA and μ-spirals, respectively. (d, e) Experimentally measured phase portrait from GA and μ-spirals, respectively. Shaded areas indicate regions where the field intensity is too low to measure the phase accurately. These regions are ignored in the FHD. Adapted from Ref. [56a] with permission from OSA.

convergents in our region of interest and allowed us to demonstrate OAM peaks up to azimuthal number 169. However, these peaks are significantly broadened by the aperturing of the beam [59]. On the other hand, we can see in Fig. 3.28 that the μ- and π-spirals have significantly fewer convergents in our region of interest. However, OAM peaks are still clearly observed at linear combinations of the convergents despite the significant broadening due to the imperfect experimental alignment.

The measured values of OAM peaks can be fully explained on the basis of the analytical far-field diffraction theory [59]. In particular, peaks in the measured OAM spectra of arbitrary Vogel spirals are always observed at numbers corresponding to rational approximations of the corresponding divergence angles. A full list of analytically calculated and experimentally measured FHD peak positions can be found in Ref. [56a].

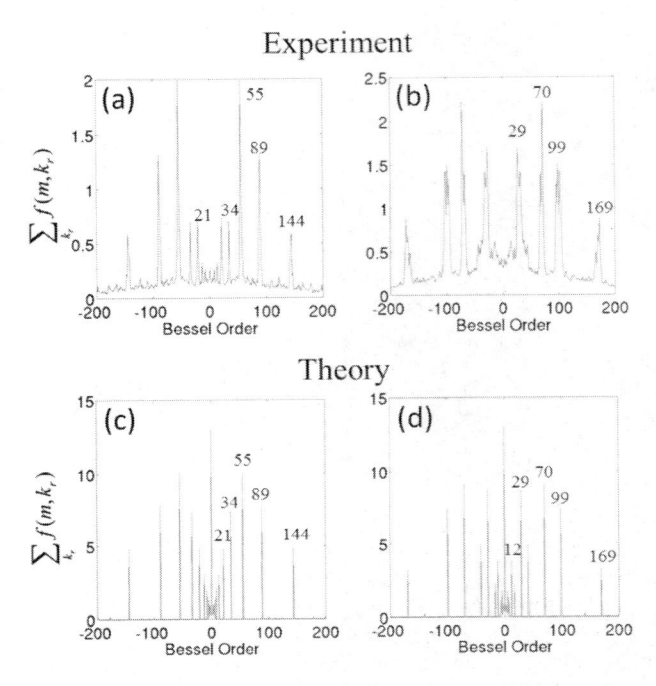

Figure 3.27 Experimentally measured Fourier–Hankel transforms of far-field scattered radiation by a (a) GA spiral and a (b) τ-spiral, respectively, summed over the radial wavenumbers k_r. The numbers in the figures indicate the azimuthal Bessel order of the corresponding FHD peaks. Analytically calculated Fourier–Hankel transforms of far-field scattered radiation by a (c) GA spiral and a (d) τ-spiral.

These results demonstrate the ability to design and generate structured light with extremely complex orbital patterns simultaneously carrying a large number of azimuthal values in the far-field zone. On the basis of the flexible platform of Vogel spiral nanophotonic structures it is possible to engineer novel light sources capable of radiating complex optical beams with programmed OAM spectra or coupling them into conventional communication systems. This unique ability promises to greatly expand the spatial multiplexing capabilities currently available in integrated optical communication systems by realizing information coding with higher-dimensional azimuthal Hilbert spaces. This

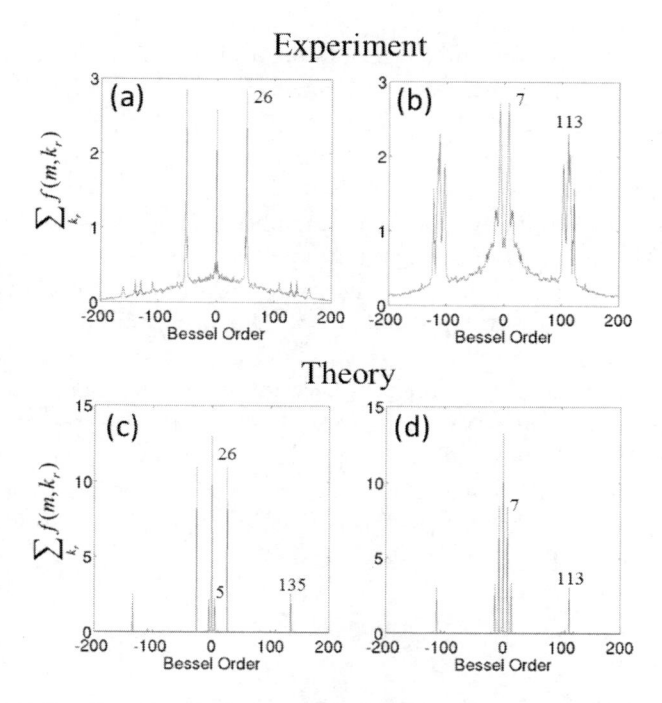

Figure 3.28 Experimentally measured Fourier–Hankel transforms of far-field scattered radiation by a (a) μ-spiral and a (b) π-spiral, respectively, summed over the radial wavenumbers k_r. The numbers in the figures indicate the azimuthal Bessel order of the corresponding FHD peaks. Analytically calculated Fourier–Hankel transforms of far-field scattered radiation by a (c) μ-spiral and a (d) π-spiral.

functionality could drastically expand the present capabilities of both classical and quantum information systems [61].

3.7 Outlook and Conclusions

In this chapter, we presented a comprehensive overview of our recent work on the engineering of optical nanostructures with Vogel spiral geometry. Specifically, we discussed the relevant background, structure–property relationships, and the light-scattering properties of arrays with different types of Vogel spiral order. These structures define a novel platform for the enhancement of light emission/absorption and the manipulation of OAM of light. Our discussion has particularly emphasized the importance of photonic diffraction phenomena in isotropic Fourier space, which could provide novel opportunities to manipulate light–matter interactions on the nanoscale.

The computational and experimental results presented in this chapter demonstrate the significance of aperiodic structures with spiral order and their potential for innovation in basic science and technological applications. In particular, the ability to create novel complex optical beams with designed OAM spectra using resonant plasmonic nanoparticle arrays suitable for planar device integration promises to impact several fields in optical trapping and sensing technologies. Moreover, light scattering by aperiodic Vogel spirals makes possible the information encoding in a larger Hilbert space azimuthal bases, which is currently of great interest for both classical and quantum cryptography.

Acknowledgments

This work was partly supported by the AFOSR programs under award numbers FA9550-10-1-0019 and FA9550-13-1-0011 and by the NSF Career Award No. ECCS-0846651.

References

1. Joannopoulos, J.D., Johnson, S.G., Winn, J.N., and Meade, R.D. (2008). *Photonic Crystals: Molding the Flow of Light*, Princeton University Press.

2. Sheng, P. (2006). *Introduction to Wave Scattering, Localization and Mesoscopic Phenomena*, Springer-Verlag.

3. Wolf, P.-E., and Maret, G. (1985). Weak localization and coherent backscattering of photons in disordered media, *Phys. Rev. Lett.*, **55**, 24, p. 2696.

4. Albada, M.P.V., and Lagendijk, A. (1985). Observation of weak localization of light in a random medium, *Phys. Rev. Lett.*, **55**, 24, p. 2692.

5. Wiersma, D.S., van Albada, M.P., and Lagendijk, A. (1995). Coherent backscattering of light from amplifying random media, *Phys. Rev. Lett.*, **75**, 9, p. 1739.

6. Lagendijk, A., van Tiggelen, B., and Wiersma, D.S. (2009). Fifty years of Anderson localization, *Phys. Today*, **62**, 8, pp. 24–29.

7. Anderson, P.W. (1985). The question of classical localization: A theory of white paint?, *Phil. Mag. Part B*, **52**, 3, pp. 505–509.

8. van Tiggelen, B.A. (1995). Transverse diffusion of light in Faraday-active media, *Phys. Rev. Lett.*, **75**, 3, p. 422.

9. Sparenberg, A., Rikken, G.L.J.A., and van Tiggelen, B.A. (1997). Observation of photonic magnetoresistance, *Phys. Rev. Lett.*, **79**, 4, p. 757.

10. Scheffold, F., and Maret, G. (1998). Universal conductance fluctuations of light, *Phys. Rev. Lett.*, **81**, 26, p. 5800.

11. Wiersma, D.S., Colocci, M., Righini, R., and Aliev, F. (2001). Temperature-controlled light diffusion in random media, *Phys. Rev. B*, **64**, 14, p. 144208.

12. Maciá, E. (2006). The role of aperiodic order in science and technology, *Rep. Progress Phys.*, **69**, 2, pp. 397–441.

13. Maciá, E. (2009). *Aperiodic Structures in Condensed Matter: Fundamentals and Applications*, CRC Press Taylor & Francis.

14. Steurer, W., and Sutter-Widmer, D. (2007). Photonic and phononic quasicrystals, *J. Phys. D: Appl. Phys.*, **40** pp. R229–R247.

15. Poddubny, A.N., and Ivchenko, E.L. (2010). Photonic quasicrystalline and aperiodic structures, *Physica E*, **42**, 7, pp. 1871–1895.

16. Dal Negro, L., and Boriskina, S.V. (2012). Deterministic aperiodic nanostructures for photonics and plasmonics applications, *Laser Photonics Rev.*, **6** pp. 178–218.

17. Kohmoto, M., Kadanoff, L.P., and Tang, C. (1983). Localization problem in one dimension: mapping and escape, *Phys. Rev. Lett.*, **50**, 23, p. 1870.

18. Kohmoto, M., Sutherland, B., and Iguchi, K. (1987). Localization of optics: quasiperiodic media, *Phys. Rev. Lett.*, **58**, 23, p. 2436.

19. Vasconcelos, M.S., and Albuquerque, E.L. (1999). Transmission fingerprints in quasiperiodic dielectric multilayers, *Phys. Rev. B*, **59**, 17, p. 11128.

20. Dulea, M., M. Severin, and Riklund, R. (1990). Transmission of light through deterministic aperiodic non-Fibonaccian multilayers, *Phys. Rev. B*, **42**, 6, p. 3680.

21. Gellermann, W., Kohomot, M., Sutherland, B., and Taylor, P.C. (1994). Localization of light waves in Fibonacci dielectric multilayers, *Phys. Rev. Lett.*, **72**, 5, pp. 633–636.

22. Merlin, R., Bajema, K., Clarke, R., Juang, F.-Y., and Bhattacharya, P.K. (1985). Quasiperiodic GaAs-AlAs heterostructures, *Phys. Rev. Lett.*, **55**, 17, pp. 1768–1770.

23. Maciá, E. (1998) Optical engineering with Fibonacci dielectric multilayers, *Appl. Phys. Lett.*, **73**, 23, pp. 3330–3332.

24. Maciá, E. (2001). Exploiting quasiperiodic order in the design of optical devices, *Phys. Rev. B*, **63**, 20, p. 205421.

25. Senechal, M. (1996). *Quasicrystals and Geometry*, Cambridge University Press.

26. Janssen, T., Chapuis, G., and De Boissieu, M. (2007). *Aperiodic Crystals: From Modulated Phases to Quasicrystals*, Oxford University Press.

27. de Graef, M., McHenry, M.E., and Keppens, V. (2008). Structure of materials: an introduction to crystallography, diffraction, and symmetry, *J. Acoustic. Soc. Am.*, **124**, 3, pp. 1385–1386.

28. Allouche, J.P., and Shallit, J.O. (2003). *Automatic Sequences: Theory, Applications, Generalizations*, Cambridge University Press.

29. Queffélec, M. (2010). *Substitution Dynamical Systems-Spectral Analysis*, Springer-Verlag.

30. Lind, D.A., and Marcus, B. (1995). *An Introduction to Symbolic Dynamics and Coding*, Cambridge University Press.

31. Schroeder, M. (2009). *Number Theory in Science And Communication: With Applications in Cryptography, Physics, Digital Information, Computing, and Self-Similarity*, Springer-Verlag.

32. Hardy, G.H., and Wright, E.M. (2008). *An Introduction to the Theory of Numbers*, Oxford University Press.

33. Miller, S.J., and Takloo-Bighash, R. (2006). *An Invitation to Modern Number Theory*, Princeton University Press.

34. Trevino, J., Cao, H., and Dal Negro, L. (2011). Circularly symmetric light scattering from nanoplasmonic spirals, *Nano Lett.*, **11**, 5, p. 2008.

35. Vogel, H. (1979). A better way to construct the sunflower head, *Math. Biosci.*, **44**, 3–4, pp. 179–189.

36. Trevino, J., Forestiere, C., Di Martino, G., Yerci, S., Priolo, F., and Dal Negro, L. (2012). Plasmonic-photonic arrays with aperiodic spiral order for ultra-thin film solar cells, *Optics Express*, **20**, pp. A418–A430.

37. Trevino, J., Liew, S.F., Noh, H., Cao, H., and Dal Negro, L. (2012). Geometrical structure, multifractal spectra and localized optical modes of aperiodic Vogel spirals, *Opt. Express*, **20**, pp. 3015–3033.

38. Shechtman, D., Blech, I., Gratias, D., and Cahn, J.W. (1984). Metallic phase with long-range orientational order and no translational symmetry, *Phys. Rev. Lett.*, **53**, 20, pp. 1951–1953.

39. Levine, D., Steinhardt, P.J. (1984). Quasicrystals: A New Class of Ordered Structures, *Phys. Rev. Lett.*, **53**, 26, pp. 2477–2479.

40. Kolmogorov, A.N., and Fomin, S.V. (1999). *Elements of the Theory of Functions and Functional Analysis*, Dover.

41. Barber, E.M., *Aperiodic Structures in Condensed Matter: Fundamentals and Applications*, CRC Press.

42. Ghosh, A., and Karmakar, S. (1999). Existence of only delocalized eigenstates in the electronic spectrum of the Thue-Morse lattice, *Physica A*, **274**, 3, pp. 555–562.

43. Iguchi, K. (1994). Theory of ternary quasiperiodic lattices: scaling-group approach, *Phys. Rev. B*, **49**, 18, p. 12633.

44. de Bruijn, N.G. (1981). Algebraic theory of Penrose's non-periodic tilings of the plane, *Kon. Nederl. Akad. Wetensch. Proc. Ser. A*, **43**, p. 84.

45. Danzer, L. (1989). Three-dimensional analogs of the planar Penrose tilings and quasicrystals, *Discrete Math.*, **76**, 1, pp. 1–7.

46. Radin, C. (1994). The pinwheel tilings of the plane., *Ann. Math.*, **139**, 3, pp. 661–702.

47. Lee, S.Y., Forestiere, C., Pasquale, A.J., Trevino, J., Walsh, G., Galli, P., Romagnoli, M., and Dal Negro, L. (2011). Plasmon-enhanced structural coloration of metal films with isotropic Pinwheel nanoparticle arrays, *Opt. Express*, **19**, 24, pp. 23818–23830.

48. Thompson, D.A.W. *On Growth and Form*, Dover.

49. Prusinkiewicz, P., and Lindenmayer, A. (1990). *The Algorithmic Beauty of Plants*, Springer-Verlag.

50. Ball, P. (2009). *Shapes. Nature's Patterns*, Oxford University Press.

51. Jean, R.V. (1995). *Phyllotaxis*, Cambridge University Press.

52. Adler, I., Barabe, D., and Jean, R.V. (1997). A history of the study of phillotaxis, *Ann. Bot.*, **80**, pp. 231–244.

53. Naylor, M. (2002). $\sqrt{2}$, and π flowers: a spiral story, *Math. Mag.*, **75**, 3, pp. 163–172.

54. Mitchison, G.J. (1977). Phyllotaxis and the Fibonacci series, *Science*, **196**, 4287, pp. 270–275.

55. Lawrence, N., Trevino, J., and Dal Negro, L. (2012). Aperiodic arrays of active nanopillars for radiation engineering, *J. Appl. Phys.*, **111**, p. 113101.

56. (a) Lawrence, N., Trevino, J., and Dal Negro, L. (2012). Control of optical orbital angular momentum by Vogel spiral arrays of metallic nanoparticles, *Opt. Lett.*, **37**, 24, pp. 5076–5078; (b) Auguié, B., and Barnes, W.L. (2008). Collective resonances in gold nanoparticle arrays, *Phys. Rev. Lett.*, **101**, p. 143902.

57. Pecora, E.F., Lawrence, N., Gregg, P., Trevino, J., Artoni, P., Irrera, A., Priolo, F., and Dal Negro, L. (2012). Nanopatterning of silicon nanowires for enhancing visible photoluminescence, *Nanoscale*, **4**, 9, pp. 2863–2866.

58. Capretti, A., Walsh, G.F., Minissale, S., Trevino, J., Forestiere, C., Miano, G., and Dal Negro, L. (2012). Multipolar second harmonic generation from planar arrays of Au nanoparticles, *Opt. Express*, **20**, 14, pp. 15797–15806.

59. Dal Negro, L., Lawrence, N., and Trevino, J. (2012). Analytical light scattering and orbital angular momentum spectra of arbitrary Vogel spirals., *Opt. Express*, **20**, 16, pp. 18209–18223.

60. Liew, S.F., Noh, H., Trevino, J., Dal Negro, L., and Cao, H. (2011). Localized photonic band edge modes and orbital angular momenta of light in a golden-angle spiral, *Opt. Express*, **19**, 24, pp. 23631–23642.

61. Simon, D.S., Lawrence, N., Trevino, J., Dal Negro, L., and Sergienko, A. V. (2013). Quantum key distribution with Fibonacci orbital angular momentum states, *Phys. Rev. A.*, **87**, p. 032312.

62. Pollard, M.E., and Parker, G.J. (2009). Low-contrast bandgaps of a planar parabolic spiral lattice, *Opt. Lett.*, **34**, 18, pp. 2805–2807.

63. Baddeley, A., and Turner, R. (2005). Spatstat: an R package for analyzing spatial point patterns, *J. Stat. Soft.*, **12**, 6, pp. 1–42.

64. Ripley, B.D. (1977). Modelling spatial patterns, *J. R. Stat. Soc. Ser. B*, **39**, 2, pp. 172–212.

65. Janot, C. (1997). *Quasicrystals: A Primer*, Oxford University Press.

66. Yang, J.K., Noh, H., Liew, S.F., Rooks, M.J., Solomon, G.S., and Cao, H. (2011). Lasing modes in polycrystalline and amorphous photonic structures, *Phys. Rev. A*, **84**, 3, p. 033820.

67. Torquato, S., and Stillinger, F.H. (2003). Local density fluctuations, hyperuniformity, and order metrics, *Phys. Rev. E*, **68**, 4, p. 041113.

68. Illian, J., Penttinen, A., Stoyan, H., and Stoyan, D. (2008). *Statistical Analysis and Modeling of Spatial Point Patterns*, Wiley.

69. Meade, R.D., Devenyi, A., Joannopoulos, J.D., Alerhand, O.L., Smith, D.A., and Kash, K. (1994). Novel applications of photonic band gap materials: low-loss bends and high Q cavities, *J. Appl. Phys*, **75**, 9, pp. 4753–4755.

70. Mekis, A., Chen, J.C., Fan, S., Villeneuve, P.R., Joannopoulos, J.D. (1996). High transmission through sharp bends in photonic crystal waveguides, *Phys. Rev. Lett*, **77**, 18, pp. 3787–3790.

71. Krauss, T.F., Labilloy, D., Scherer, A., De La Rue, R.M. (1998). Photonic crystals for light-emitting devices, *Proc. SPIE*, **3278**, pp. 306–313.

72. Notomi, M., Suzuki, H., Tamamura, T., and Edgawa, K. (2004). Lasing action due to the two-dimensional quasiperiodicity of photonic quasicrystals with a Penrose lattice, *Phys. Rev. Lett.*, **92**, 12, p. 123906.

73. Chan, Y., Chan, C., and Liu, Z. (1998). Photonic band gaps in two dimensional photonic quasicrystals *Phys. Rev. Lett.*, **80**, 5, pp. 956–959.

74. Feder, J. (1988). *Fractals*, p. 283, Plenum Press.

75. Gouyet, J.F., and Mandelbrot, B.B. (1996). *Physics and Fractal Structures*, Springer-Verlag.

76. Mandelbrot, B.B. (1982). *The Fractal Geometry of Nature*, Henry Holt and Company.

77. Stanley, H.E., and Meakin, P. (1988). Multifractal phenomena in physics and chemistry, *Nature*, **335**, 6189, pp. 405–409.

78. Mandelbrot, B.B. (1988). *An introduction to multifractal distribution functions.* In Stanley, H.E., and Ostrowsky, N. (eds.), *Fluctuations and Pattern Formation*, pp. 345–360, Kluwer Academics.

79. Frisch, U. (1980). Fully developed turbulence and intermittency, *Ann. N Y Acad. Sci.*, **357**, 1, pp. 359–367.

80. Muzy, J.F., Bacry, E., and Arneodo, A. (1994). The multifractal formalism revisited with wavelets, *Int. J. Bifurcat. Chaos*, **4**, 2, pp. 245–302.

81. Halsey, T.C., Jensen, M.H., Kadanoff, L.P., Procaccia, I., and Shraiman, B.I. (1986). Fractal measures and their singularities: the characterization of strange sets, *Phys. Rev. A*, **33**, 2, pp. 1141–1151.

82. Chhabra, A., and Jensen, R.V. (1989). Direct determination of the $f(\alpha)$ singularity spectrum, *Phys. Rev. Lett.*, **62**, 12, pp. 1327–1330.

83. Karperien, A. (1999-2013). *FracLac for ImageJ version 2.5*, http://rsb.info.nih.gov/ij/plugins/fraclac/FLHelp/Introduction.htm.

84. Rasband, W.S. (1997–2014). ImageJ. U.S. National Institutes of Health, http://imagej.nih.gov/ij/.

85. Albuquerque, E.L., and Cottam, M. (2003). Theory of elementary excitations in quasiperiodic structures, *Phys. Rep.*, **376**, 4, pp. 225–337.

86. Jiang, X., Zhang, Y., Feng, S., Huang, K.C., Yi, Y., and Joannopoulos, J.D. (2005). Photonic band gaps and localization in the Thue-Morse structures, *Appl. Phys. Lett*, **86**, 20, pp. 201110–201113.

87. Mallat, S. (1999). *A Wavelet Tour of Signal Processing*, Academic Press

88. van den Berg, J. (2004). *Wavelets in Physics*, Cambridge University Press.

89. Buckheit J., Chen S., Donoho D., Johnstone I., and Scargle J. (2005). *WaveLab850*, http://www-stat.stanford.edu/~wavelab.

90. Ling, Y., Cao, H., Burin, A.L., Ratner, M.A., Liuu, X., and Chang, R.P.H. (2001). Investigation of random lasers with resonant feedback, *Phys. Rev. A*, **64**, 6, p. 63808.

91. Song, D., Cho, E-C., Conibeer, G., Flynn, C., Huang, Y., and Green, M.A. (2008). Structural, electrical and photovoltaic characterization of silicon nanocrystals embedded in SiC matrix and Si nanocrystals/crystalline silicon heterojunction devices, *Sol. Energ. Mater. Sol. C*, **92**, 4, pp. 474–481.

92. Green, M.A. (2002). Third generation photovoltaics: solar cells for 2020 and beyond, *Physica E*, **14**, p. 65.

93. Haase, C., and Stiebig, H. (2007). Thin-film silicon solar cells with efficient periodic light trapping texture, *Appl. Phys. Lett.*, **91**, p. 061116.

94. Shir, D., Yoon, J., Chanda, D., Ryu, J.-H., and Rogers, J.A. (2010). Performance of ultrathin silicon solar microcells with nanostructures of relief formed by soft imprint lithography for broad band absorption enhancement, *Nano Lett.*, **10**, 8, pp. 3041–3046.

95. Mallick, S.B., Agrawal, M., and Peumans, P. (2010). Optimal light trapping in ultra-thin photonic crystal crystalline silicon solar cells, *Opt. Express*, **18**, pp. 5691–5706.

96. Biswas, R., Bhattaacharya, J., Lewis, B., Chakravarty, N., and Dalal, V. (2010). Enhanced nanocrystalline silicon solar cell with a photonic crystal back-reflector, *Sol. Energ. Mater. Sol. C*, **94**, 12, pp. 2337–2342.

97. Ferry, V.E., Verschuuren, M.A., Claire van Lare, M., Schropp, R.E.I., Atwater, H.A., and Polman, A. (2011). Optimized spatial correlations for broadband light trapping nanopatterns in high efficiency ultrathin film a-Si:H solar cells, *Nano Lett.*, **11**, 10, pp. 4239–4245.

98. Ferry, V.E., Sweatlock, L.A., Pacifici, D., and Atwater, H.A. (2008). Plasmonic nanostructure design for efficient light coupling into solar cells, *Nano Lett.*, **8**, 12, pp. 4391–4397.

99. Rockstuhl, C., Fahr, S., and Lederer, F. (2008). Absorption enhancement in solar cells by localized plasmon polaritons, *J. Appl. Phys.*, **104**, p. 123102.

100. Catchpole, K.R., and Polman, A. (2008). Design principles for particle plasmon enhanced solar cells, *Appl. Phys. Lett.*, **93**, p. 191113.

101. Atwater, H.A., and Polman, A. (2010). Plasmonics for improved photovoltaic devices, *Nat. Mater.*, **9**, p. 205.

102. Eisele, C., Nebel, C.E., and Stutzmann, M. (2001). Periodic light coupler gratings in amorphous thin film solar cells, *J. Appl. Phys.*, **89**, p. 7722.

103. Nakayama, K., Tanabe, K., and Atwater, H.A. (2008). Plasmonic nanoparticle enhanced light absorption in GaAs solar cells, *Appl. Phys. Lett.*, **93**, p. 121904.

104. Beck, F.J., Polman, A., and Catchpole, K.R. (2009). Tunable light trapping for solar cells using localized surface plasmons, *J. Appl. Phys.*, **105**, p. 114310.

105. Stuart, H.R., and Hall, D.G. (1998). Island size effects in nanoparticle-enhanced photodetectors, *Appl. Phys. Lett.*, **73**, pp. 3815–3817.

106. Ostfeld, A.E., and Pacifici. (2011). Plasmonic concentrators for enhanced light absorption in ultrathin film organic photovoltaics, *Appl. Phys. Lett.*, **98**, 11, p. 113112.

107. Forestiere, C., Miano, G., Rubinacci, D., and Dal Negro, L. (2009). Role of aperiodic order in the spectral, localization, and scaling properties of plasmon modes for the design of nanoparticle arrays, *Phys. Rev. B*, **79**, 8, p. 085404.

108. Forestiere, C., Miano, G., Boriskina, S.V., and Dal Negro, L. (2009). The role of nanoparticle shapes and deterministic aperiodicity for the design of nanoplasmonic arrays, *Opt. Express*, **17**, 12, pp. 9648–9661.

109. Lin, D., Tao, H., Trevino, J., Mondia, J.P., Kaplan, D.L., Omenetto, F.G., and Dal Negro, L. (2012). Direct transfer of sub-wavelength plasmonic nanostructures on bio-active silk films, *Adv. Mater.*, **24**, 45, pp. 6088–6093.

110. Weismann, C., Bergenek, K., Linder, N., and Schwartz, U.T. (2009). Photonic crystal LEDs: designing light extraction, *Laser Photonics Rev.*, **3**, 3, pp. 262–286.

111. Allen, L., Padgett M.J., and Babiker, M. (1999). The orbital angular momentum of light, *Prog. Opt.*, **39**, pp. 291–372.

112. Allen, L., Beijersbergen, M.W., Spreeuw, R.J.C., and Woerdman, J.P. (1992). Orbital angular momentum of light and the transformation

of Laguerre-Gaussian laser modes, *Phys. Rev. A*, **45**, 11, pp. 8185–8189.

113. Grier, D.G. (2003). A revolution in optical manipulation, *Nature*, **424**, 6950, pp. 810–816.

114. Mair, A., Vaziri, A., Weihs, G., and Zeilinger, A. (2001). Entanglement of the orbital angular momentum states of photons, *Nature*, **412**, 6844, pp. 313–316.

115. Wang, J., Yang, J.Y., Fazal, I.M., Ahmed, N., Yan, Y., Huang, H., Ren, Y., Yue, Y., Dolinar, S., Tur, M., and Willner, A.E. (2012). Terabit free-space data transmission employing orbital angular momentum multiplexing, *Nat. Photonics*, **6**, 7, pp. 488–496.

116. Berkhout, G.C.G., Lavery, M.P.J., Courtial, J., Beijersbergen, M.W., and Padgett, M.J. (2010). Efficient sorting of orbital angular momentum states of light, *Phys. Rev. Lett.*, **105**, 15, p. 153601.

117. Padgett, M., Courtial, J., and Allen, L. (2004). Light's orbital angular momentum, *Phys. Today*, **57**, 5, pp. 35–40.

118. Heckenberg, N.R., McDuff, R., Smith, C.P., and White, A.G. (1992). Generation of optical phase singularities by computer-generated holograms, *Opt. Lett.*, **17**, 3, pp. 221–223.

119. Yu, N., Genevet, P., Kats, M.A., Aieta, F., Tetienne, J., Capasso, F., and Gaburro, Z. (2011). Light Propagation with phase discontinuities: generalized laws of reflection and refraction, *Science*, **334**, 6054, pp. 333–337.

120. Chavez-Cerda, S., Padgett, M.J., Allison, I., New, G.H.C., Gutierrez-Vega, J.C., O'Neil, A.T., MacVicar, I., and Courtial, J. (2002). Holographic generation and orbital angular momentum of high-order Mathieu beams, *J. Opt. B: Quant. Semiclass. Opt.*, **4**, pp. S52–S57.

121. Intonti, F., Caselli, N., Lawrence, N., Trevino, J., Wiersma, D.S., and Dal Negro, L. (2013). Near-field distribution and propagation of scattering resonances in Vogel spiral arrays of dielectric nanopillars, *New J. Phys.*, **15**, 085023.

122. Ozaktas, H.M., Zalevsky, Z., and Alper Kutay, M. (2001). *The Fractional Fourier Transform with Applications in Optics and Signal Processing*, Wiley.

123. West, B.J., Bologna, M., Grigolini, P. (2003). *Physics of Fractal Operators*, Springer.

124. Mendlovic, D., and Ozaktas, H.M. (1993). Fractional Fourier transforms and their optical implementation I, *J. Opt. Soc. Am. A*, **10**, 9, pp. 1875–1881.

125. Mendlovic, D., and Ozaktas, H.M. (1993). Fractional Fourier transforms and their optical implementation II, *J. Opt. Soc. Am. A*, **10**, 12, pp. 2522–2531.

126. Narayanan, V.A., Prabhu, K.M.M. (2003). The fractional Fourier transform: theory, implementation and error analysis, *Microproc. Microsyst.*, **27**, pp. 511–521.

127. Bultheel, A., and Martínez-Sulbaran, H. (2004). Computation of the fractional Fourier transform, *Appl. Comput. Harmon. Anal.*, **16**, 3, pp. 182–202.

128. http://nalag.cs.kuleuven.be/research/software/FRFT/.

Chapter 4

Disordered Photonics

Francesco Riboli

Department of Physics, University of Trento, via Sommarive 14, 38123 Trento, Italy
f.riboli@unitn.it

4.1 Introduction

Electromagnetic waves are all around us. The extreme large-frequency spectrum allows us to perceive the world on different length scales, from a few hundreds of nanometers for visible light to centimeters for radar waves. Any source of visible light, the sun or a simple lamp, emits light rays that are reflected, refracted, diffracted, absorbed, re-emitted, and scattered from every single piece of matter or even from isolated atoms or molecules encountered on its path. Light rays that reach our eyes have undergone a huge amount of such events and store a lot of information. Trying to decode information from the scattered light that reaches our eyes is something that our brain does continuously in our daily lives, allowing us to perceive and have an image of the surrounding world.

Nevertheless there exist situations where the scattering events that light rays perform from the source to the sensor are many and

Nanodevices for Photonics and Electronics: Advances and Applications
Edited by Paolo Bettotti
Copyright © 2016 Pan Stanford Publishing Pte. Ltd.
ISBN 978-981-4613-74-3 (Hardcover), 978-981-4613-75-0 (eBook)
www.panstanford.com

randomly distributed in space and time. This situation occurs, for example, when we look at an object placed behind an opaque screen or when we try to read a neon sign on a very foggy day. In both cases, the directions of the light rays coming from the source are scrambled in space and consequently our eyes are not able to resolve the exact shape or profile of the source. Anyhow the lack of ability to see through the fog or through an opaque screen does not mean that the light rays are not carrying the information. More in general, encoding the information carried by multiple scattered light rays remains a formidable task to deal with.

Diffusion theory proves to be a very useful model to describe the multiple scattering regime that take place in opaque media [1]. In these systems light undergoes a large amount of independent scattering events from randomly positioned particles and therefore interference between different paths is smoothed out by the disorder. The final transport properties become mostly independent, both from the wave nature and from the particular nature of the scatterers.

The first scientific description of diffusion-like mechanisms is due to Robert Brown, who studied the motion of pollen in water [2]. To have a mathematical and physical description of such a phenomenon we had to wait until 1905, the year in which Albert Einstein published his famous paper about the Brownian random walk [3]. Particles that undergo Brownian-like motion, change their direction randomly in every single step, but their average position after a large number of steps is always the same. Nevertheless, as time goes by, a particle will explore all the available space, although the average distance explored increases slowly. In one or two dimensions a particle will eventually experience all the space, visiting the same point a large number of times. This is not true in three dimensions, where a particle will never come back to a point where it already passed.

The evolution of the probability to find, at a given point and at a given time, a particle that undergoes a Brownian random walk is the solution of the diffusion equation. This equation describes the average motion of particles and it has nothing to do with the microscopic properties of the system, that is, the irregular motion of each particle. One simple example to picture diffusion-like processes is to follow the evolution of a small drop of coffee dropped in a glass of milk. After a transient time where coffee filaments and droplets evolve in a very complicated way, the coffee cloud expands slowly. This is roughly the same picture that we have to figure out when we think about a light stream that propagates within a system composed of a disordered arrangement of scattering centers. In this case the multiple scattering path can be described as a random walk. In such systems the light intensity diffuses in much the same way as coffee in milk.

The first study to use the diffusion equation to study propagation of visible light was done in the field of astrophysics. The target was to extract information from interstellar light passing through the dust present in outer space. The huge steps forward made in these last years in the comprehension of the multiple scattering processes allow us now to understand better the mesoscopic physics of waves in disordered materials and thus to conceive new techniques to increase the overall efficiency of solar cells [4], to see through opaque media [5, 6], or to better understand and reproduce biomimetic-based materials [7]. Moreover, from a broader point of view, the study of multiple scattering of light also allows us to understand how other kinds of waves, like mechanical or matter waves, behave in disordered media [8–10]. Indeed, regardless of the waves we are considering, the equations that describe their evolution can be, formally, written in a similar way. The main advantage to using light is to understand the basic physics of multiple scattering besides the near independence of the transport properties from temperature effects and also in the absence of mutual interactions between photons (in the linear regime). These features make experiments with light easier to interpret.

A complete different scenario is realized when light propagates through a strongly scattering medium, a regime that is achieved when the degree of disorder is very strong, eventually inhibiting the diffusion of light through the material itself. This regime, known as the *localization regime* goes beyond the diffusion picture, because the diffusion constant goes to zero, and it is characterized by the interference mechanism between multiple scattered closed paths that are not connected with the boundary of the system. The first step in the localization phenomena in disordered materials has been made by P. W. Anderson in the context of disordered semiconductor materials, thus dealing with electron wave packets [11]. Indeed, in 1958, Anderson published the famous paper entitled "Absence of Diffusion in Certain Random Lattices," which despite initial skepticism from the scientific community earned him the Nobel Prize in 1977. In the eighties, the original idea of Anderson, that a sufficient amount of structural disorder in a perfect lattice could lead to spatial localization of the propagating wave, was extended to the case of light waves. This allowed us to understand novel optical phenomena in disordered photonic systems, like the coherent back-scattering mechanism, and the possibility to confine light very efficiently in extremely small modal volume [12–14]. Only at the end of the nineties and in the first year of the new century, the localization of light in 1D, 2D, and 3D disordered photonic systems has been proved experimentally [15–19]. Nowadays the knowledge gained of the physics of disordered photonic systems allows us to design artificial materials for applications in the field of renewable energy [4], as well as materials for the study of new transport regimes of light, which are observed in peculiar realization of disorder [20].

Finally the opposite situation of localization of waves is the superdiffusion mechanism. Superdiffusion is characterized by the fact that light propagates at a velocity faster than the velocity set by the diffusion. It can be observed in inhomogeneous disordered photonic systems, like Levy glasses, and it has attracted a lot of attention in this last years because it pave the way for the right optical system to study and characterize superdiffusive phenomena that occur in other fields, like foraging in animals, the flow in a turbulent fluid, and the flow of banknotes all around the world. Superdiffusion of light has been recently demonstrated in artificially created photonic systems [21]. In the following section we will provide some key concepts about the basic principles underlying the phenomenon of diffusion and localization of electromagnetic waves. We will provide also some practical examples of devices that make use of such phenomena.

4.2 Diffusion

The standard approach to studying propagation of waves in disordered media is to use the diffusion equation. This approach neglects any kind of interference effects, but it is able to grasp the main physics of propagation of waves in disordered media.

The materials with which we deal every day are disordered on optical length scales and are constituted of a random packing of small units, each of these units scattering light. For example, biological tissues contain cells, whose typical sizes are between 1 µm and a few tens of microns; concrete is made of grains, microns to millimeters in size; and paint is made from submicron grains. The transport of light in these materials is driven by a multiple scattering process that gives the opaque aspect to objects. Inside these materials the transport of light is the result of the interference between the incident wave and the wave scattered by every single constituent. This intricate situation makes that the wave intensity distribution inside the system is extremely complex, and depends on the exact position of every scatterer in that specific disorder realization. The most striking effect resulting from this complexity is the speckle pattern (Fig. 4.1).

The speckle pattern is the spatial profile of a coherent wave transmitted (or reflected) though an opaque material. It is constituted of a random intensity distribution of bright spots and dark

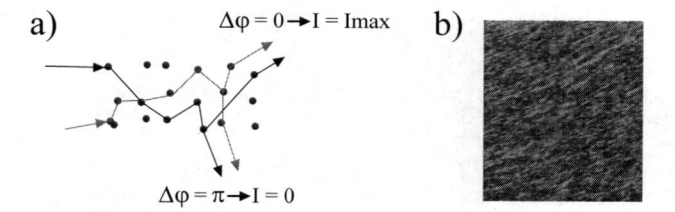

a) $\qquad \Delta\varphi = 0 \rightarrow I = I_{max}$ b)

$$\Delta\varphi = \pi \rightarrow I = 0$$

Figure 4.1 (a) The electromagnetic field incident on a disordered medium accumulates a random phase during its propagation. In each point, the intensity of the field is the sum of all these random-phased contributions, which implies a random intensity distribution. (b) Example of a speckle pattern obtained by shining a sheet of paper with laser light. The bright and dark spots are the speckles.

spots, where the intensity goes to zero. A typical speckle pattern can be observed by shining an opaque material with a laser pointer and looking at the transmitted light on the wall. Nevertheless in daily life we are used to seeing the transmission and reflection from opaque materials as a smooth and diffuse illumination. This is due to the common light sources that are incoherent and broadband, thus providing a mechanism to smooth out the speckle fluctuations by averaging the intensity distribution over wavelength. Moreover, movements of the scatterers, in fluids, glasses, or moving media, provide another mechanism that averages out interference effects and provides the familiar diffuse aspect to light coming from opaque systems. By contrast the description of the averaged transport of light intensity is less complicated than the full description of wave transport. Indeed, by averaging over disorder the wave equation, it is possible to describe the transport of the intensity of light in terms of a random walk that leads to the diffusion equation. Given the complexity of the topic, in the following sections we will only show the main steps of how to derive the diffusion equation in a rigorous way, starting from the equation for electromagnetic waves. The full treatment with further details can be found in the book by E. Akkermans and G. Montambaux [22] or in the book by P. Sheng [23]. There are also many reviews that deal with the propagation of waves in disordered systems in the mesoscopic regime. Among all these we suggest the review by T. M. Nieuwenhuizen and M. C. W. van Rossum [24] that treats in a deeper way the problem of the multiple scattering of classical waves.

4.3 Light Transport: Microscopical Description

To describe the transport of an electromagnetic wave in a disordered medium, we have to solve the wave equation, taking into account the random permittivity of the material. For the sake of simplicity we consider light as a scalar wave, neglecting all effects of polarization, but it is possible to generalize the approach for vector waves. The vectorial model shows that the scalar wave approach is a good approximation if the scattering is isotropic on average. More in general, the mean free path and diffusion constant become tensorial quantities. In the following section we will study multiple scattering in isotropic materials, where a scalar approximation is valid. The wave equation thus reads

$$[\nabla^2 + k_0\varepsilon(\mathbf{r})]E(\mathbf{r}, k_0) = j(\mathbf{r}) \tag{4.1}$$

where $E(\mathbf{r}, k_0)$ is the electromagnetic field, k_0 is the free-space wavevector, $j(\mathbf{r})$ is the source current density, and $\varepsilon(\mathbf{r})$ is the position-dependent permittivity. We now make the assumption that the disordered material is composed by a set of randomly placed scatterers, having a certain polarizability $\alpha(\mathbf{r})$. In this case, the permittivity is given by

$$\varepsilon(\mathbf{r}) = 1 + \Sigma_i\alpha(\mathbf{r} - \mathbf{r}_i). \tag{4.2}$$

For simplicity it is possible to assume that the size of the scatterers is much smaller than the wavelength. This implies that each scatterer can be described as a point dipole, which allows us to simplify $\alpha(\mathbf{r} - \mathbf{r}_i) = \alpha\delta(\mathbf{r} - \mathbf{r}_i)$. One of the most useful approaches to solve Eq. 4.1 is to use the Green's function method, which allows us to get the solution for any source distribution inside the system. This approach describes multiple scattering by decomposing light transport in a sequence of scattering events with free-space propagation in between.

Green's function $g(\mathbf{r}, \mathbf{r}_0; k_0)$ describes the electric field at the point $\mathbf{r}$ due to a point source at the point $\mathbf{r}_0$:

$$[\nabla^2 + k_0\varepsilon(\mathbf{r})]g(\mathbf{r}, \mathbf{r}_0; k_0) = \delta(\mathbf{r} - \mathbf{r}_0) \tag{4.3}$$

To derive the general solution of the wave equation we can use the linearity of Maxwell's equations with respect to the electric field. The electromagnetic field due to a sum of sources distributed with a source density $j(\mathbf{r})$ is given by

$$E(\mathbf{r}) = \int dr' \, j(\mathbf{r}')g(\mathbf{r}, \mathbf{r}'; k_0). \tag{4.4}$$

Because the Green's function formalism allows us to decompose the propagation of light within a disordered medium in a sequence of scattering events, it makes sense to solve Eq. 4.3 in free space and in the single scattering configuration. In free space $\varepsilon(\mathbf{r}) = 1$ and in 3D space Green;s function is

$$g^{R,A}(\mathbf{r}, \mathbf{r}_0; k0) = -\frac{1}{|\mathbf{r} - \mathbf{r}_0|}e^{\pm ik_0|\mathbf{r}-\mathbf{r}_0|}, \tag{4.5}$$

where the subscripts R and A correspond to the anticasual ($-$ minus sign) and casual ($+$ plus sign) solutions. The two solutions are the advanced Green's function (g_0^A) and the retarded Green's function (g_0^R). They represent a spherical field distribution in free space that expands (causal solution) around the source or contracts onto it (anticausal solution). The two solution are related by

$$g_0^A(\mathbf{r}, \mathbf{r}_0; k0) = g_0^R(\mathbf{r}, \mathbf{r}_0; k_0)^*. \tag{4.6}$$

Because the world is causal, we have to use the retarded Green's function to calculate the electromagnetic field distribution:

$$E(\mathbf{r}) = \int g_0^R(\mathbf{r}, \mathbf{r}'; k_0) j(\mathbf{r}') d\mathbf{r}'. \tag{4.7}$$

In the following section, the retarded Green's function will be noted by g.

4.3.1 Single Scattering

Let's suppose we have one single scattering center in empty space. The wave equation becomes

$$[\nabla^2 + k_0^2] g(\mathbf{r}, \mathbf{r}_0; k_0) = \delta(\mathbf{r} - \mathbf{r}_0) + V_1(\mathbf{r}) g(\mathbf{r}, \mathbf{r}_0; k_0), \tag{4.8}$$

where $V_1(\mathbf{r}) = -k_0^2 \alpha(\mathbf{r})$ is the single scattering potential. Using Eq. 4.7, the wave equation can be written as follows:

$$g(\mathbf{r}, \mathbf{r}_0; k_0) = g_0(\mathbf{r}, \mathbf{r}_0; k_0) + \int d\mathbf{r}_1 g_0(\mathbf{r}, \mathbf{r}_1; k_0) V_1(\mathbf{r}_1) g(\mathbf{r}_1, \mathbf{r}_0; k_0), \tag{4.9}$$

which represents a self-consistent equation that can be solved by expanding the second term in an iterative way. The solution of the wave equation is then expressed as an expansion of the free-space Green's function

$$g(\mathbf{r}, \mathbf{r}_0; k_0) = g_0(\mathbf{r}, \mathbf{r}_0; k_0) + \int d\mathbf{r}_1 g_0(\mathbf{r}, \mathbf{r}_1; k_0) V_1(\mathbf{r}_1) g_0(\mathbf{r}_1, \mathbf{r}_0; k_0)$$

$$+ \int d\mathbf{r}_1 d\mathbf{r}_2 g_0(\mathbf{r}, \mathbf{r}_2; k_0) V_1(\mathbf{r}_2)$$

$$\times g_0(\mathbf{r}_2, \mathbf{r}_1; k_0) V_1(\mathbf{r}_1) g_0(\mathbf{r}_1, \mathbf{r}_0; k_0) + \dots \tag{4.10}$$

The above equation can be written in diagrammatic form by using the following conventions for the free-space Green's function and the scattering potential:

$$g_0 = \text{———} \tag{4.11}$$

$$V = \circ \tag{4.12}$$

With the above definitions, Eq. 4.10 can be written in a more readable form:

$$g = \text{—} + \text{—}\circ\text{—} + \text{—}\circ\text{···}\circ\text{—} + \text{—}\circ\text{···}\circ\text{···}\circ\text{—} + \dots, \tag{4.13}$$

where the dotted line ($\circ\text{···}\circ$) connects scattering events that happen on the same scattering center. This is indeed the key concept of the scattering mechanism of light. Indeed, despite the fact that we are dealing with a single scattering in empty space, the resulting Green's function (i.e., the scattered electric field) is the sum of many amplitudes, that is, the sum of the field that comes directly from the source, plus the field scattered once by the scatterer, plus the field recurrently scattered. Indeed the electromagnetic polarizability of the scatterer is driven by the total field, that is, the sum of the incident field and the field emitted by the scatterer itself. In other words the incident field polarizes the scatterer, which emits an electromagnetic field, which modifies its own polarization, which modifies consequently the scattered field, and so on. This is essentially a self-consistent process that can be further schematized by defining the t-matrix, which in diagrammatic form reads as follows:

$$t = \times = \circ + \circ\text{···}\circ + \circ\text{···}\circ\text{···}\circ + \circ\text{···}\circ\text{···}\circ\text{···}\circ + \dots \tag{4.14}$$

The t-matrix completely describes how a scatterer radiates the scattered field from the incident field. The definition of the t-matrix is very convenient also because it allows us to express Green's function as

$$g = \text{—} + \text{—}\times\text{—} . \tag{4.15}$$

The meaning of this equation in very simple. Indeed, the electromagnetic field traveling from one point in space to the other is the sum of the nonscattered field, plus the field radiated by the scatterer, taking fully into account the self-consistent contribution of the scatterer.

4.3.2 *Multiple Scattering*

Because the transport of the electromagnetic field in a multiple scattering regime is a sequence of free-space propagations and scattering events, we are now in the right condition to evaluate it. We start again from the wave equation with a potential that describes the multiple scattering $V(\mathbf{r}) = \sum_i V_1(\mathbf{r} - \mathbf{r}_i)$:

$$g(\mathbf{r}, \mathbf{r}_0; k_0) = g_0(\mathbf{r}, \mathbf{r}_0; k_0) + \int d\mathbf{r}_1\, g_0(\mathbf{r}, \mathbf{r}_1; k_0) V(\mathbf{r}_1) g(\mathbf{r}_1, \mathbf{r}_0; k_0).$$

(4.16)

Following the previous notations, and introducing the t-matrix, we then find the following diagrammatic form that summarizes the process of the multiple scattering:

$$
\begin{aligned}
g = \quad & \text{—} \quad + \quad \text{—}\times\text{—} \quad + \quad \text{—}\times\text{—}\times\text{—} \\
+ \quad & \text{—}\times\text{—}\times\text{—}\times\text{—} \quad + \quad \text{—}\times\text{—}\times\text{—}\times\text{—} \\
+ \quad & \text{—}\times\text{—}\times\text{—}\times\text{—} \quad + \ldots
\end{aligned}
$$

(4.17)

This equation describes comprehensively the process of multiple scattering of the electromagnetic field. Its meaning is the following: The transport of the field in a disordered material can be viewed as the sum of the unscattered field, plus the field scattered one time inside the material, plus the field scattered two times, plus the field scattered three timess, and so on (see Fig. 4.2). To solve exactly

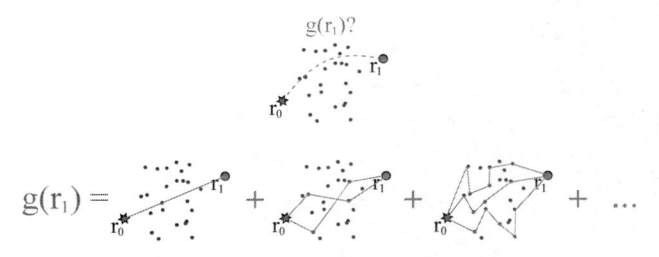

Figure 4.2 Multiple scattering expansion: the field traveling from one point in the space $\mathbf{r}_0$ to another point $\mathbf{r}$ can be viewed as the sum of the field following every possible trajectory. The total field is the sum of the unscattered fields, plus the field scattered only by one scatterer, plus the field scattered from two distinct scatterers, plus the field scattered from three scatteres, and so on.

this equation we have to know the precise position of every single scatterer. If this is the case, the solution of the multiple scattering equation gives the exact speckle patter that such arrangement of scattering center generates. In practice it is impossible to know exactly the position of every single scatterer and indeed we are used to dealing with ensemble averaged quantities. It is thus more natural to look for Green's function averaged over disorder, a quantity that is independent of the precise position of each scattering center and thus gives the average properties of the disordered medium. The averaged, or dressed, Green's function is

$$< g >= G = \text{———} = \text{—} + \text{—}\otimes\text{—} + \text{—}\otimes\text{—}\otimes\text{—}$$
$$+ \text{—}\otimes\text{—}\otimes\text{—}\otimes\text{—} + \text{—}\otimes\cdots\otimes\text{—}\otimes\text{—}$$
$$+ \text{—}\otimes\text{—}\otimes\cdots\otimes\text{—}\otimes\text{—} + \dots, \quad (4.18)$$

where $\otimes =< \times >=< t >= T$ is the averaged T-matrix. To grasp the physics behind the multiple scattering process and to simplify the diagrammatic notation, it is possible to define the self-energy Σ. This quantity represents the sum of all multiple scattering paths (all diagrams) that cannot be split in a smaller scattering sequence without neglecting a sequence that represents recurrent scattering. These diagrams are commonly called irreducible diagrams.

$$\Sigma = \otimes + \otimes\cdots\otimes\text{—}\otimes + \otimes\cdots\otimes\text{—}\otimes\text{—}\otimes + \dots \quad (4.19)$$

Thanks to the introduction of the self-energy the wave equation becomes

$$G = \text{———} = \text{—} + \text{—}\Sigma\text{———} \quad (4.20)$$

This is the well-known Dyson equation. It is valid for any multiple scattering system because it was obtained without any approximation. Taking a step back and using the definitions of diagrammatic notations, we can rewrite the Dyson equation in real space:

$$G(\mathbf{r} - \mathbf{r}_0; k_0) = g_0(\mathbf{r} - \mathbf{r}_0; k_0) + \int d\mathbf{r}_2 \int d\mathbf{r}_1 g_0(\mathbf{r} - \mathbf{r}_2; k_0)$$
$$\times \Sigma(r_2, r_1; k_0)G(\mathbf{r}_1 - \mathbf{r}_0; k_0) \quad (4.21)$$

Conceptually this equation expresses Green's function as the sum of the field following all possible sequences of single scattering events. All the possible sequences are contained in the self-energy term Σ. The solution of the Dyson equation is subject to the exact knowledge of all possible trajectories that the waves follow during the multiple scattering process. If for a moment we assume to know all the possible trajectories, by adding all the contributions it would be possible to calculate the self-energy of the system. Unfortunately, the sum of all the diagrams contributing to Σ is not known, and it is necessary to make some simplifications to go further.

The first and most simple approximation we can do is to neglect the recurrent scattering diagrams. This approximation holds in dilute systems, while for denser systems the effect of such a diagram can be included in the effective medium description. The self-energy of the system reduces to the disorder average of the t-matrix. For a homogeneous material, with a density of scatterer N, we have

$$\Sigma(\mathbf{r}_2 - \mathbf{r}_1) = \otimes = NT\delta(\mathbf{r}_2 - \mathbf{r}_1). \tag{4.22}$$

Within this approximation the Dyson equation becomes

$$G(\mathbf{r} - \mathbf{r}_0; k_0) = g_0(\mathbf{r} - \mathbf{r}_0; k_0) + \int d\mathbf{r}_1 \, g_0(\mathbf{r} - \mathbf{r}_1; k_0) N T \, G(\mathbf{r}_1 - \mathbf{r}_0; k_0). \tag{4.23}$$

The solution of the Dyson equation in real space is

$$G^R(r - r_0; k_0) = -\frac{1}{4\pi |r - r_0|} e^{\pm i K |r - r_0|}, \tag{4.24}$$

having defined

$$K = \sqrt{k_0^2 + NT} \approx \sqrt{k_0^2 + N\Re(T)} + \frac{iN}{2k_0}\Im(T), \tag{4.25}$$

where $\Re(T)$ and $\Im(T)$ are the real and imaginary parts of the t-matrix. The interpretation of the solution of the Dyson equation 4.24 in the regime of single scattering approximation has a simple meaning: the propagator (Green's function) in a multiple scattering 3D system is a spherical wave whose wavenumber K is renormalized by the disorder. The real part of the wavenumber models the modification of the refractive index due to the presence of the scattering centers, while the imaginary part induces a damping of the field oscillations, whose decay constant defines the scattering mean free path ℓ_s of the disordered system, through the relation

$$\frac{1}{\ell_s} = \frac{N}{k_0}\Im(T) = N\sigma_s.$$

Here σ_s is the scattering cross section of the scatterer. We then found the definition of one of the most important quantities in the physics of multiple scattering: the scattering mean free path ℓ_s. Practically speaking this quantity represents the maximum distance that we can observe when we look at an object on a foggy day. If, for example, visibility is a few tens of meters, this means that the mean free path of the visible light, when it propagates inside the fog, is of this order. Finally, if the size of the scatterers is comparable to the wavelength of the electromagnetic wave, we have to renormalize the scattering mean free path with the angular averaged scattering cross section of the scattering center. In this case the important quantity that identifies the ability of a disordered system to scatter the propagating wave is identified by the transport mean free path $\ell_t = \ell_s/(1- <\cos\theta>)$, where $<\cos\theta>$ is the angular average of the differential scattering cross section of the scatterer.

4.3.3 *Transport of Intensity*

Up to this point we have described the propagation of the electromagnetic field in a disordered medium. We are now interested to derive the equation that describes the propagation of intensity of the field. Because many of the steps that we have to follow to derive the equation for the propagation of intensity are very similar to those for the propagation of the field, we will show only the key steps that lead to the diffusion equation after the appropriate approximations.

The propagator of intensity $p(\mathbf{r}, \mathbf{r}_0, \Omega)$ is the product of the advanced and retarded Green's functions:

$$p(\mathbf{r}, \mathbf{r}_0, \Omega) = g^R\left(\mathbf{r}, \mathbf{r}_0, \omega + \frac{\Omega}{2}\right) g^A\left(\mathbf{r}, \mathbf{r}_0, \omega - \frac{\Omega}{2}\right), \qquad (4.26)$$

where ω is the frequency of the field, while Ω is the frequency of the modulation of the envelope of the intensity. Following the previous diagrammatic notation, the expression for the product of the advanced and retarded Green's functions is represented by coupling the diagrammatic expansion of g^R with that of g^a.

$$p(\mathbf{r}, \mathbf{r}_0, \Omega) = g^R g^A$$

$$(4.27)$$

It is now possible to calculate the average propagator $\mathscr{P}(r - r_0, \Omega) = <gg^*>$ that reads as

$$(4.28)$$

The calculations have been made by grouping all averaged diagrams into a unique operator called the reducible vertex **R**. The first term describes the transport of the unscattered intensity $P_s(r - r_0, \Omega)$, while the second term describes the transport of the multiply scattered intensity. In analogy with the definition of the self-energy it is possible to define the quantity $U(r_2, r_1)$ as the sum of all irreducible diagrams for the intensity and finally after some calculation we find the equation that describes the transport of the intensity of the electromagnetic field in a disordered medium, better known as the Bethe–Salpeter equation:

$$\mathscr{P}(\mathbf{r} - \mathbf{r}_0, \Omega) = P_s(\mathbf{r} - \mathbf{r}_0, \Omega)$$
$$+ \int dr_1 \, dr_2 \; P_s(\mathbf{r} - \mathbf{r}_2, \Omega) \, U(\mathbf{r}_2, \mathbf{r}_1) \, \mathscr{P}(\mathbf{r}_1 - \mathbf{r}_0, \Omega)$$

$$(4.29)$$

This is just a re-expression of the wave equation in a form suitable for intensity transport through a disordered medium. No hypothesis have been made so far on the type or strength of the disorder. As for the case of the Dyson equation, the solution of the Bethe–Salpeter equation requires the exact knowledge of all possible diagrams. This implies that we have to make some approximation in order to find an explicit solution. First of all we can neglect the recurrent diagrams. Indeed all the diagrams that do not follow the same sequence of scatterers for both the advanced and the retarded Green's function give a zero contribution because they have an uncorrelated phase at the output surface of the material. This means the advanced and retarded Green's functions have to follow

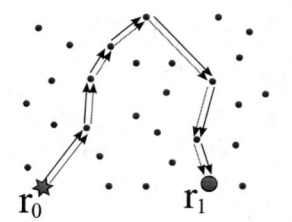
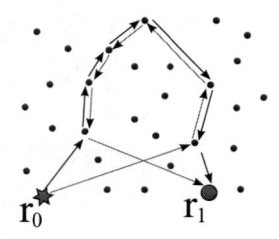

Figure 4.3 Schematization of the scattering sequence that mainly contributes to the calculation of $\mathscr{P}(r_2, r_1)$. (a) The ladder diagram where both the advanced and retarded Green's functions follow the same sequence of scatterers in the same direction. (b) The crossed diagram where the Green's function follows the sequence of scatterers in opposite directions.

the same sequence of scatterers (in the same order or in the opposite order) to give a nonzero contribution to the dressed Green's function of the system (see Fig. 4.3). The two possible paths illustrated in Fig. 4.3 are the ladder diagram and the crossed diagram.

The ladder diagram represents the schematization of the well-known random walk, while the crossed diagram represents the most important correction: the fact that the electromagnetic wave is not following a real random walk, being a wave and not a particle. So far the approximation that we did has been to neglect the recurrent scattering diagram. As mention before, this approximation is valid only for dilute system and does not work in the regime of strong scattering. With this hypothesis, the reducible vertex is given by

$$R(\mathbf{r}_2, \mathbf{r}_1, \Omega) = L(\mathbf{r}_2, \mathbf{r}_1, \Omega) + C(\mathbf{r}_2, \mathbf{r}_1, \Omega). \qquad (4.30)$$

If we are interested in the contribution of the transport that is dominated by the pure random walk, neglecting all the interference effects induced by the multiple scattering of partial waves propagating inside the system, we can also neglect the crossed diagram, and the reducible vertex reduces to $R(\mathbf{r}_2, \mathbf{r}_1, \Omega) = L(\mathbf{r}_2, \mathbf{r}_1, \Omega)$. Without going too much into detail of the calculation, it is possible to demonstrate that within the ladder approximation, the propagator $\mathscr{P}(\mathbf{r} - \mathbf{r}_0, \Omega)$ is proportional to the ladder vertex $L(\mathbf{r}, \mathbf{r}_0, \Omega)$, where the proportionality constant is the square of the scattering mean free path ℓ_s^2. Finally making the Taylor expansion of $\mathscr{P}(\mathbf{r}_1 - \mathbf{r}_0)$ around $\mathbf{r}_1 = \mathbf{r}_0$, and using symmetry arguments it is possible to obtain the diffusion equation for the propagator:

$$(-i\Omega - D_0\nabla^2)\mathscr{P}(\mathbf{r}, \mathbf{r}_0, \Omega) = \delta(\mathbf{r} - \mathbf{r}_0), \qquad (4.31)$$

where the diffusion constant D_0, called the Boltzmann diffusion constant, as we have neglected the influence of interferences, is given by

$$D_0 = \frac{c\ell_s}{d}, \qquad (4.32)$$

where c is the speed of light, ℓ_s is the scattering mean free path, and d represents the dimensionality of the system. Summarizing, we have demonstrated that the propagator of the intensity of an electromagnetic wave traveling in a disordered dielectric medium obeys the well-known diffusion equation when all the effects induced by interference are neglected. The steps that we followed to demonstrate Eq. 4.31 are:

- Write the wave equation in vacuum. Green's function formalism.
- Write the wave equation for a single scattering $\Rightarrow$ diagrammatic representation.
- Write the wave equation for many scatterers. Averaging over many realization of disorder $\Rightarrow$ Dyson equation.
- Generalize the Green's function formalism for the propagation of intensity, averaging over disorder realization $\Rightarrow$ Bethe–Salpeter equation.
- Neglect the recurrent scattering diagram, neglect the crossed diagram $\Rightarrow$ diffusion equation.

Despite all the approximations that we did that eventually neglect all the interference effects, the diffusion equation represents one milestone in the study of propagation of light through disordered media. For any system where $\lambda < \ell_s < L_c < L$, where L_c is the coherence length of the wave and L is the size of the disordered system, the diffusion equation describes the propagation of light in a very satisfactory way, provided that the diffusion constant or the scattering or transport mean free path is known a priori. A sheet of paper, fog, clouds, human tissue, and interstellar vacuum with dust are systems that can be described within the diffusion approximation. On the other hand, when the wavelength of a wave becomes comparable with the scattering mean free path, interference effects become important and we have to go beyond this approximation. This is the case of strongly scattering media, as, for example, very dense powder or artificial disordered photonic meta-materials. In these materials interference effects renormalize the diffusion constant D, eventually blocking the propagation of light. This is the regime of disorder-induced localization, better known as Anderson localization of waves. The theoretical description of localization of light is extremely complicated because we should take into considerations all the diagrams in the multiple scattering expansion. This is not possible and many alternative theories that take into consideration only part of the diagrams have been proposed, as, for example, self-consistent theory [25]. Up to today there is no theory that is able to describe the regime of strong scattering in a simple and satisfactory way. The only way is to develop ab initio numerical calculations that are very accurate but very poor from a physical point of view.

4.4 Anderson Localization of Light

Anderson localization of waves is a second-order phase transition that occurs in disordered scattering materials [8]. The transition separates the diffusive regime from the localized regime. The parameter that drives the transition is the strength of the disorder that can be identified with the $k\ell_s$ parameter, where k is the wavevector of the propagating wave. This parameter measures the ratio between the scattering (transport) mean free path ℓ_s (ℓ_t) and the wavelength of the wave λ/n_{eff} inside the material. When $tk\ell_s$ is close to unity, the wave becomes scattered every optical cycle, meaning that the interference effects become the dominant mechanism for the transport properties of the material. For $k\ell_s \leq 1$ the wave is not more able to propagate and becomes localized inside the system, making the material an insulator for light. Thus Anderson localization is a phase transition between diffusive metallic materials and localizing insulating materials. It is indeed known as metal–insulator transition (MIT). The concept of localization is very general and thus it holds for any kind of

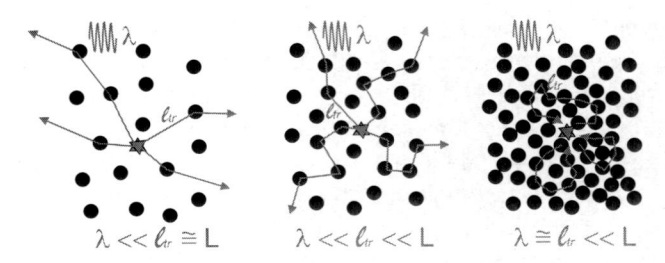

Figure 4.4 Different regime of scattering. The regime of single scattering is established when the wavelength of the wave is much shorter than the scattering mean free path that is comparable with the size of the system. In this regime, a photon generated inside the material undergoes a single scattering event before escaping from the sample (left). The diffusive regime is established when the wave is able to perform many scattering events before escaping the material (middle). The localization regime is established when the scattering strength is extremely high and the generated photons cannot escape from the material where it has been generated (right). In this case the material becomes an insulator for light.

wave, classical (electromagnetic, sound, and seismic waves) and quantum (electronic and Bose Einsteins condensate waves), that are propagating in their respective scattering medium. Originally conceived for electronic waves, functions that are propagating in a ordered lattice with a small degree of structural disorder to explain the strange insulating behavior of conductors, it has been generalized to a system where the disorder is not a perturbation of the underlying ordered lattice but it is the main characteristic of the material. The phase transition is induced by increasing the degree of structural disorder until the wave that is propagating inside the material is not more able to propagate due to the formation of closed multiple scattering loops (see Fig. 4.4).

Anderson localization being a phase transition, it depends on the dimensionality of the system. Strictly speaking only 3D disordered and infinite materials show a well-defined phase transition with a mobility edge (the energy threshold above which the material become an insulator). The opposite situation happens for 1D disordered and infinite materials, in which localization of waves is always possible, independently on the degree of disorder. The only requirement is that the size of the sample must be much larger that the localization length of the waves. In the middle between these two opposite dimensionalities we found 2D disordered systems.

Two is the critical dimension for Anderson localization. The theory and all the experiments show no phase transition in two dimensions [26]. This means that it is easy in principle to observe Anderson localization in 2D; the only requirement (as for the case of a 1D system) is to realize a sample larger than the localization length.

Experimentally, the realization of a material where transport is truly 2D can be addressed, taking advantage from the last development made in the field of photonic crystals on slab waveguides that offer an interesting and efficient method for this purpose [27].

As previously said Anderson localization is a general concept that in principle holds for any kind of wave. Even only focusing to classical waves and in particular on electromagnetic waves, the literature is extremely wide. Since the physics of 2D disordered structures is simple, intuitive, and complete, the rest of the section will be focused on 2D structures, showing how it is possible to localize light and how it is possible to experimentally characterize them. Finally we will show that single localized modes can be engineered, inducing an artificial coupling in the weak as well in the strong regime between such modes. At the end of the chapter we will show potential application of 2D disordered photonic structures in the field of photovoltaics.

4.4.1 *Localization of Light in Two Dimensions*

Up to now, 2D localization of electromagnetic waves has been observed in the microwave regime [15, 28] and in the near-infrared regime [18]. Also the localization of the transverse component of a light wave propagating in an optical fiber has been observed [17]. One of the simplest systems that shows light localization in two dimensions is based on disordered photonic crystals on slab waveguides (see Fig. 4.5).

The transport of light in this class of systems is driven by modes characterized by resonances of finite spectral width and spatial extent. Such resonances are generated by the interference between

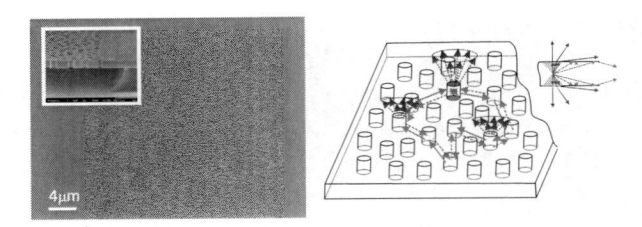

Figure 4.5 (a) Scanning electron microscopy images of disordered samples with filling fraction $f = 0.35$ and hole diameters $hd = 215$ nm. The inset of panel (a) shows the suspended photonic membrane clamped at its edges. (b) Schematic representation of the multiple scattering of guided waves. Each scattering event produces guided and radiative wavevectors. The balance between the amount of guided and radiative components determines the lifetime of localized modes.

multiple scattered light waves [23, 29] whose directions and phases are randomized in a complex manner, yet giving the possibility to define photonic modes with a characteristic spatial profile and lifetime. Figure 4.6 (left) shows the intensity profile of a delocalized mode, the envelope of which is only determined by the boundary of the sample and the typical intensity profile of a localized mode composed of a collection of speckles whose average amplitude decreases exponentially (Fig. 4.6, right).

Figure 4.5 shows the experimental realization of a sample that is able to localize light [20]. The samples is a 320 nm thick GaAs planar dielectric waveguide, optically activated by the inclusion of three layers of InAs quantum dots (QDs) in the middle plane of the slab and patterned with disordered distributions of circular holes. The membrane is suspended in air and clamped at its edges in order to obtain a symmetric high-contrast planar waveguide. The lateral size of the square pad is $L = 25$ μ. To obtain light localization in the infrared regime (the wavelength range where the QDs are emitting ranges from 1.1 μ to 1.4 μ), we have to choose the right structural parameter. The size of the holes is around 200 nm, while the filling fraction (the ratio between the area occupied by the holes and the total area) is around $f = 0.35$. The high scattering cross section of air holes, together with the high value of the filling fraction, makes that the sample is deeply in the localization regime and the recurrent

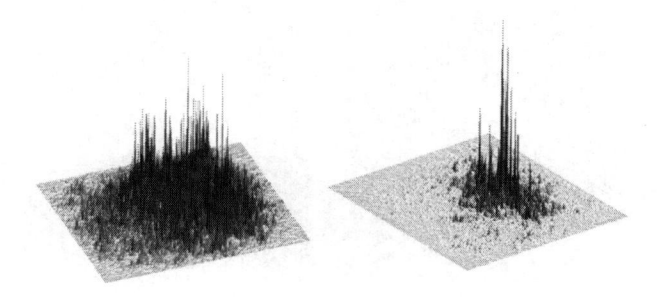

Figure 4.6 Numerical calculation of the intensity profile of a delocalized mode (left) and a localized mode (right). The delocalized mode is able to connect two different regions of the sample providing the metallic behavior, while the localized mode is well confined within a small portion of the sample, isolating such a region from the rest of the sample.

scattering diagrams become fundamental in determining the optical properties of the material.

This system can be easily studied by using a near-field scanning optical microscope (SNOM) used in an illumination/collection geometry at room temperature. The QDs are excited through the SNOM tip and the photoluminescence (PL) signal is recorded by the same tip and it is dispersed in a spectrometer and detected by an InGaAs array. This allows one to record point by point (with a combined spatial and spectral resolutions of 200 nm and 0.5 nm, respectively) the local PL emission spectrum of the QDs. These kinds of measurements provide simultaneous spectral information and spatial distributions of the optical modes.

Figure 4.7 shows the typical intensity distribution of two localized modes that are characterized by a main intensity peak surrounded by speckles with lower intensities. This is indeed the typical feature of a localized mode, that is, a collection of many speckles with amplitudes that decrease as a function of the distance from the ideal center of mass of the mode. The finite spatial extent is the main characteristic that implies the insulator characteristic of the disordered material. In other words, a photon orbiting within a localized mode does not feel the sample outside its localization length, defined as the distance over which the amplitude of the intensity decreases by a factor $1/e$. The typical localization length

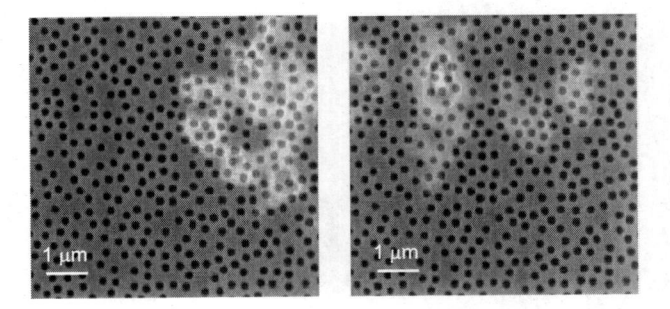

Figure 4.7 Example of two localized modes with a small degree of spatial overlap. Each mode is composed of many speckles with different intensities. The probability to find overlapping localized modes is very low, locally breaking the insulator regime of the material. These modes are better known as "necklace states" of the second order, being created by the coupling between two localized modes.

of modes shown in Fig. 4.7 is of the order of $\xi \simeq 1\mu$. It is important to remember that the localization length ξ is a statistical parameter, and it make sense when averaged over a large number of localized modes.

4.4.2 Hybridization of Localized Modes

Even if the two modes shown in Fig. 4.7 are well separated, they share a small portion of their tails, so the two modes can in principle "talk" to each other by forming an open transmission channel that connects two different regions of the sample. This is exactly the case of the two modes shown in Fig. 4.7. To demonstrate that two localized modes are able to talk to each other, we have to find a way to change the resonant frequency of one of these two modes and see what happens to the energy of the other mode. In the case where there is no coupling, the energy of the unperturbed mode will be unaffected by the perturbation, while in the opposite case we have to see some modification also in the energy of the unperturbed mode.

To control the energy of one of the localized modes in a broad spectral range, it is possible to exploit a postfabrication technique based on laser-assisted micro-oxidation. This technique is able to permanently and locally ($\simeq 1\mu^2$) modify the dielectric

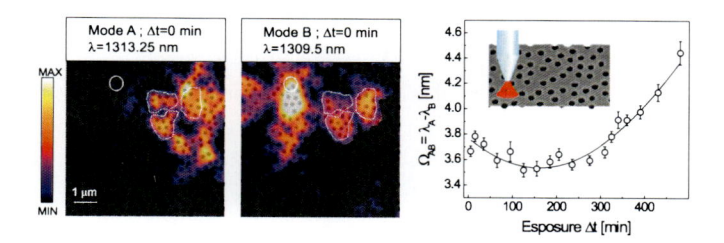

Figure 4.8 The two localized modes shown in Fig. 4.7, having highlighted the region of spatial overlap. Thanks to this overlap the two modes are able to interact in the strong coupling regime. The white circle identifies the region where the local perturbation is performed. The interaction is demonstrated by looking at the separation of the two resonant frequencies that follows the typical behavior of an avoided crossing between the two resonant frequencies.

environment by reducing the effective GaAs membrane thickness and by increasing the effective pore diameter [22]. It has been successfully used to gently blue-shift the resonance frequency of photonic-crystal-based cavities and has also been theoretically proposed as a way to change the nature of the modes in a strongly disordered system [19]. The inset of the right panel of Fig. 4.8 shows the cartoon of the laser-assisted oxidation technique, implemented by illuminating the sample with an SNOM tip, that is able to selectively identify a precise region of the sample, eventually where one of the localized modes has its maximum intensity. The central panel of Fig. 4.8 shows the intensity profile of The perturbed mode (Mode B), and the white circle identifies the region where the laser-assisted oxidation is performed. Indeed this region is in correspondence with the maximum of Mode B but is far from Mode A. The right panel shows the splitting between the mode resonant wavelengths as a function of the exposure time ($\Omega_{AB} = \lambda_A - \lambda_B$) that initially decreases, reaches the minimum value around $\Delta t \simeq 170$ min, and then increases again. The black solid line is the result of the fit of the experimental splitting with the expected strong coupling behavior [30]. The minimum experimental splitting Ω_{AB}^0 gives the interaction strength g from $\Omega_{AB}^0 = 2g = (3.55 \pm 0.25)$ nm that, compared with the average full width and half maximum (FWHM) of the two modes $\Delta\lambda = (3.00 \pm 0.08)$ nm means that the two modes are just at the onset of strong coupling. It is important to underline that the occurrence of overlapping localized modes in a disordered photonic material is very rare. These special modes that are born from the hybridization of two localized modes are better known as necklace states and have been theoretically predicted in 1D, 2D, and 3D disordered systems. Their occurrence is rare because they are formed by two or more modes with a reasonable spatial and spectral overlap. Despite their scarcity, necklace states play a fundamental role in the transport properties of a disordered material because their occurrence is the signature of the opening of a new transmission channel inside the insulator material. They has been observed in 1D and 2D disordered systems, but no experimental evidence has been reported so far in 3D disordered systems.

4.5 Light-Trapping Mechanism for Photovoltaic Applications

Disordered photonic materials in the regime of strong scattering are able to confine light very efficiently. As the mode becomes confined in space, its lifetime increases, thereby increasing the residence time of the light within the material. The trapped light performs many oscillations before escaping from the sample, thus increasing the probability to be absorbed. This is the basic principle that leads to the possibility to use disordered photonic materials in the field of thin-film photovoltaics by substituting thick and homogeneous active layers, where photons propagate freely, with ultrathin engineered disordered active layers, where photons remain trapped for a longer time.

This basic and simple idea dates back to the half of the eighties [31], when S. John demonstrated that the effect of a small imaginary part to the dielectric constant ϵ, on the propagation of waves in a disordered medium near the Anderson transition, induces a near divergence of the absorption coefficient. From a broader point of view that includes also disordered materials in the diffusive regime, the effect of the increased absorption coefficient can be attributed to the extremely slow diffusive propagation of energy in the disordered medium.

A coherent light-trapping mechanism can thus improve cell efficiency, since it allows it to absorb light in ultrathin layers, which in turn provide better collection of photogenerated charge carriers [32]. At the same time it is possible to decrease the time needed to grow the active layer, thus also reducing the amount of raw materials, and finally decrease the ratio between the production cost of the cell with respect to the photogenerated electric power. Generally speaking, among all the loss mechanisms that decrease the efficiency of a solar cell, the losses due to incomplete light trapping account for a 5% roughly, thereby giving rise to a great interest in the field of optical density of state engineering of photonic materials. Disordered materials play a fundamental role in this case because they have a large degree of freedom that allows optimal engineering of the optical density of states within a large bandwidth of operation [4].

4.5.1 *Early Designs and the Problem of Collection Efficiency*

Despite the basic idea to trap light by exploiting interference effects in the strong scattering regime, thus increasing light–matter interaction, up to today the most practical solution is one that makes use of disordered photonic materials, besides the use of disordered gratings or rough surfaces [33, 34]. These approaches have nothing to do with the mechanism of light trapping. The integration of rough surfaces just above the active material allows one to diffuse the incident light, maximizing the optical path of each light ray and therefore the probability to be absorbed. Other advantages come from the decrease of the reflection coefficient of the incident light, thanks to better refractive index matching between the absorber and the external world. Moreover this approach does not directly modify the active materials and does not influence their electronic properties. The upper limit for the absorption enhancement that can be reached by using the diffusing mechanism has been calculated by E. Yablonovitch [35]. Within the optic ray approximation, the limit sets the maximum absorption enhancement of a rough surface (in double-pass configuration), respect to flat (single-pass) geometry. This limit is $4n^2 \sin^2 /\theta$, where n is the refractive index of the active layer and θ is the angle of the emission cone in the medium surrounding the cell. By roughening a slab of silicon and putting on the bottom of it a back reflector it is theoretically possible to enhance the absorption by a factor of 50 in the visible range. Despite the fact that this limit does not take into account interference effects, it represents a reference value for absorption enhancement in a photovoltaic material.

One possible implementation of a solar cell that exploits the diffusion-enhanced mechanism can be achieved, for example, by smearing the surface of the active material with a disordered arrangement of scatterers, for example, small dielectric spheres, or more simply by gently roughening its surface. Figure 4.10 shows schematic representations of solar cells that exploit the principle of multiple scattering.

The standard architecture is a flat surface (Fig. 4.10a) where a possible ARC is deposited on its top. The optical path length is not altered and the absorption is limited by the single or double pass of the photon through the active layer. Only increasing its thickness it is possible to enhance the absorption efficiency. Figure 4.10b,c shows two possible implementations to use random scattering and thus enhance the absorption efficiency. In the first case the presence of disordered arranged dielectric spheres on the top of the active layer is able to spread the light through the active material [36], while in the second case the same result is achieved by roughening its surface [30, 37–40]. In both cases it is possible to decrease the amount of reflected light and increase the optical path length within the active layer.

To exploit the potential of disordered photonic systems and benefit from the coherent light-trapping mechanisms, we have to heavily perturb the active material by including dielectric scatterers inside it. This allows us to completely manage the flow of light within the material. Figure 4.10d shows the basic principle of the coherent light-trapping mechanism, depending on modal coupling. In this case the wave nature of light is completely exploited being the thickness of the active layer and the typical pitch of the photonic material comparable with the incident wavelength. The

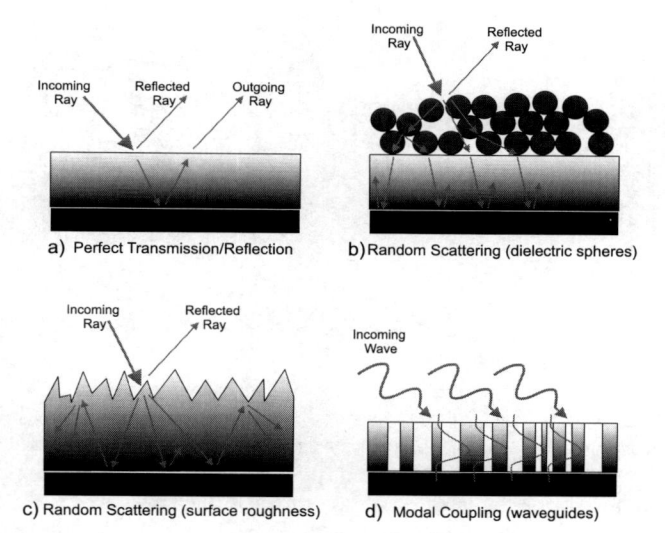

Figure 4.9 (a) The ability of an antireflection coating (ARC) to maximize optical absorption is limited because the optical path length is not altered. (b) Incoherent or geometric light trapping where random scattering is used to increase the optical path length inside the photoactive region. (c) Coherent light trapping where interference, diffraction, and optical electric field concentration are exploited to enhance the likelihood that an incident photon is absorbed in the photoactive region.

amount of coupled and reflected light, the acceptance angles, and the polarization response can be engineered by playing with the size, etch depth, and positional correlation of the scattering centers, changing to all intents and purpose the optical density of states of the disordered material [4]. Recently it has been theoretically demonstrated that by fully exploiting the potential of coherent light-trapping mechanisms, it is possible to overcome the Yablonovitch limit. This is possible if one is able to confine the guided mode within the active material to a deep-subwavelength scale [41].

One of the drawbacks of the light-trapping mechanism is the presence of scattering centers within the active materials. This is detrimental for the overall cell efficiency because the surface of each scatterer acts a recombination center for the charged carriers, decreasing the amount of photogenerated current that reach the p-n junction. One possible way to overcome this problem is to reduce

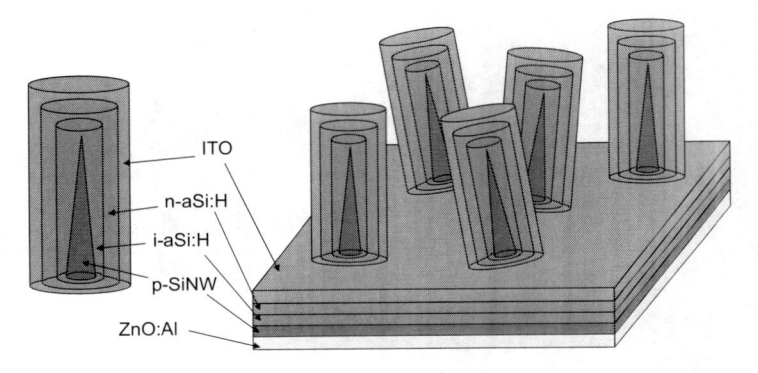

ITO
n-aSi:H
i-aSi:H
p-SiNW
ZnO:Al

Figure 4.10 Scheme of the approach to integrate the p-n junction inside each single scattering center. The radial geometry of the junction allows a short carrier diffusion, minimizing the probability of an unwanted recombination mechanism. ZnO:Al: aluminum-doped zinc oxide; p-SiNW: p-doped crystalline silicon nanowire; i-aSi:H: intrinsic hydrogenated amorphous silicon; n-aSi:H: n-doped amorphous silicon; ITO: indium tin oxide [42].

as much as possible the distance that the electron–hole pairs have to diffuse in order to avoid an unwanted recombination mechanism [42]. Figure 4.10 shows the scheme of a possible solution. The p-n junction is radially integrated directly inside the scattering center, allowing a a very short carrier diffusion length. Taking inspiration from this innovative architecture of p-n junctions, it is possible to find solutions for each specific design that involve the light-trapping mechanism.

4.5.2 Recent Designs to Coherently Manage and Trap the Flow of Light in Ultrathin Dielectric Films

The most appropriate dielectric architecture to exploit the light-trapping mechanism for photovoltaics driven applications is to conceive them in planar waveguide configurations. In these materials propagation and localization of light exploit the 2D multiple scattering of index-guided waves 4.5, thus leaving one of the planar surfaces available for the in-coupling of incident light. The advantages of the 2D architecture with respect to 3D arrangements of scattering centers are many. The most important is that the probability to trap light and create long living modes in two dimensions is much higher than in three dimensions [26]. Other advantages are related to the intrinsic two-dimensionality of the system that leaves the third degree of freedom available for a direct optical access to, for example, couple and detect light and for local engineering of the material [20]. Finally planar architectures are compatible with standard growing processes and can be integrated with electronic components. All these requirements are fundamental to conceive a solar-cell-based device. A coherent light-trapping mechanism has been developed in 2D ordered photonic crystals, better known as photonic crystal slabs [27]. This class of materials is able to confine light in a region of a few cubic wavelengths, with lifetimes up to nanoseconds, giving an opportunity to extremely enhance light–matter interaction [43]. Nevertheless, such an ability comes at the cost that they are optimized for a limited set of operating conditions, that is, an extremely small frequency bandwidth, a narrow **k**-vector acceptance cone, and a single polarization operation. By introducing a small amount of structural disorder, the ability to trap light decreases but the operating working conditions increase. Indeed the lifetime of localized modes decreases to hundreds of picoseconds or even less, while density of modes increases, and the total acceptance angle widens. By further increasing the amount of structural disorder the lifetime of the modes decreases drastically and the spectral response becomes almost flat [4]. Figure 4.11 shows the dispersion of the resonances in two different random media, with a different degree of spatial correlation between scatterers' positions [44]. The 3D architecture of such a system is similar to that of Fig. 4.5, that is, a dielectric slab perforated with cylindrical air holes that act as scattering centers.

For low frequencies around $t/\lambda \simeq 0.6$ (t is the film thickness), the resonance distribution exhibits a well-defined peak, corresponding to a propagating mode with an effective refractive index. Increasing the normalized frequency we observe a progressive broadening of the peak distribution in the reciprocal space that is associated with a stronger light scattering by the holes. When the effective wavelength of the disordered mode is comparable to the length scale over the which the material refractive index fluctuates, the resonance broadly spreads over the reciprocal space, leading to an increase of the **k**-vector density above the light line. The number of **k**-vector

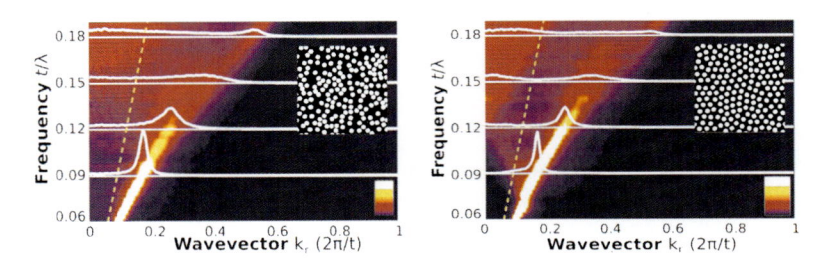

Figure 4.11 Resonance distribution of a disordered photonic material realized on a slab waveguide. The dashed line is the light line. The component of the dispersion above the light line radiates outside the slabs and at the same time allows the coupling of the light inside the material. The left panel shows the dispersion of an uncorrelated disorder material, while the right panel shows the dispersion for a slab waveguide where the positions of the scattering centers are correlated. The positional correlation redistributes the amount of guided components of the modes with respect to the radiative components, allowing a full density of state engineering of the material.

components above the light line is directly related to the lifetime of the localized modes, and for reciprocity, the same components allow one to couple light inside the dielectric membrane by illuminating it from the third dimension. The almost uniform distribution of the resonances between $k_{||} = 0$ and the light line indicates that the coupling efficiency depends only weakly on the angle of incidence of light. By comparing the two panels it is easy to see that the amount of resonance lying above the light line depends on the normalized energy t/λ and on the degree of correlation. Indeed by playing with structural parameters it is possible to engineer and redistribute the amount of leaky resonance, thus enhancing or depleting the probability to couple light inside the dielectric film. By further increasing the degree of structural correlations between scattering elements, the resonance distribution gradually changes to that of a 2D photonic crystal. The benefit provided by the coherent light-trapping mechanism in such disordered structures is quite surprising. The calculated integrated absorption efficiency of nanopatterned thin amorphous silicon (a-Si) films in the 600–800 nm wavelength range of the solar spectrum is 53% for a completely random pattern of scattering centers and 58% for the correlated case. These values have to be compared with the same quantity for a bare (unpatterned) film that is 16% and for the completely ordered counterpart (2D photonic crystal) that is 53% like in the random case. This is the main result that drives the research around disordered materials to enhance absorption efficiency. Indeed the value of the integrated efficiency suggests that the disordered structure absorbs at least the same quantity of incident light with respect to the perfect ordered case, with the advantage that disordered materials are more attractive from a large-scale production view, the structural disorder, always present in the production phase, being part of the physical process that increases the performance of the material. Realistic samples, made of a single layer of a-Si of thickness = 930 nm, deposited on top of 100 nm of silver shows absorption enhancement of a factor of 1.5 and a weak dependence on the angle of incidence and polarization of light [45].

4.5.3 *Conclusions*

The research field about the coherent light-trapping mechanism has made a huge step forward in these last years, discovering that it is possible to take advantage from the structural disorder on the wavelength scale. Experimental results made on a-Si disordered films are in agreement with the expected trend. On the other hand there are huge steps to do to minimize unwanted recombination mechanisms of electron–hole pairs, mechanisms that drastically decrease the final efficiency of the solar cell. The proposed solution of the radial junction architecture embedded inside dielectric rods could be the right way to face and solve this problem. Neglecting electronics-related issues, the theoretical efficiency of a solar cell made by a single junction of silicon could increase by 5%, only exploiting at its best light-trapping mechanism. This value should represents a huge step forward in this field. To cover this gap a careful understanding of the role that disorder plays in in respect of local density of states and mode coupling is of crucial importance to design photonic materials with better light-trapping behaviors.

References

1. Ishimaru, A. (1978). *Wave Propagation and Scattering in Random Media*, Academic Press.

2. Brown, R. (1828). A brief account of microscopical observations made in the months of June, July and August, 1827, on the particles contained in the pollen of plants; and on the general existence of active molecules in organic and inorganic bodies, *Philos. Mag.*, **4**, pp. 161–173.

3. Einstein, A. (1905). Über die von der molekularkinetischen Theorie der Wärme geforderte Bewegung von in ruhenden Flüssigkeiten suspendierten Teilchen, *Ann Phys.*, **17**, p. 549.

4. Vynck, K., Burresi, M., Riboli, F., and Wiersma, D.S. (2012). Photon management in two-dimensional disordered media, *Nat. Mater.*, **11**, pp. 1017–1022.

5. Bertolotti, J., van Putten, E.G., Blum, C., Lagendijk, A., Vos, W.L, and Mosk, A.P. (2012). Non-invasive imaging through opaque scattering layers, *Nature*, **491**, pp. 232–234.

6. Katz, O., Heidmann, P., Fink, M., and Gigan, S. (2014). Non-invasive single-shot imaging through scattering layers and around corners via speckle correlations, *Nat. Photonics*, doi:10.1038/nphoton.2014.189.

7. Finnemore, A., Cunha, P., Shean, T., Vignolini, S., Guldin, S., Oyen, M., and Steiner, U. (2012). Biomimetic layer-by-layer assembly of artificial nacre, *Nat. Commun.*, **3**, 966.

8. Lagendijk, A., van Tiggelen, B., and Wiersma, D.S. (2009). Fifty years of Anderson localization, *Phys. Today*, **62**, 8, pp. 24–29.

9. Roati, G., D'Errico, C., Fallani, L., Fattori, M., Fort, C., Zaccanti, M., Modugno, G., Modugno, M., and Inguscio, M. (2008). Anderson localization of a non-interacting Bose–Einstein condensate, *Nature*, **453**, pp. 895–898.

10. Roati, G., D'Errico, C., Hu, H., Strybulevych, A., Page, J.H., Skipetrov, S.E., and van Tiggelen, B.A. (2008). Localization of ultrasound in a three-dimensional elastic network, *Nat. Phys.*, **4**, pp. 945–948.

11. Anderson, P.W. (1958). Absence of diffusion in certain random lattices, *Phys. Rev.*, **109**, pp. 1492–1505.

12. Van Albada, M.P., and Lagendijk, A.D. (1985). Observation of weak localization of light in a random medium, *Phys. Rev. Lett*, **55**, p. 2692.

13. Wolf, P.E., and Maret, G. (1985). Weak localization and coherent backscattering of photons in disordered media, *Phys. Rev. Lett*, **55**, p. 2696.

14. John, S. (1985). Strong localization of photons in certain disordered superlattices, *Phys. Rev. Lett*, **58**, p. 2486.

15. Dalichaouch, R., Armstrong, J.P., Shultz, S., Platzman, P.M., and McCall, S.L. (1991). Microwave localization by two-dimensional random scattering, *Nature*, **354**, pp. 53–55.

16. Wiersma, D.S., Bartolini, P., Lagendijk, A., and Righini, R. (1997). Localization of light in a disordered medium, *Nature*, **390**, pp. 671–673.

17. Schwartz, T., Bartal, D., Fishman, S., Segev, M. (2007). Transport and Anderson localization in disordered two-dimensional photonic lattices, *Nature*, **446**, pp. 52–55.

18. Riboli, F., Barthelemy, P., Vignolini, S., Intonti, F., De Rossi, A., Combrie, S., and Wiersma, D.S. (2011). Anderson localization of near-visible light in two dimensions, *Opt. Lett.*, **36**, pp. 127–129.

19. Segev, M., Silberberg, Y., and Christodoulides, D.N. (2011). Anderson localization of light, *Nat. Photonic*, **7**, pp. 197–204.

20. Riboli, F., Caselli, N., Vignilini, S., Intonti, F., Vynck, K., Barthelemy, P., Gerardino, A., Balet, L., Li, L.H., Fiore, A., Gurioli, M., and Wiersma, D.S. (2014). Engineering of light confinement in strongly scattering disordered media, *Nat. Mater.*, **13**, pp. 720–725.

21. Barthelemy, P., Bertolotti, J., and Wiersma, D.S. (2008). A Lévy flight for light, *Nature*, **453**, pp. 495–498.

22. Akkermans, E., and Montambaux, G. (2001). *Mesoscopic Physics of Electrons and Photons*, Cambridge University Press.

23. Sheng, P. (1995). *Introduction to Wave Scattering, Localization, and Mesoscopic Phenomena*, Academic Press, San Diego, CA.

24. van Rossum, M.C., Nieuwenhuizen, W., and Th. M. (1999). Multiple scattering of classical waves: microscopy, mesoscopy, and diffusion, *Rev. M. Phys.*, **71**, 1, pp. 313–371.

25. Vollhardt, D., and Wölfle, P. (1992). Self-consistent theory of Anderson localization. In Hanke, W., and Kopaev, Yu. V. (eds.), *Electronic Phase Transitions*, Vol. 32 of the series *Modern Problems in Condensed Matter Sciences*, pp. 1–77, North-Holland, Amsterdam.

26. Abrahams, E., Anderson, P.W., Licciardello, D.C., and Ramakrishnan, T.V. (1979). Scaling theory of localization: absence of quantum diffusion in two dimensions, *Phys. Rev. Lett*, **42**, pp. 673–676.

27. Joannopoulos, J.D., Johnson, S.G., Winn, J.N., and Meade, R.D. (2008). *Photonic Crystals: Molding the Flow of Light Second Edition*, Princeton University Press.

28. Laurent, D., Legrand, O., Sebbah, P., Vanneste, C., and Mortessagne, F. (2007). Localized modes in a finite-size open disordered microwave cavity, *Phys. Rev. Lett*, **99**, 253902.

29. Genack, A.Z., and Zhang, S. (2009). Wave interference and modes in random media. In *Tutorials in Complex Photonic Media*, SPIE Publications.

30. Bessonov, A., Cho, Y., Jung, S.J., Park, E.A., Hwang, E.S., Lee, J.W., Shin, M., and Lee, S. (2011). Nanoimprint patterning for tunable light trapping in large-area silicon solar cells, *Sol. Energ. Mater. Sol. Cells*, **95**, 10, pp. 2886–2892.

31. John, S., (1985). Localization and absorption of waves in a weakly dissipative disordered medium, *Phys. Rev. B*, **31**, pp. 304–309.

32. Polman, A., and Atwater, H.A. (2012). Photonic design principles for ultrahigh-efficiency photovoltaics, *Nat. Mater.*, **11**, pp. 174–177.

33. Rockstuhl, C., Fahr, S., Bittkau, K., Beckers, T., Carius, R., Haug, F.J., Söderström, T., Ballif, C., and Lederer, F. (2010). Comparison and optimization of randomly textured surfaces in thin-film solar cells, *Opt. Express*, **18**, pp. A335–A341.

34. Sheng, X., Johnson, S.G., Michel, J., and Kimerling, L.C. (2011). Optimization-based design of surface textures for thin-film Si solar cells, *Opt. Express*, **19**, pp. A841–A850.

35. Yablonovitch, E., (1982). Statistical ray optics, *J. Opt. Soc. Am. A*, **71**, 7, pp. 899–907.

36. Hung, Y.J., Hsu, S.S., Wang, Y.T., Chang, C.H., Chen, L.Y., Su, L.Y., and Huang, J.J. (2011). Polarization dependent solar cell conversion efficiency at oblique incident angles and the corresponding improvement using surface nanoparticle coating, *Nanotechnology*, **22**, 48, p. 485202.

37. Janthong, B., Moriya, Y., Hongsingthong, A., Sichanugrist, P., and Konagai, M. (2013). Management of light-trapping effect for a-Si:H/c-Si:H tandem solar cells using novel substrates, based on {MOCVD} ZnO and etched white glass, *Sol. Energ. Mater. Sol. Cells*, **119**, pp. 209–213.

38. Cho, J.S., Baek, S., Park, S.H., Park, J.H., Yoo, J., and Yoon, K.H. (2012). Effect of nanotextured back reflectors on light trapping in flexible silicon thin-film solar cells, *Sol. Energ. Mater. Sol. Cells*, **102**, pp. 50–57.

39. van Lare, M., Lenzmann, F., and Polman, A. (2013). Dielectric back scattering patterns for light trapping in thin-film Si solar cells, *Opt. Express*, **18**, pp. 20738–20746.

40. Yu, X., Yu, X., Zhang, J., Hu, Z., Zhao, G., and Zhao, Y. (2014). Effective light trapping enhanced near-UV/blue light absorption in inverted polymer

solar cells via sol-gel textured Al-doped ZnO buffer layer, *Sol. Energ. Mater. Sol. Cells*, **121**, pp. 28–34.

41. Yu, Z., Raman, A., and Fan, S. (2011). Nanophotonic light-trapping theory for solar cells, *Appl. Phys. A*, **105**, pp. 329–339.

42. Yu, L., Misra, S., Wang, J., Qian, S., Foldyna, M., Xu, J., Shi, Y., Johnson, E., and Cabarrocas, P.R. (2014). Nanophotonic light-trapping theory for solar cells, *Nat. Comm.*, **4**, 4357, doi:10.1038/srep04357.

43. Akahane, Y., Asano, T., Song, B.S., and Noda, S. (2003). High Q photonic nanocavity in a two-dimensional photonic crystal, *Nature*, **425**, pp. 944–947.

44. Gaurasundar, M.C., Burresi, M., Pratesi, F., Vynck, K., and Wiersma, D.S. (2014). Light transport and localization in two-dimensional correlated disorder, *Nature*, **112**, p. 143901.

45. Burresi, M., Pratesi, F., Vynck, K., Prasciolu, M., Tormen, M., and Wiersma, D.S. (2013). Two-dimensional disorder for broadband, omnidirectional and polarization-insensitive absorption, *Opt. Express*, **21**, S2, pp. A268–A275.

Chapter 5

Nanowire Architecture for Fast Electronic Devices

Leonardo Viti,[a] Alessandro Pitanti,[a,b] and Miriam S. Vitiello[a]

[a] *NEST, CNR-Istituto Nanoscienze and Scuola Normale Superiore, Piazza San Silvestro 12, I-56126 Pisa, Italy*
[b] *California Institute of Technology, 1200 E. California Blvd., Pasadena, CA 91125, USA*
leo.viti.14@gmail.com, alessandro.pitanti@sns.it, miriam.vitiello@sns.it

5.1 Introduction

5.1.1 *Terahertz Detection*

Terahertz (THz) radiation, approximately defined in the $30-300$ μm range, lies in the region of the electromagnetic spectrum that is often called the *THz gap*. The word "gap" originates from the lack of compact, solid-state, coherent radiation sources. However, this problem has started to be seriously addressed with the development of quantum cascade lasers (QCLs) operating in the THz range [1]. These semiconductor-based sources—though still limited in operating temperatures—hold the promise of being a core component of practical THz systems devoted to spectroscopy,

Nanodevices for Photonics and Electronics: Advances and Applications
Edited by Paolo Bettotti

sensing, and, more recently, imaging in a wide variety of application areas (biomedical diagnostics, security, cultural heritage, quality and process controls, etc.), mainly due to the fact that many materials such as paper, plastics, and ceramics, which are opaque at visible frequencies, are highly transmissive across the THz and microwave ranges [2]. The other "THz gap," which needs to be overcome for imaging-oriented applications is the one concerning detectors.

THz and sub-THz radiation detection systems can be classified into two main categories, (i) incoherent detection systems (with direct detection sensors), which are only sensitive to signal amplitude and which, as a rule, are inherently broadband, and (ii) coherent detection systems, which allow simultaneous detection of signal amplitude and phase. Coherent detection is usually achieved by employing heterodyne systems in which detected signals are transferred, via mixers, to much lower frequencies (10–30 GHz) and then amplified. Basically these systems are inherently selective (narrow band). Frequently used mixers are devices having a strong electric field quadratic nonlinearity. Examples are forward-biased Schottky barrier diodes (SBDs), superconductor-insulator-superconductor (SIS) tunnel junctions, semiconductor and super-conducting hot-electron bolometers (HEBs), and superlattices (SLs) [3]. Among the advantages of heterodyne detection techniques there is (i) the possibility to be sensitive on both frequency modulation and phase modulation; (ii) the negligible sensitivity to background radiation noise, usually dominant in the THz range; and (iii) the possibility to detect much weaker radiant signal powers compared to direct detection. Clear disadvantages are (i) the need of beams equally polarized, coincident, and equal of diameter and (ii) the difficulty of producing large-format arrays. One figure of merit conventionally employed to compare millimeter and submillimeter detector performances is the noise-equivalent power (NEP). It is a function of noise and responsivity and is defined as the value of root-mean-square (rms) input radiant signal power W required to produce an rms output signal that is equal to the rms noise N value (signal-to-noise ratio [SNR] = 1). The lower the NEP, the more sensitive the detector.

In the case of thermal detectors intrinsic temperature fluctuation noise (Johnson–Nyquist noise) defines the upper NEP limit as

$$N = \sqrt{\frac{4k_B T}{R_{th}}},\qquad(5.1)$$

where k_B is the Boltzmann constant, T is the temperature of the thermistor, and R_{th} is the resistance between the detector and the heat sink. Another critical aspect of what concerns THz application requirements is the small photon energies of millimeter and submillimeter detectors compared to the thermal energy (26 meV) at room temperature. Also the Airy disk diameter (diffraction limit) defined by

$$A_D = 2.44\frac{\lambda f}{d}\qquad(5.2)$$

is large, dictating a small spatial resolution for THz systems. Here f is the focus length of the optical system and d its input diameter. Finally, detector performances are usually characterized by means of the specific detectivity (D^*), which is equal to the reciprocal of NEP, normalized per square root of the sensor area and frequency bandwidth (Δv), or alternatively via the detector responsivity $\mathcal{R}_v$ directly related to D^* by means of the noise spectral density N:

$$D^* = \frac{\sqrt{A v}}{\text{NEP}} = \frac{\mathcal{R}_v \sqrt{A}}{N}\qquad(5.3)$$

Commercially available THz detectors are based on thermal sensing elements that are either very slow (10–400 Hz modulation frequency for Golay cells or pyroelectric elements, with NEPs in the 10^{-10} W/$\sqrt{\text{Hz}}$ range) or require deep cryogenic cooling (4 K for superconducting HEBs), while those exploiting fast nonlinear electronics (Schottky diodes) are usually limited to sub-THz frequencies for best performances [3, 4].

From a technological point of view, Schottky diodes [4, 5] are the prototypical electronic components for room-temperature detection of high-frequency radiation. The operating principle exploits the nonlinearity and asymmetry of the current–voltage characteristic to generate, through rectification, a continuous signal out of the oscillating incident electromagnetic field. Despite this simple idea, practical implementations are often sophisticated, since electric transport in the device must be able to "follow" the radiation frequency, requiring minimization of the RC time constant. State-of-the-art technologies are based on vertical transport, featuring metallic air bridges for contacting. However, these structures are typically delicate and very demanding for array geometries. Furthermore, performances drop rapidly with a frequency above 1 THz. More recently, electronic devices based on the gate modulation of the channel conductance by the incoming radiation have been realized in high-electron-mobility transistor (HEMT), field-effect transistor (FET), and silicon metal-oxide semiconductor field-effect transistor (Si-MOSFET) architectures, showing fast response times and high responsivity ($\mathcal{R}_v$) [6, 7], as well as the possibility of implementing multipixel focal-plane arrays [8]. These technologies, in principle easily scalable to even large arrays, are still limited by detection cutoff frequencies of a few hundred gigahertz, above which responsivity drops and cryogenics is required.

5.1.2 *Detection Mechanism in FETs*

The basic idea behind the operation of a FET detector is that the channel of a FET can in principle act as a resonator for plasma waves, whose frequency depends on the channel dimension. In the presence of gate lengths in the µm or nanometer range, the plasma frequency could cover the THz range. The operating mechanism of a FET detector is not trivial [9] but can intuitively be interpreted as deriving from the nonlinear dependence of the FET channel current on the gate voltage near the pinch-off point. These devices have the advantage that the responsivity can be maximized with the gate bias V_G, while measuring the output at the drain with no source–drain bias applied, thus dramatically improving the SNR.

THz detection in FETs is mediated by the excitation of plasma waves in the transistor channel (Fig. 5.1). The plasma wave velocity in the gated region is typically noticeably larger compared to the electron drift velocity. A strong resonant photoresponse is usually predicted in materials having plasma-damping rates lower than both the frequency ω of the incoming radiation and the inverse of the wave transit time τ in the channel. To reach mobilities of at least several thousand cm^2/Vs at frequencies > 1 THz low temperatures

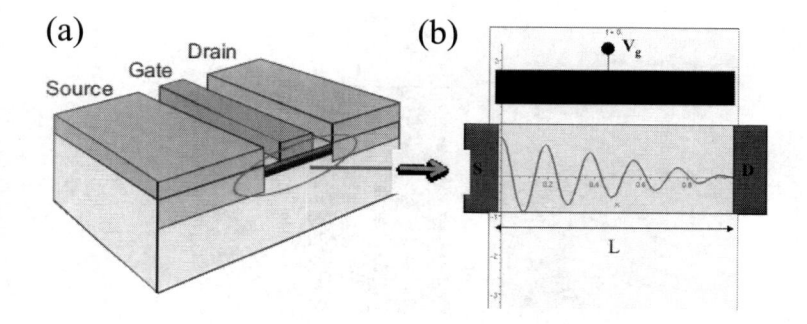

Figure 5.1 (a) Schematic of a field-effect transistor (FET). (b) Plasma oscillations in a FET.

are required. Under these conditions, stationary states arising from the quantization of plasma waves over the gate width are excited whenever V_G is such that $n\pi s/2L_g = \omega$, where n is an odd integer, s the plasma wave velocity, and L_g the gate length. On the other hand, when plasma oscillations are overdamped, that is, they decay on a distance smaller than the channel length, broadband THz detection is predicted [6]. In this case, the oscillating electric field of the incoming radiation induced between source and gate electrodes produces a modulation of both charge density and carrier drift velocity [9]. Carriers travelling toward the drain generate a continuous source–drain voltage, Δu, controlled by the carrier density in the channel that can be in turn maximized by varying V_G. High mobility at room temperature is therefore crucial to take full advantage of resonant detection [10].

5.2 Nanowire THz Detectors

One-dimensional nanostructure devices are at the forefront of studies on future electronics, although issues like massive parallelization, doping control, surface effects, and compatibility with silicon industrial requirements are still open challenges. On the other hand, significant progress is recently being made in the atomic- to nanometer-scale control of materials' morphology, size,

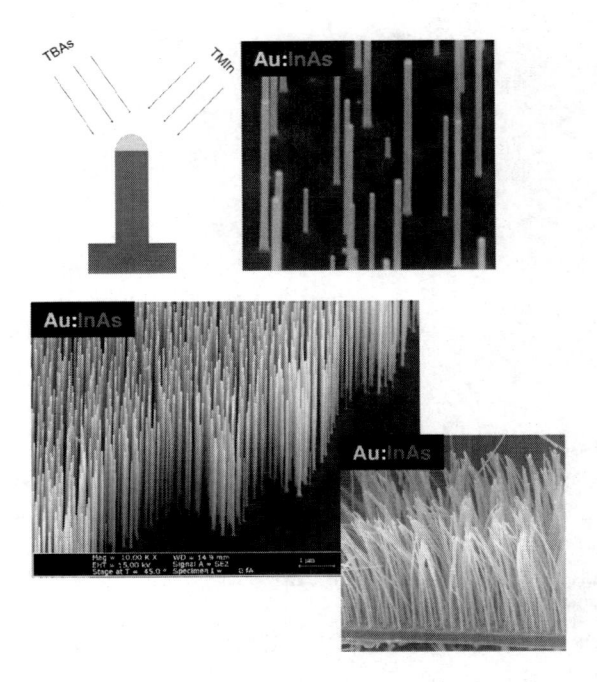

Figure 5.2 Top panel: Schematics of gold-assisted chemical beam epitaxy (CBE) growth of InAs nanowires; scanning electron micrographs of a forest of nanowires.

and composition [11], including the growth of axial [12], radial [13], and branched [14] nanowire-based heterostructures.

Nanowires can be easily removed from the host substrate (Fig. 5.2) and placed on top of a new functional one for individual contacting, even in relatively large numbers, with a simple planar technology very suitable for low-capacitance circuits. Therefore, they represent, in principle, an ideal building block for implementing rectifying diodes [3] or plasma wave detectors [6] that could be well operated into the THz, thanks to their typical attofarad-order capacitance. Surprisingly, despite the strong effort in developing these nanostructures for a new generation of complementary metal-oxide semiconductors (CMOS), memory, and/or photonic devices,

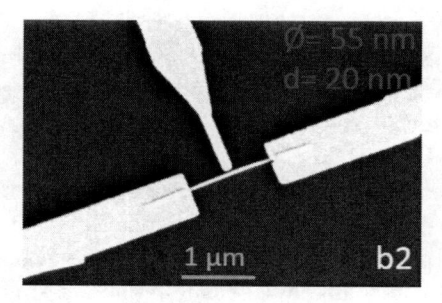

Figure 5.3 Scanning electron micrograph of a nanowire-based FET: a THz single-pixel detector.

their potential as THz radiation sensors has not been explored so far.

FETs based on InAs nanowires (Fig. 5.3) [15, 16] have attracted considerable attention in the last years mainly due to the excellent properties of InAs for electron transport. InAs has indeed high electron mobility and saturation velocity and a potentially long electron mean free path, enabling high transconductance at low drive voltages. Mobility values as high as 6000 cm^2/Vs have been indeed demonstrated at room temperature in InAs nanowire FETs [17].

The narrow bandgap and degenerate Fermi-level pinning further allow for easy formation of excellent ohmic contacts, which becomes increasingly important as the transistor is scaled down. In addition, InAs nanowires can be grown epitaxially on silicon without the use of gold seeding, thus making the process viable also for low-cost silicon technology integration [18] where deep Au levels in the silicon bandgap must be avoided. However, one of the open issues in nanowire-based transistors is the achievement of a high on-current when downscaling the nanowire diameter, which has been identified as a potential roadblock for this technology. The nanowire diameter must follow the scaling of the gate length due to electrostatic considerations, which leads to a rapid increase in the series resistance of the ungated source and drain regions. This series resistance can be reduced through doping, metal diffusion, or deposition of a highly conductive material around the wire in the source–drain regions.

The THz detection principle in a FET was first explained by the Dyakonov–Shur plasma wave theory [9]. The nonlinear properties of plasma wave excitations in nanoscale FET channels enable their response at frequencies appreciably higher than the device cutoff frequency. In the ballistic regime of operation, the momentum relaxation time is longer that the electron transit time and the FETs can also be used for resonant (with peak response at a certain wavelength directly tunable by changing the gate voltage) THz detection [19]. Alternatively a broadband, nonresonant response is obtained [20]; when THz radiation is coupled to the FET—between gate and source—the AC THz voltage modulates simultaneously the carrier density and the carrier drift velocity. As a result, the THz signal is rectified and leads to a DC signal Δu between source and drain proportional to the received power. The value of this voltage (or current, depending to the readout circuit) depends on the carrier density in the channel, which may be controlled by the gate voltage, and on their drift velocity. The device operates as a square law detector in which the largest nonlinearity, leading to the highest responsivity, is achieved around the channel pinch-off. The gate voltage value ensuring the best sensitivity is also determined by the transistor load [21]. In addition, some asymmetry between source and drain is needed to induce Δu; this can originate from the difference in the source and drain boundary conditions due to some parasitic capacitance but usually stems from the asymmetry in feeding the incoming radiation, which can be achieved either by using a special antenna or by an asymmetric design of source and drain contact pads. The radiation then predominantly creates the THz AC voltage between the source (or drain) and gate contacts. Finally, the asymmetry can naturally arise if a DC current is passed between source and drain, creating a depletion of the electron density on the drain side of the channel.

Ideally, the described detection scheme should ensure the best performance in 1D geometry, when only one plasmon mode can be excited by the incoming electromagnetic field. Furthermore, very low channel-to-gate capacitance values are needed to extend the detecting range at progressively higher frequencies. These requirements must be combined with high electron mobility to avoid strong damping of the induced plasma wave before reaching the drain contact, which would lower significantly the detector responsivity. Nanowire FETs are then very appealing candidates to fulfil the above requirements.

5.2.1 *Contacting the Nanowire*

Nanowires are semiconductor structures with small dimensions and high aspect ratios. Average dimensions of 100 nm diameters and 1–2 μm lengths are well suitable for the realization of nanotransistors, even if the bottleneck in miniaturizing the full device is essentially represented by the metallic electrodes used to contact the nanowire itself, whose lateral size is usually larger than 50 nm. Even if often neglected, the metallic regions surrounding the nanowires are certainly perturbing the fields operating the device, in terms of both gate screening and asymmetry effects. Let's consider a nanowire-based FET geometry as the one sketched in Fig. 5.4a. Here a side electrode, physically disconnected from the nanowire, is used to gate the device. The other two metallic pads are used to contact the nanowires and get a current flow through the semiconductor. The screening effect is well depicted in the electrostatic finite-element method (FEM) simulations of Fig. 5.4a,b; the electric potential generated by the lateral gate is strongly deformed when the presence of the metallic contacts is taken into account. In most experiments, the contacts are either grounded or possess a small potential bias and therefore contribute in the determination of the shape of the electric field generated by the gate. Assuming a homogeneous, nonscreened gate field makes the estimation of nanowire electrical parameters, such as mobility or charge density, a delicate matter [22] when simple charge transport experiments are taken into account. As an example, we can show how the pinch-off threshold voltage (V_{th}) is influenced by the presence of electrical contacts. The threshold voltage is defined as the gate voltage at which the transistor channel completely closes and no source–drain current is flowing. This can be seen considering a very simple capacitive coupled model where the nanowire charge Q is given by [23]

$$Q = C_{wG} V_{th},$$
(5.4)

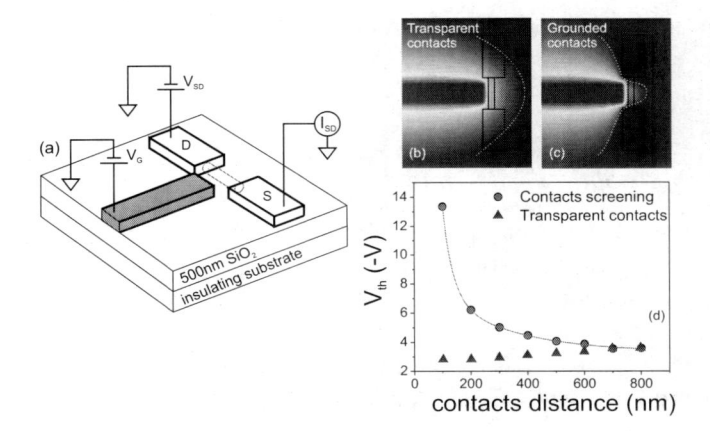

contacts distance (nm)

Figure 5.4 (a) Sketch of the model system for a lateral gate nanowire FET. A small bias is applied to the source contact (S), the drain (D) is grounded, and the gate bias (V_G) can be used to inject/deplete charge in the nanowire (red cylinder). When a 1 V potential is applied to the gate and considering transparent contacts (b), the nanowire feels an almost homogeneous field and we can consider a Fermi-level shift almost constant on the whole nanowire. By simply grounding the contacts, the potential is pinned and deformed (b) and the nanowire experiences a dishomogeneous gating field. The dotted curve represents a potential isoline to better stress the effect. (c) Modification of threshold voltages as a function of contact distances for transparent and not-transparent contacts. A detailed description of the parameters employed in the simulation is reported in the main text. Partially redrawn with permission from Ref. [22]. Copyright 2012, AIP Publishing LLC.

with C_{wG} the wire gate capacitance. Assuming a constant charge density within the nanowire, shorter channels should have smaller threshold voltages; however, the gating field being more screened when the contacts are very close, a counterintuitive increase of V_{th} is produced. The charge is directly linked to the source–drain current as

$$I_{SD}\mu \frac{Q}{L_{ch}^2} V_{SD} = \frac{\mu C_{wG}}{L_{eff}^2}(V_G - V_{th})V_{SD}, \qquad (5.5)$$

where V_{SD} is the source–drain bias, L_{eff} the transistor channel effective length, and μ the carrier mobility. It's easy to see from Eq. 5.5 that I_{SD} is zero when the gate bias is exactly V_{th}. By using 3D charge transport simulations, it is possible to extract V_{th} by

Table 5.1 Electrical parameters for doped semiconductor InAs nanowires. m_e indicates the electron mass

m^*	$0.023m_e$	E_{pinn}	10 meV
ρ_{bg}	5e6 C/m^3	μ	1e3 cm^2/(V·s)

solving Eq. 5.5 for $I_{SD} = 0$. The last ingredient we need is the link between the nanowire charge and the gate potential. By shifting the Fermi energy, the gate field can inject (positive voltage) or deplete (negative voltage) n-type carriers into/from the nanowire. We can consider the total charge density at zero temperature as composed by a background (ρ_{bg}) component due to doping and by a gate-dependent charge:

$$\rho_{\text{tot}}(V_G) = \rho_{\text{bg}} + \frac{2}{3}e^{\frac{5}{2}}\frac{\sqrt{2}m^{*\,3/2}}{\hbar^3\pi^2}(E_{\text{pinn}} + V_G)^{3/2}, \qquad (5.6)$$

where m^* is the effective carrier mass and E_{pinn} the pinning energy. A typical n-doped InAs nanowire has the electrical characteristics reported in Table 5.1 [22].

By solving numerically Eqs. 5.5 and 5.6, using the FEM, we can evaluate the threshold voltage as a function of the contact distance. The nanowire device is schematized according to the sketch in Fig. 5.4a, considering a small bias on the source (S) electrode, a grounded drain (D) electrode, and different voltages on the gate finger, placed about in the center of the nanowire at a lateral distance of 200 nm. The simulation result is reported in Fig. 5.4d. By decreasing the nanowire length, considering a constant charge density, a reduction of the charge, and thus of the threshold voltage, is expected (triangles in Fig. 5.4c). However, when the electrodes are added to the simulation, the threshold voltage rises with decreasing contact distance, due to the more prominent screening. Therefore, the presence of metallic contacts in the core of the device makes the assessment of the device electrical parameters a nontrivial task, which can be partially addressed by considering full numerical simulations of the gate geometry.

The same numerical tool can be useful to infer further effects of having asymmetrical gating fields. One of the most interesting features of transistors is the nonlinearity of the source–drain current as a function of gate bias. The gate field modulates the charge density in a nonlinear fashion, producing a nonlinear current as in Eq. 5.5. This characteristic being one of the figures of merit for transistor-based detectors, we can define the charge nonlinearity using the concept of responsivity, defined as

$$\rho_{resp} = \int \frac{\partial^2 \rho(x, y, z)}{\partial V_G^2} \cdot \left(\frac{\partial \rho(x, y, z)}{\partial V_G} \right)^{-1} \rho_{tot}(x, y, z) \, dxdydz,$$

$$(5.7)$$

where ρ_{resp} is the charge density corresponding to the photocurrent. The lateral gate influences the nanowire in the center, while the nanowire edges are mostly affected by the contacts, which pin locally the nanowire bands at their potential voltage. Therefore, even when reverse-biased, a charge accumulation region will be present close to the contacts. For nanowires with a large background charge (doping), the main contribution to charge responsivity is found in these regions, for negative gate bias, as expected for n-type FET devices. On the other hand, if small or no external doping is present, the maximum contribution to ρ_{resp} is produced when charge is injected in the nanowire (positive voltage), in the spatial region in proximity of the gate electrode finger. The shift of the ρ_{resp} maximum from positive to negative gate bias with increasing doping is a peculiar behavior that can be well verified with FEM simulations. Considering the same model of Fig. 5.4, we report in Fig. 5.5 the results of charge responsivity simulations for different background doping levels. As discussed, when no doping is present, the maximum charge responsivity is found at positive voltages, coming from charge injection into the nanowire. Increasing the charge level, a second maximum at negative voltages starts to appear, becoming more and more intense with the background charge level, until the positive one becomes negligible. The fact that the nanowire transport is strongly affected by the presence of the contacts is an important feature to be taken into account when designing nanowire-based devices. Interestingly, while most of the nanowire-based devices reported in the literature can be fairly described using our simulations for the large doping case, we will show in Section 5.3.2 that undoped InAs/InSb nanowire FET detectors show this reported effect, having responsivity contributions both at positive and negative gate biases.

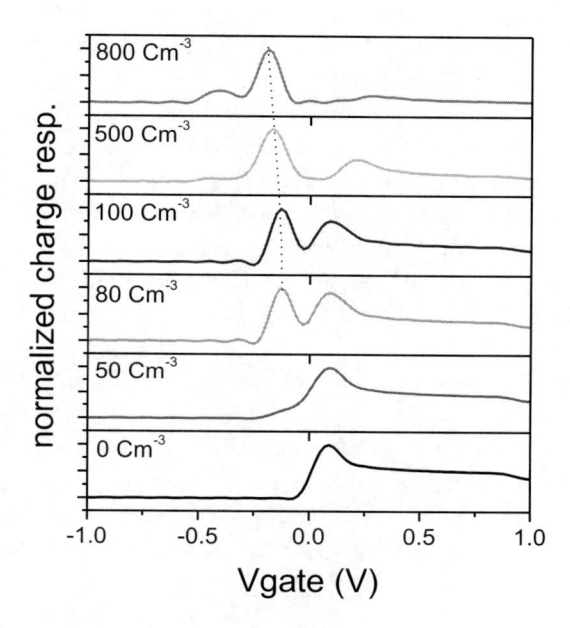

Figure 5.5 FEM simulation of the photoresponse charge density (ρ_{resp}) for different nanowire doping levels. The maximum responsivity moves from positive voltages to negative ones when the doping level is increased. Reprinted with permission from Ref. [22]. Copyright 2012, AIP Publishing LLC.

5.2.2 *Radiation Coupling*

When a THz beam is impinging on a 1D FET the effective beam spot area is significantly larger than the active channel area, meaning that a proper strategy to collect radiation should be adopted. The conventional scheme implemented to enhance the detection efficiency for submillimetric electromagnetic waves is the realization of planar metallic antennas directly patterned on the substrate and electrically connected to the device. Broadband dipole antennas, such as spirals and bowties (BTs), are commonly used [24]. In the case of THz FET detectors with an inherent symmetry in the active channel, as, for example, InAs nanowire FETs, the asymmetry can be easily induced by connecting one lobe of the dipole to the source electrode and the other to the gate. The use of an antenna has three main advantages: (i) Impedance matching of the detector to the free-space wave can be achieved over a large bandwidth, avoiding reflection losses; (ii) the effective detection area is increased; and (iii) the wavelength-dependent behavior of the antenna can be in principle employed to enhance detector response in a selected range of frequencies, once a proper resonant or quasiresonant antenna configuration is chosen. Despite these nominal benefits, a full impedance matching between antennas and feed circuit is very unlikely to be achieved in our cases, mainly because the antenna impedance is orders of magnitude lower than the nanowire resistance.

Moreover it is worth noticing that every metallic structure in the surroundings of the FET can in principle act as a radiation collector at a determined wavelength. In the case of nanostructures, like 1D FETs, electrical connections are conventionally provided by metallic *bonding pads*, which through *bonding wires* give access to the macroscopic connections of a chip (Fig. 5.6). This scheme

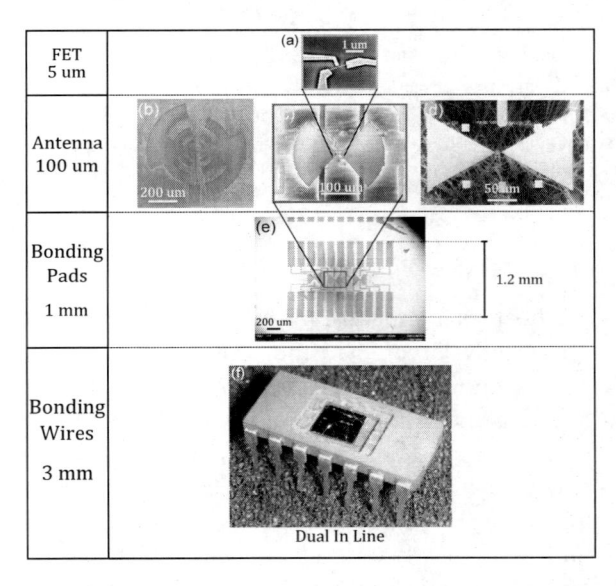

Figure 5.6 Dimensions of metallic elements in our devices. SEM images of (a) nanowire FET, (b) log periodic (LP) antenna, (c, d) bowtie (BT) antennas, and (e) on-chip electrical connections layout. (f) Picture of a *dual in-line* package.

intrinsically provides an additional coupling of the incoming radiation, with an efficiency strictly related to the wavelength of the incoming beam. Bonding pads are indeed $100 \times 200\,\mu m$ metallic rectangles placed about 1 mm away from the active area that can act as a receiving antenna for radiation of corresponding frequency; moreover the ~ 3 mm long bonding wires can couple electromagnetic waves with a frequency $\nu \approx 100$ GHz.

When the radiation wavelength is reduced below 1 mm ($\nu > 300$ GHz) the role of planar antenna coupling becomes crucial to transform electromagnetic free-space waves into a high-frequency current. Among the most effective solutions, BT and log periodic (LP) antennas offer a broad band impedance matching and better radiation characteristics compared to the classic dipole. These are often called self-complementary structures because the metallic and nonmetallic areas are of the same shape and can be transferred into each other by rotation.

5.2.2.1 Bowtie antenna

The planar BT antenna derives from the double conical antenna: it is formed by two isosceles triangles or by two circle arcs. Having a bandwidth of about two octaves it's well suited for broadband detection. The frequency response is determined by two geometrical parameters, the overall length L and the flare angle θ. The lower bound of the spectral band corresponds to a wavelength that is twice the overall length: $\lambda_{\max} = 2L$.

The BT antenna is polarization sensitive: the detector response vanishes if the electric field of the incoming beam is orthogonal to the bow axis (Fig. 5.7).

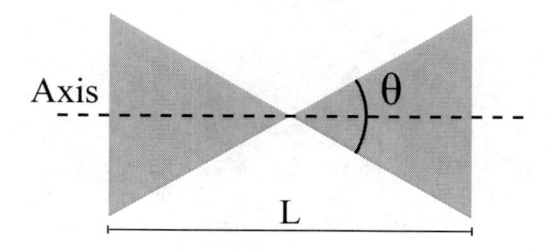

Figure 5.7 Schematic view of a bowtie antenna.

5.2.2.2 Log periodic antenna

A circular-toothed LP antenna is a broadband antenna designed to allow resonances repeated periodically with the logarithm of the frequency. This peculiarity represents a difference in comparison to BT and spiral antennas because the spectral response is not frequency independent in the LP case.

In a planar configuration an LP can be obtained combining a BT with metallic teeth, added on each lobe side, as shown in Fig. 5.8. The size and position of the concentric teeth determine the resonant wavelengths. The lobes of the LP antenna are asymmetric: the teeth on the right side of the lobe define empty spaces (*antiteeth*) that are identical to the teeth on the left side.

There is an analytical way to determine the resonances of an LP antenna [25]. As a general rule the ratio between two adjacent resonance frequencies is given by the ratio between the external radii of two consecutive teeth:

$$\frac{f_n}{f_{n+1}} = \frac{R_{n+1}}{R_n} = \frac{r_{n+1}}{r_n} = \tau, \tag{5.8}$$

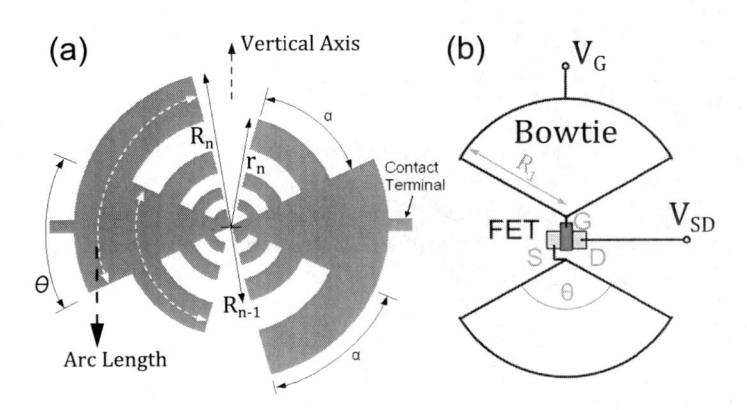

Figure 5.8 (a) Planar log periodic antenna with four teeth on the right side of the lobe and three teeth on the left side (with respect to the center). The angle θ corresponds to the *flare angle* of the bowtie and the angle α sets the length of the teeth; all the reported devices have $\alpha = \theta = 50°$. (b) Scheme of the central region of the device: the two lobes are connected to the source and the gate electrode, respectively.

where R_n and r_n are, respectively, the external and internal radii of the nth tooth on the right side of the lobe; note that r_n is also the external radius of the $(n-1)^{th}$ tooth on the left side and that the smallest radius is R_1. The ratio τ is the scaling factor of the antenna and defines the parameter $\sigma = \sqrt{\tau}$, which corresponds to the ratio between the external radii of a tooth and of the inner *anti*tooth.

Thus, by defining the number of teeth N, the innermost radius R_1, the angles θ and α and the scaling factor τ, the geometry of an LP antenna is completely determined. The external radii can be inferred from R_1 using this expression:

$$R_n = \tau^n R_1; \qquad R_n = \sigma r_n \tag{5.9}$$

The analytical calculation of the resonance wavelength can be done following a model based on the idea that an *anti*tooth behaves like a $\lambda/2$ resonator: the current induced on the antenna by the incoming electromagnetic wave has two minima and one maximum along the perimeter of the empty space between two consecutive teeth. The wavelength that resonates between the nth and $(n-1)^{th}$ teeth is then calculated as

$$\lambda_n = 2l_n = 2\left[r_n\left(1+\frac{1}{\sigma}\right)\frac{\alpha\pi}{180°} + r_n\left(1-\frac{1}{\sigma}\right)\right]. \tag{5.10}$$

From the above expression can be clearly deduced that angle θ plays no role in determining the resonant frequencies. However, as in the BT case, it defines the angular acceptance of the antenna with respect to the polarization of the electric field. It has been demonstrated that the response of LP antennas is a slowly varying function of the polarization direction.

5.2.3 *Nanofabrication*

To fabricate nanowire FET THz detectors, 1–2 µm long InAs nanowires having diameters in the range of 30–100 nm were grown bottom-up on InAs ⟨111⟩B substrates by chemical beam epitaxy (CBE) in a Riber Compact-21 system by Au-assisted growth using trimethylindium (TMIn) and tertiarybutylarsine (TBAs) as metal-organic (MO) precursors and ditertiarybutyl selenide (DtBSe) as a selenium source for n-type doping [26]. Because of its high decomposition temperature, TBAs was precracked in the injector at 1000°C. A 0.5 nm thick Au film was first deposited by thermal evaporation on the InAs wafer in a separate evaporator chamber and then transferred to the CBE system. The wafer was then annealed at 520°C under TBAs flow in order to remove the surface oxide and generate the Au nanoparticles by thermal dewetting. The InAs segment was grown for 90 min at a temperature of (430 ± 10)°C with MO line pressures of 0.3 and 1.0 Torr for TMIn and TBAs, respectively. For n-type doping, during the growth the DtBSe line pressure was varied between 0 and 0.4 Torr to achieve a Se doping level in the range of $1 - 140 \times 10^{17}$ cm^{-3}, respectively. Se doping can be used both to control the charge density and to optimize source–drain and contact resistance, while ensuring sharp pinch-off in the transconductance. Switching TBAs during growth with tertiarybutylphosphine (TBP) or tris(dimethylamino) antimony (TDMASb), it is possible to create heterostructured nanowires, growing, respectively, InP and InSb material. This allows an additional degree of freedom in engineering the nanowire electrical properties.

Nanowires were then mechanically transferred to a 350 µm thick high-resistivity Si substrate with a 500 nm SiO$_2$ insulating surface layer. Microscopic connection strips defining a lane with 10 fields, and nanometric markers defining a coordinate system in each field, were predefined on each substrate by employing ultraviolet (UV) lithography and electron beam (e-beam) lithography, respectively. The samples were then spin-coated with an e-beam-sensitive resist and contact patterns were exposed by e-beam lithography. After development of the exposed resist, residual polymer on the nanowire contact areas was removed using oxygen plasma. In addition, to remove surface oxides, InAs contact areas were passivated before metal deposition to prevent reoxidation. The passivation step before evaporation was performed using a highly diluted ammonium polysulfide $(NH_4)_2S_x$ solution. This treatment proved to be crucial for an optimal electrical behavior of the devices due to the high surface-to-volume ratio of the nanowires. Ohmic contacts were then realized, taking care of making the potential barrier between InAs and metal contacts negligible. To this aim, the metal layer to be evaporated was carefully selected in order to make its work function equal or less than the electron affinity of the bottom semiconductor. In the case of InAs, having an electron affinity of $\sim$4.9 eV, titanium is the most proper choice since its work function is 4.33 eV. A Ti(10 nm)/Au(90 nm) layer was then thermally evaporated onto the samples, and lift-off was made in heated acetone, followed by rinsing in isopropanol. The devices were then glued on a selected chip via an electrically insulating adhesive, and electrical connections were established via 25 μm Au wires using an electronic wedge bonder.

5.2.4 *Methods*

5.2.4.1 Transport measurements

The electrical characterization of the fabricated devices has been performed by independently driving at room temperature, in air, the source-to-drain (V_{SD}) and the gate (V_G) voltages, in the range of [−25 to 25 mV] and [−10 to 10 V], respectively. The drain contact has been connected to a current amplifier, also acting as a virtual ground, converting the current flowing through the nanowire into a voltage signal with an amplification of $\sim$10^4 V/A. The latter signal has been recorded through a voltmeter reader.

Figure 5.9 shows a typical $I_{SD}(V_G)$ $(I-V$ transcharacteristic) curve for homogeneous InAs nanowires, obtained by varying V_G while keeping constant V_{SD}, from which the most fundamental FET parameters can be inferred.

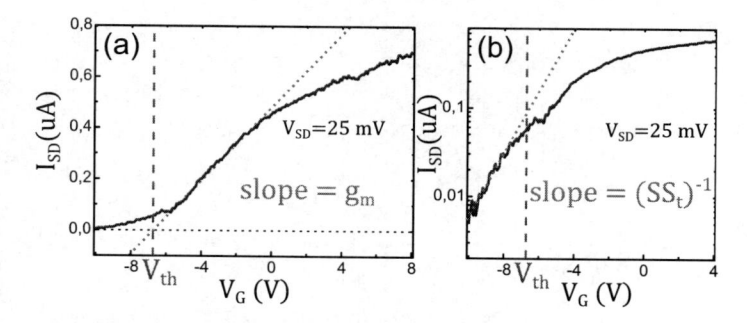

Figure 5.9 Room temperature current (I_{SD})-voltage(V_G) characteristic measured at 25 mV of applied V_{SD}. Solid lines: experimental data. Dashed lines: *pinch-off* voltage. Dotted lines: linear fits to the data used to calculate the transconductance and the subthreshold slope.

- Peak transconductance ($g_m = [A/V]$): Defined as the maximum value of the first derivative of the transfer characteristic (I_{SD} vs. V_G), it quantifies the efficiency of the *on/off* current ratio.
- Threshold voltage (V_{th}): It is the gate potential above which the transistor channel starts to be conductive. It's also known as the *pinch-off* voltage.
- Subthreshold slope (SS_t): In the region corresponding to $V_g < V_{th}$ (subthreshold region) a diffusive current still flows from drain to source. This current between D and S contacts decreases exponentially as gate bias is reduced. The slope of the logarithmic plot of I_{SD} versus V_g represents the inverse of the subthreshold slope (SS_t^{-1}). This parameter, expressed in millivolts per decade, estimates the gate (*in*)efficiency in closing the transistor channel: an efficient gate corresponds to a lower SS_t, that is, a smaller residual current flow.

At room temperature the electrical conductance through InAs nanowires is dominated by the collisions of electrons with lattice impurities and phonons. This conduction regime, called *diffusive*, arises when the length of the semiconductor specimen (in this case the transistor channel length l_{ch}) is much larger than the electron mean free path: $\bar{l} = v_F \tau_{coll}$ where v_F is the Fermi velocity and τ_{coll} is the momentum relaxation time of the electron. This situation can be easily described using the Drude model [27].

Applying a bias at the nanowire ends, an electric field **E** accelerating the electrons originates. At a specific speed the force exerted by the electric field is balanced by the viscous friction arising from the interaction with the lattice. The equilibrium velocity, called *drift velocity* ($\mathbf{v}_d$), is given by ($e > 0$)

$$\left[\frac{dp}{dt}\right]_{coll} = -\left[\frac{dp}{dt}\right]_E \Rightarrow -\frac{m^* \mathbf{v}_d}{\tau_{coll}} = e\mathbf{E} \Rightarrow \mathbf{v}_d = -\frac{e\tau_{coll}}{m^*}\mathbf{E} \quad (5.11)$$

The proportionality constant that relates $\mathbf{v}_d$ to **E** is defined as electron *mobility*:

$$\mu = \frac{e\tau_{coll}}{m^*} \quad (5.12)$$

In the case of nanowire FETs the mobility can be easily inferred from the transcharacteristic curve [28]:

$$\mu = \frac{g_m W_g^2}{C_{wG} V_{SD}}, \tag{5.13}$$

where W_g is the gate width and C_{wG} is the electrical capacitance between the nanowire and the gate. Once μ is known, the carrier density n in the nanowire can be estimated. The motion of the electrons from source to drain (or vice versa) is identified with the current density $\mathbf{j} = -ne\mathbf{v}_d$. This current is related with the electric field via *Ohm's law*:

$$\mathbf{j} = \sigma \mathbf{E} = \frac{\mathbf{E}}{\rho}, \tag{5.14}$$

where σ and ρ are the conductivity and resistivity of the channel, respectively. From the above equations' results

$$ne\mu = \sigma \quad \Rightarrow \quad n = (\mu e \rho)^{-1} \tag{5.15}$$

Huang et al. proposed an alternative method for the calculation of the carrier density in a nanowire FET [29]. Although this model provides a still rough estimation of n, it is more appropriate for the description of lateral gate nanowire FETs. Starting from the already reported relation $Q = C_{wG} V_{th}$, n can be simply calculated dividing Q by the volume of the depletion region in the nanowire:

$$n = \frac{C_{wG} V_{th}}{e \pi r^2 W_g}, \tag{5.16}$$

where r is the radius of the nanowire. With Eqs. 5.13 and 5.16 the mobility and carrier density in the nanowire can be independently estimated once the parameter C_{wG} is preventively known.

A common approach for the calculation of C_{wG} is the so-called *metallic approximation*. The latter assumes that the electrostatic potential cannot penetrate inside the nanowire, whose surface behaves like a metallic layer [30]. This model considerably simplifies the actual system but is known to be still valid for nanowires having a doping concentration $\geq 10^{17}$ cm^{-3}. Under the metallic approximation the following assumptions should be valid: (i) The carrier density is enough to screen the electrostatic potential out of the semiconductor, (ii) the electron mobility is uniform along the wire, and (iii) the nanowire itself is an equipotential surface.

We performed 3D FEM electrostatic simulations to approximate C_{wG}, designing a geometry that exactly resembles the actual devices. The nanowire FETs are placed on a silicon wafer with a 500 nm SiO_2 top layer and the InAs nanowires have been approximated as metal wires. The software computes the Poisson equation for the electrostatic potential $\phi(\mathbf{r})$ and calculates the capacitance between the gate electrode and the surroundings [31]:

$$C = \varepsilon_0 \int_V \varepsilon_r(\mathbf{r}) \, |\nabla\phi(\mathbf{r})|^2 \, d\mathbf{r}, \qquad (5.17)$$

where the integral is taken over the whole geometry volume. Figure 5.10 explains how to estimate C_{wG} from the simulation results: C_{wG} is given by the difference between C_{tot} and $C_{electrodes}$, C_{tot}

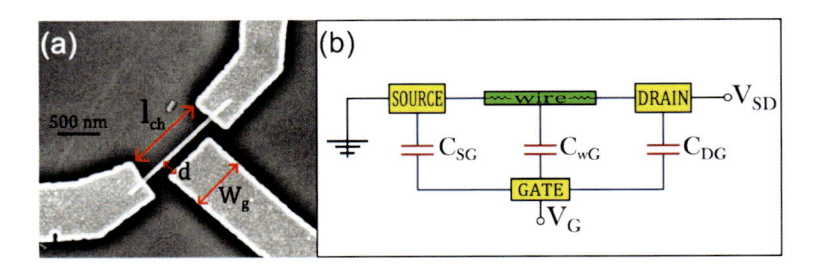

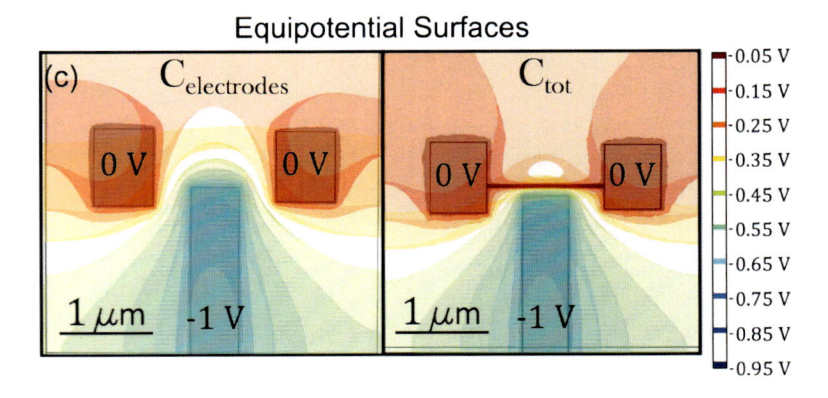

Figure 5.10 3D FEM model for the calculation of C_{wG}. (a) SEM image of a lateral gate nanowire FET; relevant lengths are reported: channel length l_{ch}, gate width W_g, and gate-to-wire distance d. (b) Equivalent DC circuit. (c) Equipotential surfaces for the estimation of $C_{electrodes}$ and C_{tot}.

being the capacitance between the gate electrode and the conductor formed by the union of S, D, and the nanowire and $C_{electrode}$ being the capacitance between the gate and the other electrodes without considering the presence of the nanowire. Although the model takes into account the deformation of the static field induced by the presence of the source and drain contacts, the nanowire is here approximated with a metal wire, meaning that the extracted μ and n values should be here considered as lower and upper values, respectively.

5.2.4.2 Optical measurements

The optical characterization of nanowire FETs was performed employing two different sources, a fixed-frequency (0.292 THz) electronic source and a 1.5 THz QCL. In both cases antennas matched with the frequency of the incoming radiation have been properly designed.

The gate and source electrodes were connected to the antenna lobes (the source lead is grounded), while the drain output was sent to the readout apparatus. As a first approximation, the rectified photocurrent is given by

$$j_{dc} = e\langle n(t)v_d(t)\rangle, \tag{5.18}$$

where e is the electron charge and the temporal average is taken over one period of oscillation. Under an open circuit configuration, an electric field along the channel compensates for the rectification current and a DC photovoltage Δu arises. A simple approach to describe the photoresponse of the FET-based detectors is based on a diffusive model; Δu directly depends on the derivative of the conductance σ with respect to the gate voltage [32]:

$$\Delta u \propto \frac{1}{\sigma}\frac{d\sigma}{dV_G} \tag{5.19}$$

By Eq. 5.19 it is possible to predict the best-performing devices, starting from electrical measurements: the higher the transconductance, the higher the photoresponse.

Photoresponse experiments (Fig. 5.11) at 292 GHz [33, 34] were performed by using an electronic source based on frequency multipliers. The radiation was collimated and focused by off-axis

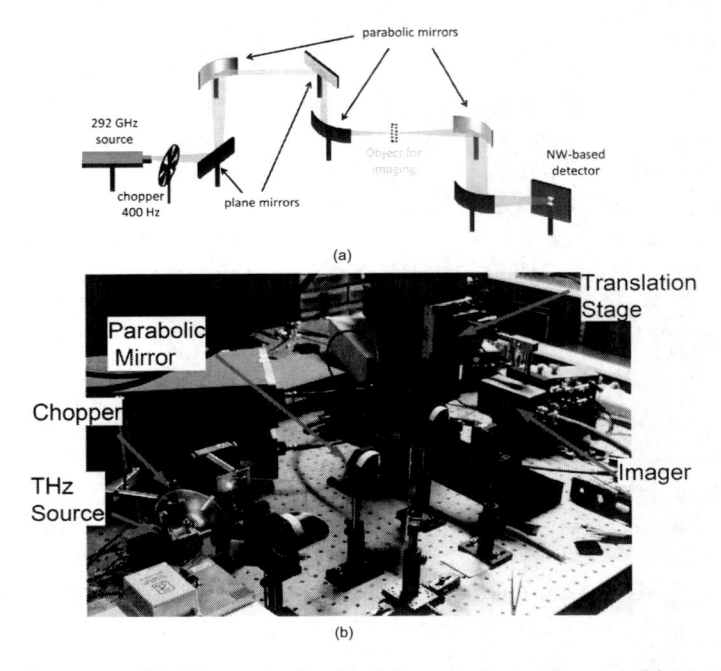

(a)

(b)

Figure 5.11 (a) Schematic sketch of the experimental setup. (b) Picture of the optical bench with main instruments. From Ref. [34]. © IOP Publishing. Reproduced with permission. All rights reserved.

parabolic mirrors in a 4 mm spot diameter beam; the intensity was mechanically chopped at 475 Hz and the photoinduced source–drain voltage was measured by using a lock-in connected with a low-noise voltage preamplifier having an input impedance of 10 $M\Omega$ and an amplification factor equal to 25. A miniature light-emitting diode (LED) in combination with an indium tin oxide (ITO) mirror helped with the alignment of the source. The detector was moved with a motorized X-Y translation stage. The output power P_t of the electronic source was measured using a large-area calibrated detector. The vertically polarized incoming radiation impinged from the free space onto the nanowire devices with an optical power $P_t = 2.3$ mW. The response to the sub-THz radiation was measured at zero applied source–drain bias, as a DC voltage at the drain contact, while the source is grounded. Δu can be inferred from the signal measured by the lock-in amplifier (LIA) using the following relation:

$$\Delta u\,(V_G) = \frac{\frac{\pi}{4}2\sqrt{2}\,\mathrm{LIA}}{G},\tag{5.20}$$

where 2 is due to peak-to-peak magnitude, $\sqrt{2}$ originates from the LIA *rms* amplitude, and $\pi/4$ is the fundamental sine wave Fourier component of the square wave produced by the chopper. The preamplifier gain (G) was set to 25.

The 1.5 THz detection experiment [35] was performed using a QCL, fabricated in a double-metal waveguide and operating at $T = 10$ K in pulsed mode, with a train of 2168 pulses (1.7 A amplitude, 435 ns pulse width, 62.8% duty cycle) repeated at a modulation frequency of 333 Hz. The QCL active region design is based on a combined bound-to-continuum and LO-photon depletion scheme [36]. The radiation was collimated and focused on the antenna by a set of two $f/1$ off-axis parabolic mirrors, and the photoinduced source–drain voltage was measured by using a lock-in without any preamplification stage. The vertically polarized incoming radiation impinges from the free space onto the nanowire devices through a ≈ 1 mm diameter pinhole with an optical power $P_t \approx 200\,\mu$W. The latter was measured with a pyroelectric detector and compared with the P_t values extracted with a calibrated absolute THz power meter (Thomas Keating Instruments). The detector was moved with a motorized X-Y translation stage. The photoresponse Δu can be extracted from the LIA signal using Eq. 5.20 with $G = 1$.

5.2.4.3 Noise measurements

Since photoresponsivity measurements were performed with no bias between source and drain, there was no static current flowing through the nanowire and the $1/f$ noise (*Flicker noise*) was expected to be negligible at the chosen modulation frequencies. The most important contribution to the noise of our devices originated from the thermal fluctuation of carriers. Assuming that the nanowire is a resistor, the Johnson–Nyquist noise spectral density is given by $N_{\mathrm{th}} = \langle V \rangle / \Delta\nu = \sqrt{4k_B T R_{\mathrm{nw}}}$, where $\langle V \rangle$ is the variance of the voltage drop across the wire.

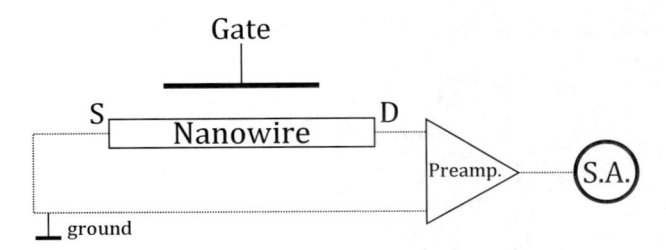

Figure 5.12 Scheme of the experimental setup for the Johnson–Nyquist noise measurement.

The noise level was extracted directly with a dynamic spectrum analyzer under the same experimental conditions of Fig. 5.12, that is, while keeping the source grounded, $V_{SD} = 0$, and while varying V_G in the range $[-10\text{ V}, 10\text{ V}]$.

5.2.4.4 Responsivity and NEP

The first figure of merit we extract from our data is the responsivity ($\mathcal{R}_v$), defined as the ratio between the photoinduced signal (in volts) and the electromagnetic power impinging on the detector. $\mathcal{R}_v$ can be directly extracted from the measured Δu by using the relation $\mathcal{R}_v = (\Delta u S_t)/(P_t S_a)$, where S_t is the radiation beam spot area, S_a is the active area, and P_t is the total power of the THz source. This formula assumes that all the power incident on the antenna is effectively coupled to the nanowire FET. In our case, owing to the relatively high nanowire impedance it is likely that a considerable fraction of the radiation field is not properly funneled onto the device due to the impedance mismatch with the antenna output that typically, for such broadband designs, is of the order of $\sim 100\ \Omega$ or below. Estimated values of S_t are 12.6×10^{-6} m^2 for the 0.3 THz experiment and 0.79×10^{-6} m^2 for the 1.5 THz one. In both cases, since the total area of our nanowire transistor, including the antenna and the contact pads, is smaller than the diffraction-limited area $S_\lambda = \lambda^2/4$ the active area was taken equal to S_λ.

The second parameter to evaluate is NEP, extracted from the ratio between the noise spectral density and $\mathcal{R}_v$ at different gate voltages.

5.3 Homogeneous and Heterostructured Nanowire FET Detectors

5.3.1 *InAs-Based Nanowire FETs*

As stated at the beginning of this chapter, the potentiality of InAs-based nanowire FETs stems from the electrical properties of the material, above all the high electron mobility due to the very low effective mass ($m^* = 0.023\,m_e$) and to the long electron mean free path. Beside this, InAs nanowires, with a narrow bandgap and degenerate Fermi-level pinning, proved to be successful in preserving good detection performances, while scaling down the dimension of the device to increase the detected frequency up to the 1.53 THz range, accessible with QCL sources [35]. It is well known that in a FET the current flowing through the channel when a fixed voltage difference is applied between the source and drain contacts can be varied by setting a polarization bias on the gate electrode. In fact V_G is capable of altering the charge density across the channel, thus changing its conductance. It has been observed that the majority carriers in the nanowires are electrons and InAs-based nanowire FETs are n-type depletion-mode FETs: the threshold voltage V_{th} is negative and the drain current increases when the gate potential is raised (see Fig. 5.9).

5.3.1.1 Influence of doping level

To test the effect of selenium doping on the electrical properties of InAs nanowires we employed simple lateral gate low-capacitance FET geometry, which although not ideal for transistor characteristics allowed the realization of the first plasma wave detectors operating at THz frequencies [33].

Five individual nanowire growths have been performed in a CBE system by increasing the DtBSe (metal-organic selenium compound for n-type doping) line pressure from 0 (sample A) to 0.4 Torr (sample E) with intermediate values of 0.05 Torr (sample B), 0.1 Torr (sample C), and 0.2 Torr (sample D). The corresponding scanning electron microscopy (SEM) images are reported in Fig. 5.13 (sample A, B, C, D, and E, respectively). The statistical distributions of

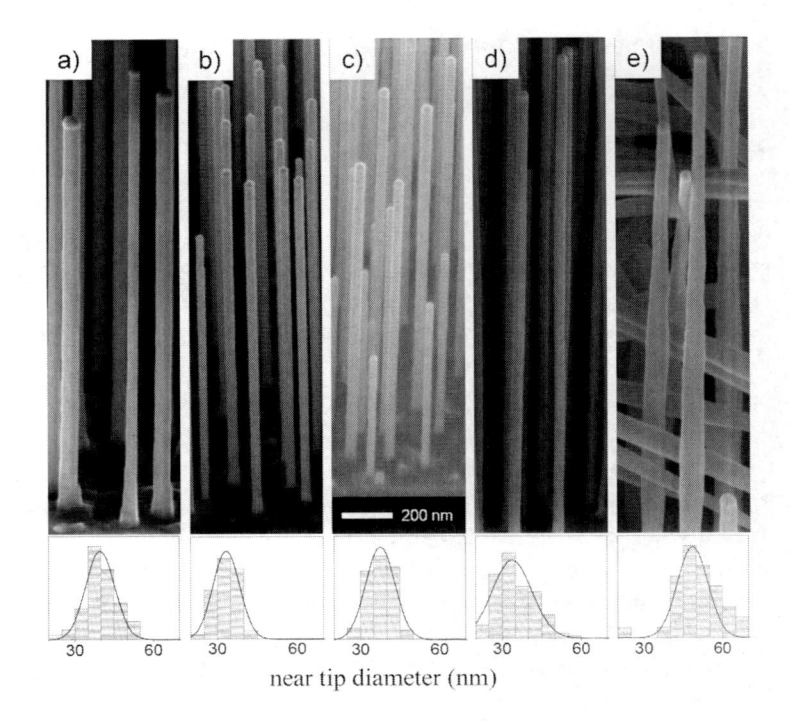

near tip diameter (nm)

Figure 5.13 (Samples A, B, C, D, and E) 45° tilted SEM image of the investigated InAs nanowires. They were 1 to 2 μm long and grown with DtBSe precursor line pressures of (a) 0 Torr, (b) 0.05 Torr, (c) 0.1 Torr, (d) 0.2 Torr, and (e) 0.4 Torr. The bottom panels show the histograms representing the as-grown nanowire near-tip diameter distribution of the corresponding nanowire samples. The solid line is a Gaussian fit of the distribution. From Ref. [26].

the nanowire near-tip diameter are reported on the bottom panels of each SEM image.

To electrically characterize the room-temperature transport of the five classes of InAs nanowires we measured I_{SD} as a function of V_{SD} by independently varying V_{SD} and V_G in the range of -0.025 to 0.025 V and -10 to 10 V, respectively (Fig. 5.14, left column). By driving the lateral gate at positive voltage values, the nanowire resistance decreases by about 2 orders of magnitude. Moreover, regardless of the nanowire carrier density, a linear increase of the nanowire current was observed as a function of the source–

drain bias with slopes progressively larger at higher n-doping concentrations. Indeed, at a gate voltage $V_G = 10$ V, the nanowire resistance (R) varies from 470 to 3 kΩ, while the DtBSe line pressure is increased from 0 to 0.4 Torr. By approximating the nanowire with a cylindrical geometry, we calculated the nanowire resistivity ρ as $\rho = R\pi D^2/4l$, where the nanowire diameter D has been obtained by averaging the diameter of the nanowire over its free length l, between the contacts. Resistivity values in the range of 300 to 8 $\Omega \times \mu$m have been measured for nanowires grown with increasing precursor line pressure.

The carrier density has also a dominant role on the FET "gating effect" that is based on the manipulation of the charge distribution. The applied gate bias indeed modifies the charge distribution in the nanowire, namely the electron carrier density in the n-channel of the FET.

The right column of Fig. 5.14 shows the change of the I_{SD} as a function of V_G. A gate voltage sweep was applied to the nanowire transistor, keeping a fixed $V_{SD} = 0.025$ V value. In all cases, at negative gate voltages, the electric field is able to produce a depletion region narrowing the channel and turning off the current flowing through the nanowire when V_G approaches the pinch-off voltage value. In the last case (samples D and E), the gate potential is indeed not able to remove the electrons and a current still flows through the channel even at $V_G = -10$ V.

The threshold voltages V_{th} of our nanowire FETs were determined by the intercepts with the horizontal axis of the linear fit of the I_{SD} vs. V_G characteristics in the region of maximum transconductance (g_m). The comparison between the five samples shows that V_{th} progressively decreases from -0.5 to -250 V as a function of the precursor pressure, that is, by increasing the carrier density. However, the analysis of the extracted slopes shows that the peak transconductance, normalized to the gate length, varies from 10 to 100 mS/m, reaching a maximum when the precursor pressure of 0.1 Torr (sample C) is employed. The strong increase of the Se doping suppresses the peak transconductance, probably due to the screening of the gate by holes or the more ionized donors for electrons to scatter from. Under the same experimental conditions, a minimum inverse subthreshold slope of ≈ 11 V/dec was found.

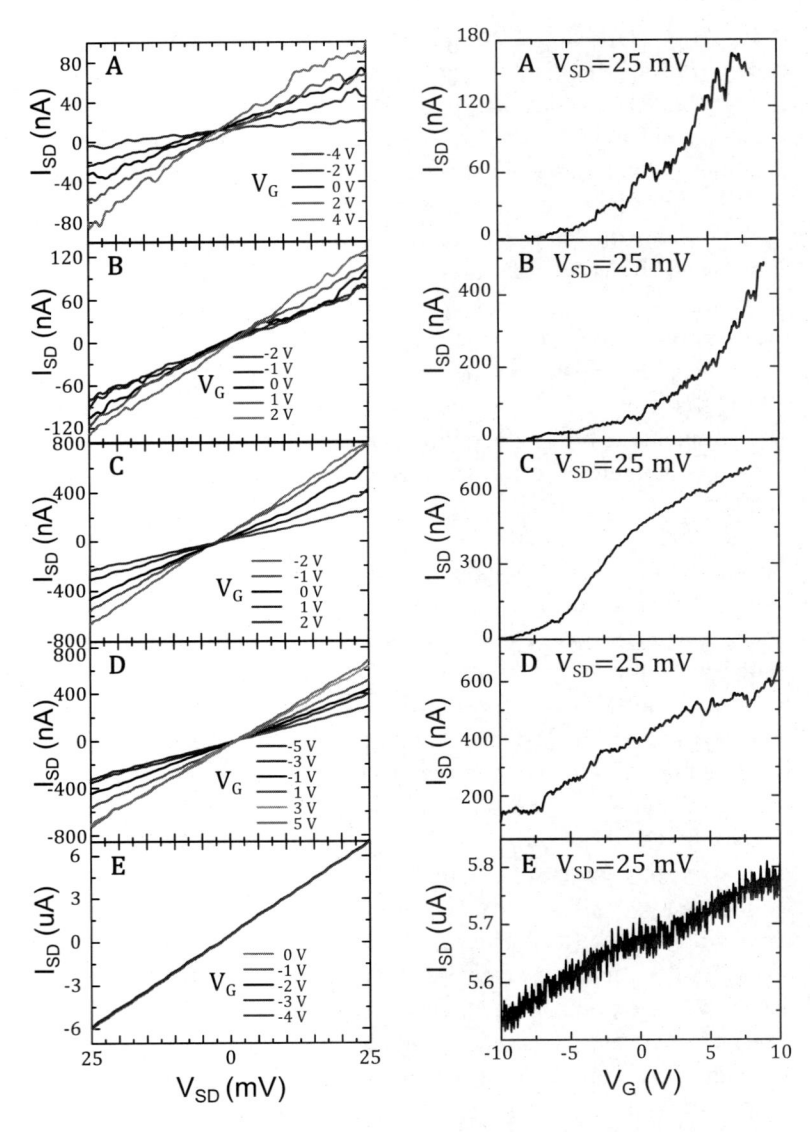

Figure 5.14 Transport characterization of the five categories of homogeneous devices. (Left) $I_{SD}(V_{SD})$ curves at different gate voltages: linearity ensures the ohmicity of metal-semiconductor contacts. (Right) Transcharacteristic curves $I_{SD}(V_G)$ at fixed $V_{SD} = 25$ mV.

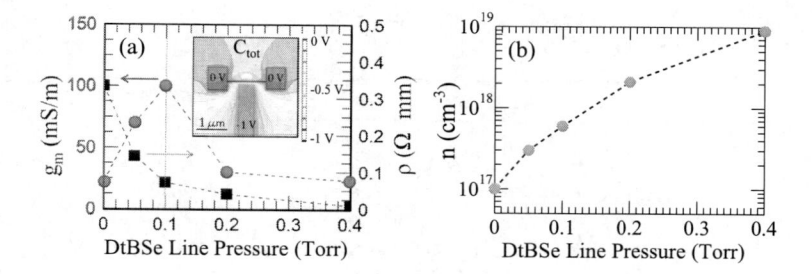

Figure 5.15 (a) Peak transconductance (red circles) normalized to the gate length and resistivity (black squares). Values are plotted as a function of the DtBSe precursor line pressure. Inset: 3D finite-element electrostatic potential simulation calculated by applying a gate voltage $V_G = -1\ V$, while keeping $V_{SD} = 0$. Continuous boundary conditions are applied to the air–Si interface. Isosurfaces corresponding to different values of the electrostatic potential are shown on the graph. (b) Extracted carrier density (green circles) in the transistor channel as a function of the DtBSe precursor line pressure.

Figure 5.15a shows the change of the transconductance and resistivity values measured for samples A, B, C, D, and E as a function of the growth conditions. The comparison between the g_m curve and the corresponding ρ values shows that to optimize the transport properties of the FETs, a compromise should be found between high transconductances and sufficiently low nanowire resistivities. By using the strategy described in Section 5.2.4 it is possible to extract the electron mobility (μ) of our devices. Employing 3D finite-element simulations C_{wG} values were estimated for every device, spanning the range of 2.8–7.5 aF. Mobility values in the range of 10^3–10^4 cm^2/Vs were extracted for samples A–E. From the mobility and resistivity data, the carrier concentration (n) at $V_G = 0\ V$ was calculated from the relation $n = (\mu e \rho)^{-1}$. This allows us to correlate the precursor line pressure during Se doping with the effective carrier density through the nanowire. The results plotted in the figure show that the carrier density in our devices increases significantly for DtBSe line pressures above 0.1 Torr and that the best compromise in the FET transport performances (dashed line in the figure) is found at $n \approx 5 \times 10^{17}$ cm^{-3}.

Table 5.2 Comparison of average electrical properties of FETs for different doping levels of the InAs nanowire

Class	A	B	C	D	E
Resistivity ρ (10^{-5} $\Omega \cdot$ m)	30	25	7	5	1
Conductivity σ (kS/m)	3	4	14	25	120
Transconductance g_m (nA/V)	20	30	50	20	10
Subthreshold Slope SS_t (V/dec)	14.7	13.0	11.1	16.8	–
Threshold Voltage V_{th} (V)	−0.5	−1.4	−3.1	−15.1	−250
Carrier Density n (10^{17} cm^{-3})	< 1	< 3	6	21	90
Mobility (cm^2/Vs)	1200	1350	1700	750	500

In Table 5.2 the estimated average electrical properties of the measured devices are reported.

The nanowires belonging to the **C** category showed the best electrical performances in terms of transconductance and mobility. For this reason they were selected to be integrated in the channel of a FET for the development of THz detectors. As explained in Section 5.2.4 the electrical properties of the devices strongly affect the sensitivity of the detectors since the mechanism of photodetection is based on the fact that when radiation is funneled onto the antenna, a direct photoinduced current is created from the modulation of both the drift velocity v_d and the carrier density n.

5.3.1.2 Photodetection

The homogeneous InAs-based nanowire FETs have been optically tested in both the experiments described in Section 5.2.4 using planar broadband antennas with arms connected to the gate and source electrodes.

In the 300 GHz measurements we designed FETs with channel length $l_{ch} \approx 1$ μm, gate width $W_g = 500$ nm, and gate-to-wire distance $d \leq 100$ nm. A set of devices exploiting different antenna geometries was realized. (i) We patterned a broadband BT equiangular dipole antenna with a length of 220 μm (sample a), (ii) a LP circular-toothed structure with an outer diameter of 645 μm (sample b), and (iii) an LP antenna with identical dimensions and with the substrate thickness reduced to 120 μm (sample c). A further sample of type *a* featured a silicon hyperhemispherical

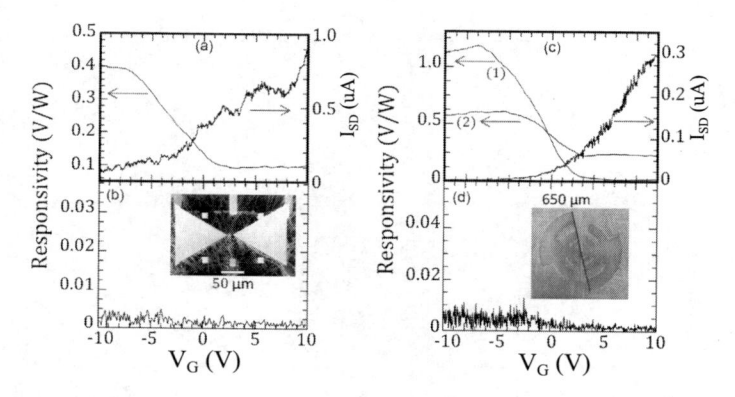

Figure 5.16 (a) Room-temperature responsivity and transfer characteristic (at $V_{SD} = 0.01$ V) as a function of V_G for sample a. (b) $\mathcal{R}_v$ measured when the 292 GHz source is switched off; inset: SEM image of the detector geometry; each bowtie arm is an equilateral triangle ($\theta = 60°$) of 110 μm height. (c) $\mathcal{R}_v$ and I_{SD} vs. V_G curves for samples b (2) and c (1). (d) $\mathcal{R}_v$ measured when the 292 GHz source is switched off; inset: SEM image of the patterned log periodic antennas: $\alpha = \theta = 50°$, the ratio of the radial size of successive teeth is $\tau = 2$, and the size ratio of tooth and antitooth is $\sigma = \sqrt{2}$. Reprinted with permission from Ref. [33]. Copyright (2008) American Chemical Society.

lens having a diameter $D = 6$ mm mounted on the back of the Si substrate so that the radiation beam is properly focused on the nanowire after crossing the Si/SiO$_2$ layers. This makes the 220 μm bow antenna perfectly resonant with the 292 GHz source available for the measurements and reduces the beam spot area S_t by a factor of ~3. The device was then inserted in a compact package and wire-bonded, taking care that the source contact of the bow antenna was grounded on the metal package itself (sample d). Geometries and responsivity plots are depicted in Figs. 5.16 and 5.17.

Responsivity values of ~1 V/W were reached, which is significantly high if one considers the small detecting element and the fact that no real matching of the beam-focusing optics to the actual expected antenna lobe pattern was performed. As expected from Eq. 5.19, the responsivity decreases as a function of the gate voltage and remains roughly constant and low when $V_G > V_{th}$, while a huge increase of the conductance is observed, as shown from the transconductance characteristics. It is worth noticing that in the case

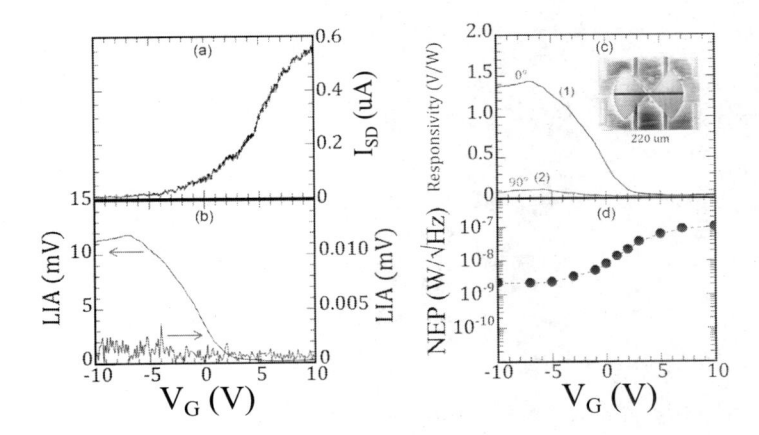

Figure 5.17 Sample d. (a) Transfer characteristic measured at room temperature and at a drain-to-source voltage $V_{SD} = 0.01$ V. (b) Lock-in amplifier (LIA) signal as a function of the gate voltage measured at $T = 300$ K, while sample d is irradiated at 292 GHz (left vertical axis) or when the 292 GHz source is off (right vertical axis). (c) Calculated responsivity while the polarization of the incoming beam is parallel (curve 1) or orthogonal (curve 2) to the bowtie antenna axis; inset: SEM image of the employed 220 µm bowtie antenna ($\theta = 120°$). (d) NEP(V_G) for sample d. Reprinted with permission from Ref. [33]. Copyright (2008) American Chemical Society.

of sample a, although the antenna dimensions were not perfectly matched with the frequency of the incoming sub-THz beam, the bonding wires might also act as an antenna, helping funnel the incoming beam power.

From the comparison between Fig. 5.16a and Fig. 5.16c (curve 2), it appears that the use of LP antenna geometry, perfectly matched with the frequency of the incoming beam, allows a 25% increase of the measured responsivity. It is important to mention that by thinning the silicon substrate from 350 to 120 µm, the responsivity value further increases by a factor of 2, as shown from the comparison between curves 1 and 2 in Fig. 5.16c. Furthermore, it is worth noticing that the noise is only 2 orders of magnitude lower than the signal (Fig. 5.16b,d) due to the absence of the collection silicon lens, instead of what shown in Fig. 5.17b.

Figure 5.17c shows the responsivity values extracted directly from sample d, while the 292 GHz beam impinges with a polarization parallel (1) and orthogonal (2) to the antenna on the Si substrate through the silicon lens. In this case, the bow length was perfectly matched with the pumping wavelength through the Si/SiO$_2$ layers. This allows increasing the nanowire FET responsivity by a factor of ~4. Furthermore, the use of focusing optics directly mounted in contact with the device substrate allowed a significant increase of the collection efficiency of the optical system, resulting in a signal-to-noise ratio (SNR) improvement of about 2 orders of magnitude. It is worth noticing that when the antenna orientation was perpendicular to the polarization of the incoming beam, the detector responsivity was drastically reduced by more than 1 order of magnitude, confirming the effectiveness of the employed antenna geometry.

To characterize the sensitivity of the nanowire FETs we measured the noise level with the technique described in Section 5.2.4.3. The extracted noise value $N \approx 4 \times 10^{-9}$ $V/\sqrt{\text{Hz}}$ leads to an NEP value of ~1 $\times$ 10^{-8} W/$\sqrt{\text{Hz}}$ at zero gate bias, that is, roughly comparable To the Johnson–Nyquist one. The noise figure is dominated by that of the voltage amplifier (nominally precisely 4 nV/$\sqrt{\text{Hz}}$, meaning that our NEP estimate should indeed be considered as an upper limit. The NEP values as a function of the gate voltage are plotted in Fig. 5.17d for sample d. A minimum NEP value of ~2.5 $\times$ 10^{-9} W/$\sqrt{\text{Hz}}$ is obtained in the subthreshold regime, demonstrating the good performance of our nanowire FETs.

In the experiment with the 1.5 THz QCL source, a broadband BT antenna with a flare angle of 105° and an arm length of 100 μm, perfectly resonant with the radiation wavelength, was patterned with the nanowire FET. To keep the parasitic capacitances low, W_g was reduced below 100 nm. The Si substrate was lapped down to 80–100 μm, and the device was mounted on a dual in-line (DIL) package after being preventively kept one day under vacuum. This treatment improved the electrical performance of the FET up to a peak transconductance value ~5 μA/V. As shown in Fig. 5.18a the nanowire FET detectivity is, as expected from electrical considerations, dramatically improved in this case. Figure 5.18a shows the responsivity plotted as a function of the gate voltage V_G by varying the antenna orientation angle (φ) with respect to the polarization of the incoming beam. The detector responsivity

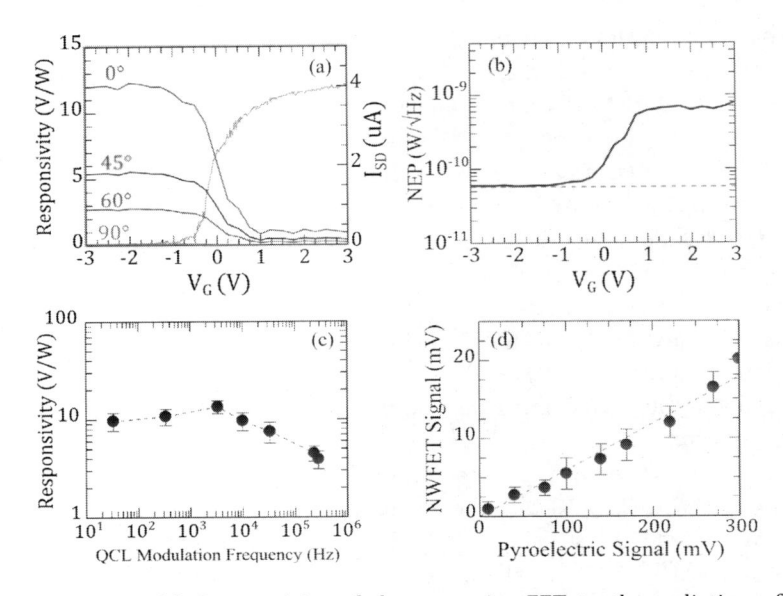

Figure 5.18 (a) Responsivity of the nanowire FET to the radiation of a 1.5 THz QCL, modulated at 333 Hz, as a function of the gate voltage measured at $T = 300$ K and at zero applied V_{SD}, while the angle φ between the polarization of the incoming beam and the bowtie antenna axis is varied. The right vertical axis shows the current–voltage (I_{SD}–V_G) transfer characteristic measured at room temperature and at a drain-to-source voltage $V_{SD} = 0.005$ V. (b) NEP as a function of the gate voltage. The dashed line shows the minimum recorded NEP value. (c) Detector responsivity plotted as a function of the QCL modulation frequency. The dashed line is a guide to the eye. (d) Photoresponse signal Δu recorded while varying the bias/current of the QCL modulated at 333 Hz, plotted as a function of the corresponding pyroelectric response. The dashed line is a linear fit to the data. Reprinted with permission from Ref. [33]. Copyright (2008) American Chemical Society.

reaches up to 12 V/W for $\varphi = 0$, allowing a significant improvement of our detector performances with respect to previous experimental results at 0.3 THz, despite the 5 times higher operating frequency. This is mostly due to three main factors: (i) the 1 order of magnitude reduction of the nanowire resistance, (ii) the antenna geometry perfectly resonant with the 1.5 THz QCL, and (iii) the narrow gates employed. It is worth noticing that while φ increased, the detector responsivity was drastically reduced by more than 1 order of magnitude, confirming the efficiency of the employed antenna geometry.

Figure 5.18b shows the NEP as a function of the gate voltage, measured from the ratio $N/\mathcal{R}_v$, where N has been considered as the Johnson–Nyquist noise $N = \sqrt{4k_B T R_n w}$. A minimum NEP value of $\sim 6 \times 10^{-11}$ $W/\sqrt{Hz}$ was reached in the subthreshold regime, confirming the dramatic improvement of our detector sensitivity levels, significantly better than the commercial thermal uncooled detection system operating at frequencies > 1 THz [3].

To test the response times of our nanowire detectors we modulated the QCL beam in a frequency range spanning from 33 Hz up to ≈ 300 kHz. The detector responsivity as a function of the modulation frequency is displayed in Fig. 5.18c. Within a four-decade bandwidth $\mathcal{R}_v$ decreases by a factor of 3, still remaining significantly higher than 1 V/W, meaning that the response speed of our room-temperature detectors is really competitive with any cooled detection systems operating in the far-infrared region. It is worth noticing that the employed experimental setup poses a limit on the maximum modulation frequency at ≈ 300 kHz. The corresponding response time (≈ 3 μs) has then to be considered as an upper limit. Eventually, to investigate the dynamic range of our devices, we checked their linearity against a standard THz pyroelectric system. Figure 5.18d shows the nanowire FET photoinduced voltage Δu measured while varying the QCL drive current in its operating regime and by simultaneously recording the QCL optical power with a pyroelectric detector at a fixed modulation frequency of 333 Hz. The trend is basically linear up to the highest QCL power.

5.3.2 *InAs-/InSb-Based Nanowire FET*

As previously discussed in Section 5.2.4.2, in a simple scheme for light revelation with FETs, the detection current density can be written, taking the temporal average on an oscillation cycle of the radiation $2\pi/\omega$, as [6]

$$j_{det} = e\langle \rho_1(t) \cdot v_{d,1}(t) \rangle_{2\pi/\omega}, \qquad (5.21)$$

where e is the electron charge and ρ_1 and $v_{d,1}$ are the oscillating components of charge density and drift velocity, respectively. While the first depends on the I-V_G characteristic, the second one mainly originates from the asymmetry between S and D contacts. In the previous paragraphs, we have shown how the nonlinearity can be controlled and tuned considering both gating geometry and nanowire doping, being directly dependent on how the charge can be efficiently depleted from or injected into the transistor channel. On the other hand, S–D asymmetry was merely imposed by unbalanced contacts, where in particular the S contact was linked to one arm of the coupling antenna, whereas the D was kept grounded. In this way, the drift velocity can be modulated and the device used for radiation detection. Indeed, any additional asymmetry in the nanowire FET contacts can be beneficial to the detection mechanism. The versatility in nanowire growing allows one to play with the constituting semiconductor materials; in particular, an asymmetric heterojunction could enhance the drift velocity modulation and be useful to obtain particular electrical transport characteristics. Indium-based III–V semiconductors can be easily incorporated in a nanowire: interesting, among them, InAs and InSb material electronic bands form a broken-gap alignment (type III) at the heterointerface. Though being practically unattainable in bulk due to lattice constant mismatch, the almost monodimensional nanowire structure allows to grow this kind of device with small residual stress. Through the formation of an asymmetric triangular barrier, type III band alignment results in S–D current rectification, making the nanowire an electrical diode along its growth axis [37]. The nanowire cross section being extremely small, this heterostructured system is one of the best candidates for ultrafast diodes and, when employed in a FET, for radiation detection. The S–D assymmetry can be explored by means of transport characterization. Contacting the nanowires with four probes, as in Fig. 5.19a, and using a back gate realized with heavily degenerate doped Si, we can assess the transport properties of InAs and InSb segments and of their heterojunction. Note that the configuration for nanowire-based detectors requires only two contacts and a lateral gate, the Si substrate underneath the device being almost intrinsic.

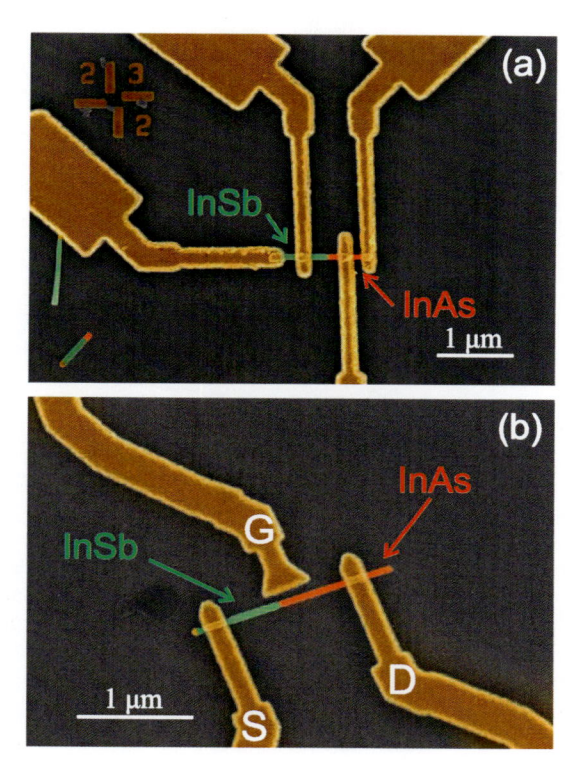

Figure 5.19 False color SEM pictures of contacted nanowire devices. The red (green) segments are made of InAs (InSb). (a) Four-probe device for electrical characterization. (b) Lateral gate device for radiation detection. The lateral gate and InSb contact terminates in the two arms of the log periodic antenna.

5.3.2.1 Electrical properties

Room-temperature $I - V$ characterization shows that Ti-Au electrodes form an ohmic contact on the nanowire passivated surface, producing linear $I-V$ curves with resistance of tens to a few hundreds of kΩ for the investigated nanowires. Though no external doping was used, both nanowires segments show n-type majority carriers characteristics, the conductivity being reduced at negative gate biases and increased at positive ones, as reported in Fig. 5.20. This excess charge is essentially believed to be originated from

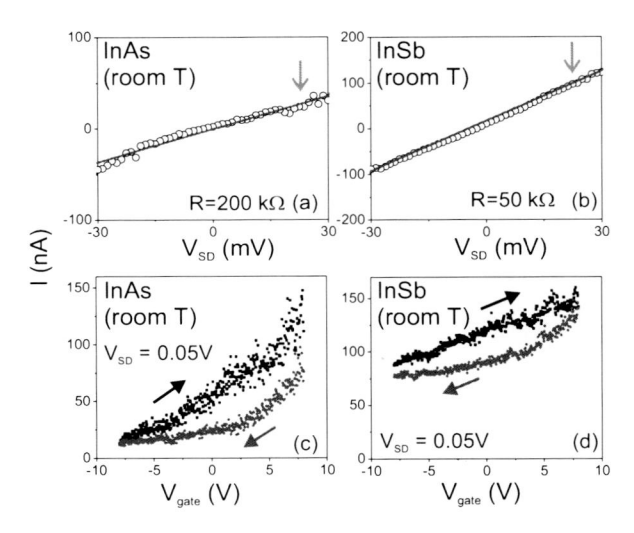

Figure 5.20 (a, b) Room-temperature $I-V$ characteristic of InAs and InSb segments at zero gate bias. (c, d) $I_{SD}-V_G$ characteristic for a source–drain voltage of 50 mV. Note that despite both materials being clearly n-type, the InSb channel does not completely close due to the complicated mixing of majority and minority carriers close to the heterointerface. Reprinted with permission from Ref. [37]. Copyright 2008, AIP Publishing LLC.

the pinning of the Fermi level at the nanowire surface (see, for example, [38]). Interestingly, at room temperature, we are able to totally deplete the charge with a resulting zero conductivity in InAs segments, while in the InSb ones we still have a saturated, residual S–D current, even at negative bias. This is due to the strong band bending at the heterointerface, which influences the transport in the homogeneous segments where a complicated interplay between majority (electrons) and minority carriers (holes) takes place, as confirmed by electron-beam-induced current (EBIC) experiments [39]. A simple picture to understand the rectifying property of the junction can be drawn by looking at the band alignment at zero bias. As reported in the 1D Schrödinger–Poisson simulations of Fig. 5.21a [40], an asymmetric triangular barrier for electrons is formed at the InAs/InSb interface. Similar to a Schottky barrier, this results in the strongly asymmetric room-temperature $I-V$ curve of Fig. 5.21c. Interestingly, by adding a small InP layer in between

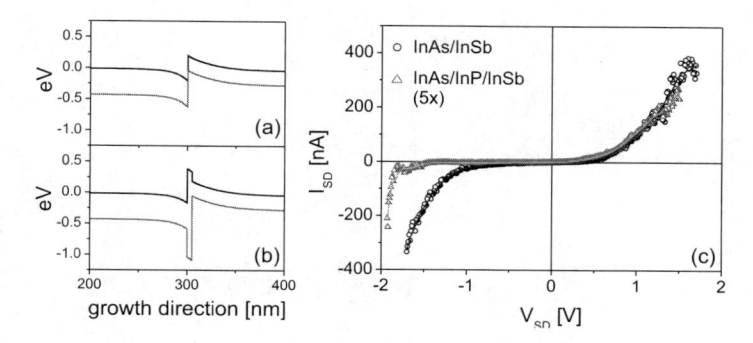

Figure 5.21 (a, b) Schrödinger–Poisson 1D simulation of the energy bands of InAs/InSb and InAs/InP/InSb nanowires, assuming an electron concentration of 10^{17} cm^{-3} (Fermi-level pinning). As expected, the presence of InP enhances the diode asymmetry, though reducing the direct bias conductivity. (c) Room-temperature measurements of current–voltage characteristic; the continuous lines are a guide for the eyes.

the two materials a higher barrier is created (see Fig. 5.21b), both reducing the leakage current due to thermionic contributions in reverse bias and, yet at the same time, increasing the resistivity under direct biasing. Figure 5.21c shows the I–V characteristic of this second nanowire, where the rectification is enhanced at the detriment of direct conductivity.

5.3.2.2 Photodetection

The strong intrinsic asymmetry in InAs/InSb nanowires is a promising effect to develop FET-based radiation detectors. Exploiting the geometry shown in Fig. 5.19b, where the gate and source electrodes were terminated with two arms of an LP antenna, we measured the photoresponse of the heterostructured nanowires under 0.3 THz illumination using the setup described in Section 5.2.4.2. As can be seen in Fig. 5.22a the photovoltage at first increases, peaking at about 15 µV, and then suddenly drops, switching sign, reaching about −50 µV. Even if not doped and not designed for perfect impedance matching, the detector performance can be compared with the basics InAs-based devices, described in the same section. On the other hand, the sign change in the voltage response is a

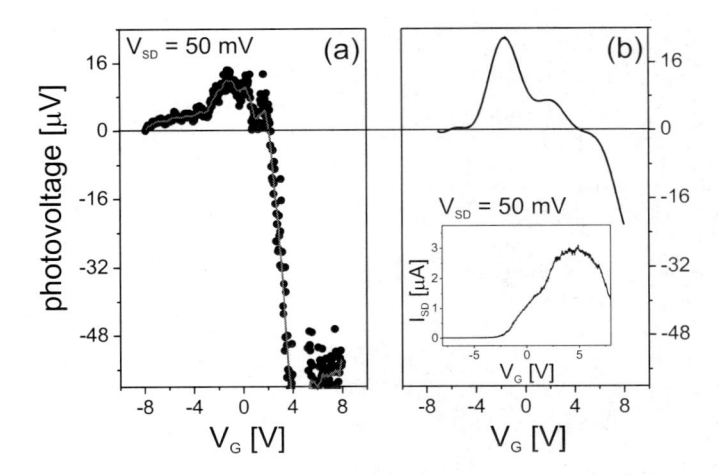

Figure 5.22 (a) Photoresponse under 0.3 THz illumination (same experimental condition of the InAs devices). (b) Responsivity obtained applying Eq. 5.22 to the room-temperature I–V characteristic, reported in the inset.

peculiar signature of this kind of device. Interestingly, this result could be expected by looking at DC charge transport measurements; the device responsivity can be obtained from the I–V characteristic, considering that is proportional to [41, 42]

$$R \propto \frac{1}{1 + R_{nw}/Z_L} \frac{1}{\sigma} \frac{d\sigma}{dV_G}, \tag{5.22}$$

where σ is the conductivity, R_{nw} the nanowire resistance, and Z_L the load impedance. By using Eq. 5.22, we obtain the expected responsivity reported in Fig. 5.22b, where the measured I–V characteristic is reported in the inset. Even though obtaining a smooth derivative is challenging when working with experimental data, the responsivity obtained is in good qualitative agreement with the measured photodetection. A first maximum is present at negative bias, while for bias slightly larger than zero, the curve changes sign, with a fast decreasing negative voltage. Interestingly, this kind of behavior is similar to what reported in Section 5.2.1 with responsivity maxima at both negative and positive gate biases. Interestingly, even in our experimental case the nanowires are undoped, falling in the right category to observe these kinds of effects. We believe the origin of this phenomenon is then related to the asymmetric gate structure realized in our device. It's worth noticing that a strong gating field, other than injecting/depleting charge to/from the device, can alter the spatial distribution of the charge itself, as reported in single-electron nanowire transistors based on quantum dots [43]. Therefore, we believe that a strong gate bias can push some of the charge within the outermost nanowire surface, where the transport properties can be very different from the nanowire core due to the presence of surface trap states, detrimental to charge conductivity. This could explain while, after an initial charge injection that raises the conductivity, this is then slowly reduced by the effect of the more resistive surface transport channel.

InAs/InSb heterostructured nanowires can represent a valid candidate to improve detection efficiency; moreover, fundamental physics investigation of the complicated electronic structure realized in this unique heterojunction can represent a notewhorthy research topic and have important technological spillovers.

References

1. Köhler, R., Tredicucci, A., Beltram, F., Beere, H.E., Linfield, E.H., Davies, A.G., Ritchie, D.A., Iotti, R.C., and Rossi, F. (2002). Terahertz semiconductor-heterostructure laser, *Nature*, **417**, pp. 156–159.

2. Tonouchi, M. (2007). Cutting-edge terahertz technology, *Nat. Photonics*, **1**, pp. 97–105.

3. Sizov, F. and Rogalski, A. (2010). THz detectors, *Prog. Quant. Electron.*, **34**, pp. 278–347.

4. Siegel, P.H. (2002). Terahertz technology, *IEEE Trans. Microwave Theory Tech.*, **50**, p. 910.

5. Hübers, H.-W. (2008). Terahertz heterodyne receivers, *IEEE J. Sel. Top. Quant. Electron.*, **14**, p. 378.

6. Knap, W., Dyakonov, M., Coquillat, D., Teppe, F., Dyakonova, N., Lusakowski, J., Karpierz, K., Sakowicz, M., Valusius, G., Seliuta, D., Kasalynas, I., El Fatimy, A., Meziani, Y.M. and Otsuji, T. (2009). Field effect transistors for terahertz detection: physics and first imaging applications, *Int. J. Infrared Millim. Waves*, **30**, p. 1319.

7. Lisauskas, A., Boppel, S., Matukas, J., Palenskis, V., Minkevicius, L., Valusis, G., Haring-Bolivar, P., Roskos, H.G. (2013). Terahertz responsivity and low-frequency noise in biased silicon field-effect transistors, *Appl. Phys. Lett.* **103**, p. 153505.

8. Öjefors, E., Pfeiffer, U.R., Lisauskas, A., and Roskos, H.G., (2009). A 0.65 THz focal-plane array in a quarter-micron CMOS process technology, *IEEE J. Solid-State Circuits*, **44**, pp. 1968–1976.

9. Dyakonov, M. and Shur, M. (1993). Shallow water analogy for a ballistic field effect transistor: new mechanism of plasma wave generation by dc current, *Phys. Rev. Lett.*, **71**, pp. 2465–2468.

10. Teppe, F. *et al.* (2005). Room-temperature plasma waves resonant detection of sub-terahertz radiation by nanometer field-effect transistor, *Appl. Phys. Lett.*, **87**, p. 052107.

11. Li, Y., Qian, F., Xiang, J. and Lieber, C.M. (2006). Nanowire electronic and optoelectronic devices, *Mater. Today*, **9**, p. 18.

12. Gudiksen, M.S., Lauhon, L.J., Wang, J., Smith, D.C. and Lieber, C.M. (2002). Growth of nanowire superlattice structures for nanoscale photonics and electronics, *Nature*, **415**, p. 617.

13. Lauhon, L.J., Gudiksen, M.S., Wang, D. and Lieber, C.M. (2002) Epitaxial core-shell and core-multi-shell nanowire heterostructures, *Nature*, **420**, p. 57.

14. Dick, K.A., Deppert, K., Larsson, M.W., Martensson, T., Seifert, W., Wallenberg, L.R. and Samuelson, L. (2004). Synthesis of branched "nanotrees" by controlled seeding of multiple branching events, *Nat. Mater.*, **3**, p. 380.

15. Thelander, C., Fröberg, L.E., Rehnstedt, C., Samuelson, L. and Wernersson, L.E. (2008). Vertical enhancement-mode InAs nanowire field-effect transistor with 50-nm wrap gate, *IEEE Electron Device Lett.*, **29**, p. 206.

16. Fröberg, L.E., Rehnstedt C., Thelander, C., Lind, E., Wernersson, L.E., and Samuelson, L., (2008). Heterostructure barriers in wrap gated nanowire FETs, *IEEE Electron Device Lett.*, **29**, p. 981.

17. Dayeh, S.A., Aplin, D.P.R., Zhou, X., Yu, P.K.L., Yu, E.T. and Wang, D. (2007), High electron mobility indium arsenide nanowire field effect transistors, *Small*, **3**, p. 326.

18. Mårtensson, T., Wagner, J.B., Hilner, E., Mikkelsen, A., Thelander, C., Stangl, J., Ohlsson, B.J., Gustafsson, A., Lundgren, E., Samuelson, L. and Seifert, W. (2007). Epitaxial growth of indium arsenide nanowires on silicon using nucleation templates formed by self-assembled organic coating, *Adv. Mater.*, 19, p. 1801.

19. El Fatimy, A., Teppe, F., Dyakonova, N., Knap, W., Seliuta, D., Valusis, G., Shchepetov, A., Roelens, Y., Bollaert, S., Cappy, A. and Rumyantsev, S. (2006), Resonant and voltage-tunable terahertz detection in InGaAs/InP nanometer transistors, *Appl. Phys. Lett.*, **89**, p. 131926.

20. Knap, W., Kachorowskii, V., Deng, Y., Rumyantsev, S., Lu, J.-Q., Gaska, R., Shur, M.S., Simin, G., Hu, X., Khan, M.A., Saylor, C.A. and Brunal, L.C. (2002). Nonresonant detection of terahertz radiation in field effect transistors, *J. Appl. Phys.*, **91**, p. 9346.

21. Lisauskas, A., Pfeiffer, U., Öjefors, E., Haring Bolìvar, P., Glaab, D. and Roskos, H.G. (2009). Rational design of high-responsivity detectors of terahertz radiation based on distributed self-mixing in silicon field-effect transistors, *J. Appl. Phys.*, **105**, p. 114511.

22. Pitanti, A., Roddaro, S., Vitiello, M.S. and Tredicucci, A. (2012). Contacts shielding in nanowire field effect transistors, *J. Appl. Phys.* **111**, p. 064301.

23. Sze, S.M., (1985). *Semiconductor Devices*, Wiley, New York.

24. González, F.J. and Boreman, G.D. (2005). Comparison of dipole, bowtie, spiral and log-periodic IR antennas, *Infrared Phys. Technol.*, **46**, pp. 418–428.

25. Scheuring, A., Wuensch, S. and Siegel, M. (2009). A novel analytical model of resonance effects of log-periodic planar antennas, *IEEE Trans. Antennas Prop.*, **57**, 11, p. 3482.

26. Viti, L., Vitiello, M.S., Ercolani, D., Sorba, L. and Tredicucci, A. (2012). Se-doping dependence of the transport properties in CBE-grown InAs nanowire field effect transistors, *Nanoscale Res. Lett.*, **7**, p. 159.

27. Datta, S. (1995). *Electronic Transport in Mesoscopic Systems*, Cambridge University Press.

28. Wunnicke, O. (2006). Gate capacitance of back-gated nanowire field-effect transistors, *Appl. Phys. Lett.*, **89**, p. 083102.

29. Huang, Y., Duan, X., Cui, Y. and Lieber, C.M. (2002). Gallium nitride nanowire nanodevices, *Nano Letters*, **2**, p. 101.

30. Vashaee, D., Shakouri, A., Goldberger, J., Kuykendall, T., Pauzauskie, P. and Yang, P. (2006). Electrostatics of nanowire transistors with triangular cross sections, *J. Appl. Phys.*, **99**, p. 054310.

31. Jackson, John David (1998). *Classical Electrodynamics*, Third Edition, John Wiley and Sons.

32. Vicarelli, L., Vitiello, M.S., Coquillat, D., Lombardo, A., Ferrari, A.C., Knap, W., Polini, M., Pellegrini, V. and Tredicucci, A. (2012). Graphene field-

effect transistors as room-temperature terahertz detectors, *Nat. Mater.*, **11**, p. 865.

33. Vitiello, M.S., Coquillat, D., Viti, L., Ercolani, D., Teppe, F., Pitanti, A., Beltram, F., Sorba, L., Knap, W. and Tredicucci, A. (2012). Room-temperature terahertz detectors based on semiconductor nanowire field-effect transistors, *Nano Lett.*, **12**, p. 96.

34. Romeo, L., Coquillat, D., Pea, M., Ercolani, D., Beltram, F., Sorba, L., Knap, W., Tredicucci, A., and Vitiello, M.S. (2012). Nanowire-based field effect transistors for terahertz detection and imaging systems, *Nanotechnology*, **24**, p. 214005.

35. Vitiello, M.S., Viti, L., Romeo, L., Ercolani, D., Scalari, G., Faist, J., Beltram, F., Sorba, L. and Tredicucci, A. (2012). Semiconductor nanowires for highly sensitive, room-temperature detection of terahertz quantum cascade laser emission, *Appl. Phys. Lett.*, **100**, p. 241101.

36. Walther, C., Fischer, M., Scalari, G., Terazzi, R., Hoyler, N. and Faist, J. (2007). Quantum cascade lasers operating from 1.2 to 1.6 THz, *Appl. Phys. Lett.* **91**, p. 131122.

37. Pitanti, A., Ercolani, D., Sorba, L., Roddaro, S., Beltram, F., Nasi, L., Salviati, G., and Tredicucci, A. (2011). InAs/InP/InSb Nanowires as low capacitance n-n heterojunction diodes, *Phys. Rev. X* **1**, p. 011006.

38. Khanal, D.R., Walukiewicz, W., Grandal, J., Calleja, E., and Wu, J. (2009). Determining surface Fermi level pinning position of InN nanowires using electrolyte gating, *Appl. Phys. Lett.* **95**, p. 173114.

39. Chen, C.Y., Shik, A., Pitanti, A., Tredicucci, A., Ercolani, D., Sorba, L., and Beltram, F. (2012). Electron beam induced current in InSb-InAs nanowire type-III heterostructures, *Appl. Phys. Lett.* **101**, p. 063116.

40. Tan, I.H., Snider, G., and Hu, E. (1990). A self–consistent solution of Schrödinger–Poisson equations using a nonuniform mesh, *J. Appl. Phys.* **68**, p. 4071.

41. Pitanti, A., Coquillat, D., Ercolani, D., Sorba, L., Teppe, F., Knap, W., De Simoni, G., Beltram, F., Tredicucci, A., and Vitiello, M.S. (2012). Terahertz detection by heterostructured InAs/InSb nanowire based field effect transistors, *Appl. Phys. Lett.* **101**, p. 141103.

42. Sakowicz, M., Lifshits, M.B., Klimenko, A., Schuster, F., Coquillat, D., Teppe, F., and Knap, W. (2011). Terahertz responsivity of field effect transistors versus their static channel conductivity and loading effects, *J. Appl. Phys.* **110**, p. 054512.

43. Roddaro S., Pescaglini, A., Ercolani, D., Sorba, L., and Beltram, F. (2011). Manipulation of electron orbitals in hard-wall InAs/InP nanowire quantum dots, *Nano Lett.* **11**, p. 1659.

Chapter 6

Nano-optomechanical Oscillators: Novel Effects and Applications

Daniel Navarro-Urrios

Catalan Institute of Nanoscience and Nanotechnology ICN2
Phononic and Photonic Nanostructures Group, Campus Bellaterra - Edifici CIN2
Bellaterra (Barcelona), 08193 Spain
daniel.navarro@icn.cat

The field of optomechanics has emerged as a new interface for light–matter interactions, paving the way to the implementation of micro- or even nano-optomechanical systems (MOMS/NOMS) in integrated circuits [1]. Functionalities such as ultrasensitive detection of small displacements or weights and possible uses in quantum information processing are some of the appealing applications driving the fast developments in this area. One of the most appealing features in the optomechanics field is the possibility of momentum exchange between photons confined within an optical cavity and mechanical devices that are either inside or part of the cavity. Photons in an optomechanical (OM) cavity can be converted into phonons in the mechanical device and vice versa. Thus, if a pumping laser wavelength is tuned to a slightly lower frequency than the resonance of the optical cavity, the light can be used to perform optically induced damping of the mechanical motion, thus effectively cooling

Nanodevices for Photonics and Electronics: Advances and Applications
Edited by Paolo Bettotti

the oscillator. Research toward this goal has rapidly progressed and culminated in the use of optical forces to cool nanoscale mechanical oscillators into their quantum ground state of motion at temperatures of about 20 K [2]. On the other hand, if the pumping laser is tuned to the other side of the optical resonance, the OM cavity could be driven into the opposite regime [3, 4]. In microscale structures a high-amplitude regime can be achieved and optically driven phonon lasing on the MHz range has been recently reported [5], though efficiencies are rather low because of the strong damping due to the interaction with the surrounding medium and the low photon–phonon coupling. Great advances along this direction have been reached and recently the first nonvolatile nanomechanical memory cell that is operated exclusively by light has been reported, exploiting both the effects of blue-detuning the pump laser and the existence of two stable mechanical states [6]. The nanoscale OM crystal (with both phononic and photonic bandgaps) regime promises stronger phonon–photon couplings, resonant mechanical frequencies scaled up to the gigahertz, and lower damping processes.

This chapter will be organized as follows. First, we will give a brief introduction to basic concepts of optomechanics. Then we will focus on a particular OM architecture among the plethora of different devices present in the literature: the OM crystals. The last section, and the larger, will be dedicated to describing a basic characterization of one of these particular devices, paying attention to the characterization techniques that are necessary for succeeding.

6.1 Basic Concepts of Optomechanics

In this section, we will follow the formalism followed in Ref. [7], where the dynamical equations of the coupled system of an optical and a mechanical oscillator are treated semiclassically. This is enough for the purposes of this chapter, since it is valid for sufficiently large photon and phonon numbers and can be usually solved numerically. The formalism using second quantization Hamiltonians can be found elsewhere [8].

Thus, we can write down the equations for the complex light amplitude $a(t)$, a^2 being proportional to the optical energy density stored in the cavity (ρ_u), and an oscillator position $x(t)$:

$$\frac{da}{dt} = i\Delta(x)a - \left(\frac{1}{2\tau_o} + \frac{1}{2\tau_{ex}}\right)a + i\sqrt{\frac{1}{\tau_{ex}}}s \qquad (6.1)$$

$$m_{\text{eff}}\frac{d^2x}{dt^2} + m_{\text{eff}}\frac{\Omega_m}{2Q_m}\frac{dx}{dt} + m_{\text{eff}}\Omega_m^2 x = \alpha a^2(t) + F_L(t) \qquad (6.2)$$

Regarding the first equation, s^2 represents the power launched inside the system by an external source and Δ the detuning of the resonance with respect to the laser frequency ($\Delta = \omega_{\text{laser}} - \omega_0$). The optical field decays at a rate given by $2\tau^{-1} = 2\tau_0^{-1} + 2\tau_{\text{ex}}^{-1}$, which is a combination of the intrinsic decay rate of the isolated cavity and the extrinsic decay rate due to the interaction with outgoing modes originated by the external source.

The second equation concerns the global amplitude $x(t)$ of the motion and is representative of a suitably normalized dimensionless mode function $\mathbf{u}(\mathbf{r}, t)$. The displacement field would be $\mathbf{u}(\mathbf{r}, t) = x(t)\mathbf{u}(\mathbf{r})$, $\mathbf{u}(\mathbf{r})$ being the spatial displacement function describing an specific mechanical eigenstate associated to a mechanical eigenfrequency Ω_m. It describes the motion of a forced mechanical oscillator with a mode effective mass (m_{eff}) and a damping given by $1/2Q_m$, where Q_m is the mechanical quality factor. The first of the forces driving the mechanical motion is the radiation pressure force, which depends on a^2 and its efficiency as applied to activate the specific mechanical mode. The exact calculation of the overall force will depend on the OM cavity under study and the optical and mechanical eigenmodes involved in the equations. The second force is the fluctuating thermal Langevin force [9]. In the absence of the optical force, the mechanical oscillations will have a randomly-time-varying amplitude and phase, where the average value of the square amplitude is given by

$$< x^2 > = \frac{k_b T}{m_{\text{eff}}\Omega_m^2}, \qquad (6.3)$$

T being the temperature of the bath and k_b the Boltzmann constant. These fully nonlinear coupled differential equations will be the basis for our discussion of nonlinear phenomena. Both equations are linked by the optical force driving the mechanical motion and by the OM coupling effect, that is, the displacement of the optical eigenfrequency due to the mechanical deformation. Indeed, the cavity resonance frequency is modulated by the mechanical amplitude and the detuning can be written as

$$\Delta(x) = \Delta(0) + \frac{d\omega}{dx} + \dots \quad (6.4)$$

For most of the cases it is enough to keep just until the linear term, where we define the optical frequency shift per displacement as

$$G = -\frac{d\omega}{dx}. \quad (6.5)$$

Frequently, it is better to express it as a function of the vacuum OM coupling strength, expressed as $g_0 = G x_{zpf}$, where x_{zpf} is the zero-point fluctuation motion, and express the spread of x in the ground state. Generally speaking, g_0 is more fundamental than G, since the latter is affected by the definition of the displacement.

Two cases will be of interest—that of a blue-detuned laser $\Delta > 0$ and that of a red-detuned laser $\Delta < 0$.

The steady-state solution of the equations is not unique on the mechanical displacement. Indeed, this can give rise to multistable situations that can be observed if the stored energy is high enough [10].

Now, let us discuss the dynamic solutions of the equations and the effects of the retarded nature of the optical force in relation with the deformation. We will work under the approximation of having a radiative lifetime much longer than the period of the mechanical oscillation. In this case, the optical force is linked instantaneously with ρ_u but its response can be developed into different orders of retardation with respect to the deformation. Then, by keeping only terms up to dx/dt, we get the following:

$$\rho_u = \rho_{u,0} + \frac{d\rho_u}{dx}x + \frac{d\rho_u}{\frac{dx}{dt}}\frac{dx}{dt} + \dots \quad (6.6)$$

The first two terms are the adiabatic response to the deformation and give rise to instantaneous variations of the stored energy. In Eq. 6.2, the second term effectively changes the effective spring constant of the oscillator and, consequently, the mechanical eigenfrequency, creating the so-called light-induced rigidity or optical spring effect. When $\Delta > 0$ the optical spring effect gives rise to an increasing of the spring constant and vice versa.

The last term of the expansion enters in that equation at parity with the damping term; therefore its effect is that of changing the damping rate of the oscillator. This effect is called dynamical back action.

When $\Delta > 0$ the damping rate is reduced and a net flow of energy passes from the optical mode to the mechanical mode. Indeed, for a high-enough ρ this term can compensate the intrinsic damping rate of the oscillator and enter into a regime where the damping is negative, that is, where there is mechanical amplification using optical pumping. This regime will be demonstrated experimentally in Section 6.4.2.2.

When $\Delta < 0$ the damping rate is enhanced, effectively cooling the mechanical mode under study, that is, the average amplitude of the mechanical mode will be lower than that measured at the same temperature without the optical force. In this case, there is a net flow of energy from the mechanical mode to the optical mode. By carefully optimizing the intrinsic optical and mechanical damping rates and the bath temperature the achievement of the lowest state of motion available has been already demonstrated several times [2, 11].

It is worth noting that the number of photons is not changed by the OM interaction. Indeed, the process can also be understood in terms of the creation of Stokes and anti-Stokes sidebands (see Fig. 6.1). In the scattering mechanism the average energy of the photons is increased when $\Delta < 0$ and decreased when $\Delta > 0$.

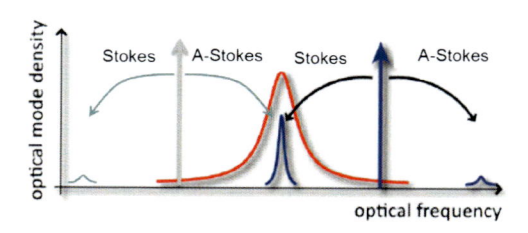

Figure 6.1 Blue-detuned (blue) and red-detuned (gray) lasers with respect to optical mode line shape (red) provide mechanical amplification and cooling, respectively, by enhancing the Stokes or anti-Stokes scattering mechanism through the optical density of states of the cavity.

6.2 Nano-optomechanical Oscillators

Photonic crystals are created by a periodic modulation of the index of refraction of some material, which leads to the formation of optical bands and, eventually, bandgaps of energy, where light cannot propagate. To form localized resonances in a central part of the photonic structures, the discrete translational symmetry of crystal is intentionally disrupted by a "defect." The created modes do not decay into the continuum inside the structure, because its propagation is forbidden outside the defect due to the presence of the gap. These structures are called photonic crystal cavities, which can localize photonic modes within the defect regions. A phononic cavity is the phononic counterpart of the photonic crystal, that is, the periodic tailoring of the elastic properties of the material can lead to the formation of phononic bands and the interruption of the periodicity to the formation of localized phononic or mechanical states. Thus if a material is engineered to be periodic so that both photons and phonons are simultaneously affected (and controlled), the previously mentioned effects and phenomena can be combined, giving rise to the so-called OM crystal. Thus, OM crystals are quasiperiodic structures in which it is possible to achieve simultaneous control of light and sound in a way that OM interactions are greatly enhanced. Indeed, since light and motion can be confined in the same region of the structure, they can interact in different ways, one of them by exchanging energy between the photon and phonon populations. The mechanical motion deforms the cavity boundaries and stresses in the material, both of which contribute to the OM coupling between the cavity photons and the mechanical modes of the structure.

The pioneer work of Eichenfield et al. [12] opened the possibility of using the interactions described in the previous section in a chip-scale platform. The small cavity dimensions and the good overlap between optical and mechanical field distributions, in combination with the small mass of the localized mechanical mode, result in values close to a few megahertz, which is 2–3 orders of magnitude higher than what offered by other common OM architectures such as microdisks or microtoroids.

There are different mechanical modes available in these structures. Usually they present extended modes, such as those of a double-clamped cantilever that appear in the range of the tens to hundreds of megahertz, depending mostly on the length of the structure. On the other hand, confined cavity modes can reach frequencies up to several gigahertz and several families can appear in different spectral ranges. In the view of the cooling mechanism, high-frequency modes are very interesting since the thermal occupation is inversely proportional to the frequency. Indeed, by exploiting all these characteristics, it was recently possible to achieve the quantum ground state of motion for a localized mode at 5 GHz in a thermal bath of 20 K [2]. This kind of structure also offers the possibility of integrating both optical and electrical actuation [5].

On the other hand, the product $\Omega_m Q_m$ is a figure of merit that reveals the degree of mechanical oscillation coherence that could be reached at a given temperature. Thus, reaching high values of Q_m is also very important to achieve strong coherent interaction mechanisms with the optical field. Usually thermal effects dominate Q_m at room temperature, but at cryogenic temperatures, they become no longer important. Several strategies can be explored in these systems toward the goal of optimizing Q_m at cryogenic temperatures, depending on creating phononic bandgaps where the localized mode could be placed.

A full phononic bandgap is defined by the absence of any phononic band in a given frequency range; meanwhile, a pseudo bandgap is defined by the absence of bands of a particular symmetry in the frequency range of definition, even if there are still bands of other symmetries at those frequencies. In Chang et al. [2], the Q_m of the confined phononic mode at cryogenic temperatures is limited by unwanted fabrication imperfections that break the perfect symmetry of the ideal structure and allow coupling among different symmetry phononic propagative bands inside the pseudo bandgaps. This loss is partly mitigated by introducing a surrounding periodic structure matched to the phonon wavelength ("phonon shield"), which is a 2D phononic crystal with a complete bandgap at the frequencies of the localized modes.

Finally, the great majority of these structures, which are mostly silicon based, are compatible with the devices of integrated photonics. Thus, they have the potential to be integrated with current nanofabrication capabilities and to enable scaling up to network sizes, which is interesting in the context of classical and quantum information processing, and for the study of collective dynamics.

6.3 Basic Properties of Optomechanical Crystals: Case Study

Here we present a basic study on a 1D OM crystal whose geometry is equivalent to that presented by Eichenfield et al. [12]. We will present standard optical and mechanical results obtained in those samples using a setup that is similar to what is widely used in the optomechanics community. With this experimental technique it is possible to transduce the mechanical motion of the OM crystal by pumping a confined optical mode. Moreover it is even possible to amplify optically the localized mechanical motion of the OM crystal.

More specifically, we will also show that high optical quality factors combined with small modal volumes make this kind of devices prone to thermal-induced effects, which strongly depend on the intensity of the electromagnetic field. Indeed, thermo-optic effects have been experimentally reported in 2D photonic crystals [13], optical and OM circular microresonators [14–16].

Monomode tapered fibers are the ideal tool to optically excite such devices by means of evanescent coupling [17, 18], since they provide high coupling efficiency with low nonresonant insertion loss. However, the accurate control of the coupling distance is not unproblematic and very often the measurements are done in contact mode to keep the coupling conditions stable in time.

We will also address the modifications of the optical and mechanical properties of the OM crystal as a function of ρ_u. Moreover, we show that the interaction between resonator and tapered fibers strongly influences the optical and mechanical properties of the resonator as well as its dynamic behavior. In this regard, two important effects of this experimental configuration are studied, the fiber loading onto the resonator and the attractive optical forces between the fiber and the resonator. Indeed, both are related since optical forces result in an effective modification of the fiber loading.

6.3.1 *Experimental Setup*

The experimental setup used to measure the optical, mechanical, and OM properties of the fabricated structures is shown in Fig. 6.2. A fiber-coupled tunable infrared (IR) laser covering the range from 1.5 μm to 1.64 μm is used as a light source. After a fiber polarization controller, light enters a microlooped tapered optical fiber, which is used to couple light within the structures under test. The light transmitted through the fiber finally reaches the IR photodiode. The polarization state of the transmitted signal can be optionally analyzed after the tapered region. Once the laser is resonant with the OM structure and light is coupled in, variations on the central resonance position due to the thermal motion will translate in variations of the transmitted signal intensity at the frequencies of the phononic modes that have a certain degree of OM coupling with the optical modes and are thermally excited.

To check for the presence of a radio frequency (RF) modulation on the transmitted signal related to the presence of optomechanically coupled phononic modes we have used an IR avalanche photodiode (APD) with a bandwidth of 12 GHz. The RF voltage is connected to the 50 ohm input impedance of a signal analyzer with a bandwidth of 20 GHz.

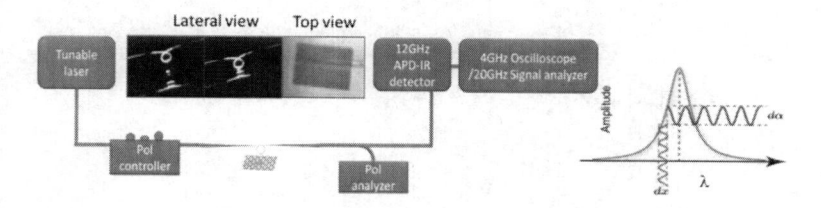

Figure 6.2 Experimental setup for the OM characterization of OM structures. The left and central photos over the scheme show a lateral view of the real microlooped tapered fiber away from the sample (left) and close to the sample (center). The mirror image of the fiber can be seen reflected in the sample. The right photo shows a top view of the tapered fiber in contact with the central zone of an OM structure.

The sample is placed below the tapered fiber and its position is controlled with nanometer-scale precision. Light is coupled into the structures, thanks to the long tail of the evanescent field of the tapered fiber mode. Indeed, the tapered fiber is approximately 1 μm thick and shaped into a microlooped form with $R \approx 10$ μm, which allows the local optical excitation of the OM structures with nanometer-scale precision on a region of less than 5 μm^2.

6.3.1.1 Tapered fiber fabrication procedure

We have used a homemade setup in which a telecom optical fiber is stretched in a controlled way using two motorized stages. The central part of the fiber is placed within a microheater on a region where the temperature is about 1180°C (Fig. 6.3). It is possible to monitor the fiber transmission at 1.5 μm while the pulling procedure is performed. To check the transition to a single-mode configuration, a short-time fast Fourier transform (FFT) algorithm is applied to the signal. This is illustrated in Fig. 6.4, where the transmitted light is represented for a reduced pulling region. It is apparent on the figure that the transmitted light before 47.3 mm is much noisier than after that distance. This is explained in terms of interference of the multiple modes that are supported by the

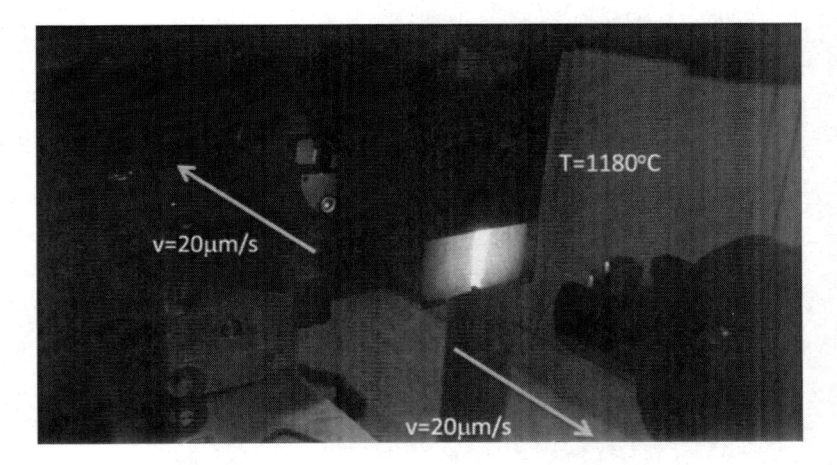

Figure 6.3 Experimental setup for fabricating tapered fibers with a microheater.

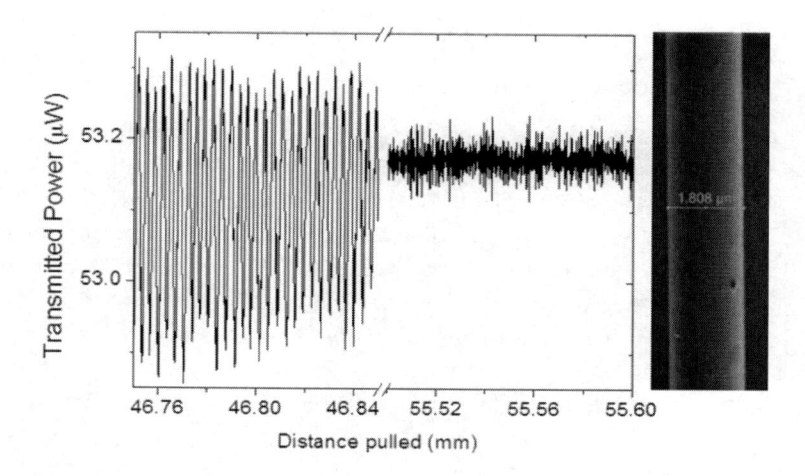

Figure 6.4 (Top) SEM image of the thinnest part of a tapered fiber. (Bottom) Transition from multimode to monomode while pulling the two extremes of the fiber.

fiber. Indeed, each supported mode experiments a different change in optical path as we pull, which is a consequence of the different effective refractive indices.

When the various frequency components that are observed during the pulling procedure totally disappear, the monomodal condition on the tapered region has been achieved and higher-order modes are radiative.

Following this procedure, we are able to fabricate tapered fibers with diameters lower than 1.5 μm, which is of the order of the wavelength of interest and ensures an evanescent field tail of several hundreds of nanometers.

Using two rotating fiber clamps, the taper is twisted by two turns on itself. The two fiber extremes are then approached for several hundreds of micrometers so that a looped structure forms in the tapered region. The two extremes are afterward pulled apart to reduce the loop size down to a few tens of micrometers, as shown in Fig. 6.2. In that process, the two parts of the fiber at the loop closing point slide smoothly in opposite senses.

6.3.2 *1D Silicon Optomechanical Crystals*

6.3.2.1 Geometry of the structures

We fabricated 1D silicon OM crystals with the design illustrated in Fig. 6.5. This geometry is equivalent to that presented by Eichenfield et al. [12]. The defect region consists of 15 central holes in which the hole-to-hole spacing (pitch) is varied quadratically from the nominal lattice constant at the beam perimeter (362 nm) to 85% of that value for the two holes in the central part. At both sides of the defect the pitch is kept constant along 30 cells more on each side acting as an effective mirror for the defect region so that the total number of cells is 75. The structures were fabricated in silicon-on-insulator (SOI) standard samples with a top silicon layer thickness of 220 nm (resistivity $\rho \approx 1 - 10 \ \Omega$ hmcm, p-doping of $\sim 10^{15}$ cm^{-3}) and a buried oxide layer thickness of 2 μm. The OM crystal cavity fabrication process was based on electron beam (e-beam) direct writing on a coated 170 nm of poly(methyl methacrylate) (PMMA) 950 K resist layer. This e-beam exposure was optimized with an acceleration voltage of 10 keV and an aperture size of 30 μm with a Raith150 tool. After developing, the resist patterns were transferred into the SOI samples employing by inductively coupled plasma-reactive ion etching. To release the silicon membranes, a second photolithography process based on UV exposure with a mask-aligner tool was employed to open a window in the UV resist just where the structures were placed. Finally, the silicon dioxide under the membranes was removed by using a buffered hydrofluoric acid (BHF) bath.

Two designs were fabricated with the nominal parameters summarized in Table 6.1, the main difference among both being the width.

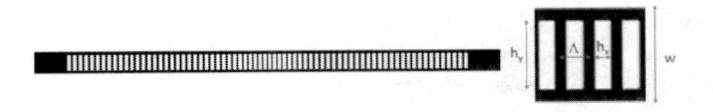

Figure 6.5 Geometry of the 1D optomechanical structure as seen from the top. On the right we show a zoomed region illustrating the different geometric parameters characteristic of this particular design.

Table 6.1 Values of the parameters determining the geometry of the structures

Parameter	Device 1	Device 2
Ndefect	15	15
Ntotal	75	75
Λ (nm)	362	365
w (nm)	1396	864
hy (nm)	992	575
hx (nm)	190	183
thickness (nm)	220	220

Within the wafer, a total of 18 columns with devices were produced, each one associated to a slightly different e-beam dose. Increasing the latter value enlarges the actual sizes of the holes and reduces the total widths of the nanobeam structures but does not affect the pitch.

6.4 Optical and Mechanical Simulations

The localized optical modes of the finite OM structure have been found by means of finite-difference time domain (FDTD) simulations. We have used MEEP, [19], a free software package developed at MIT. The optical quality factors (Q_0) are extracted with the HARMINV subpackage, which uses the filter diagonalization method [20]. This particular design allows the localization of five photonic transverse electric (TE)-polarized eigenmodes with different energy and spatial distribution along the long axis of the OM structure, all of them with quality factors well above 10^5. The spectral dependence of the intensity stored in the structure is represented in the left part of Fig. 6.6, each peak corresponding to a localized mode illustrated in the right part of Fig. 6.6. When the fiber is in contact with the OM crystal there is a significant overlap between the evanescent field of the mode supported by the tapered fiber and the localized mode within the OM crystal (Fig. 6.7c), thus allowing an efficient excitation of the latter. In the same way, the resonant mode extends within the tapered fiber glass region,

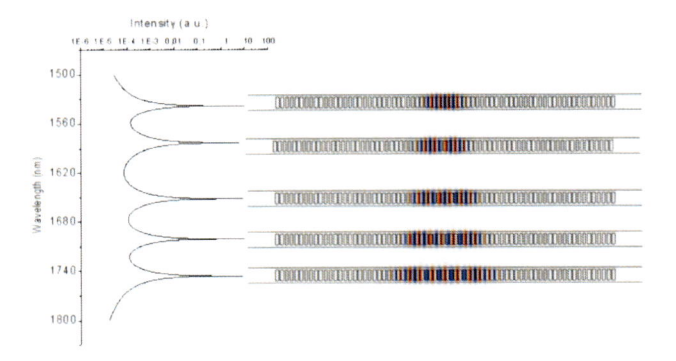

Figure 6.6 Optical localized modes calculated for the experimentally fabricated (Device 1).

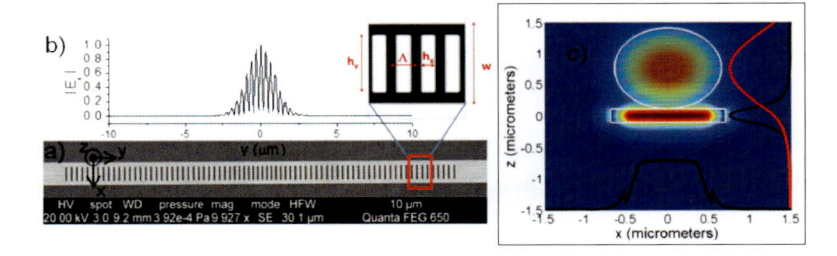

Figure 6.7 (a) SEM micrograph of Device 1. (b) $|Ex|$ along the y axis for the fundamental optical mode. (c) Color contour plot of $|Ex|$ in the $x - z$ plane for the modes supported by the OM crystal and the tapered fiber. The black (red) lines correspond to the mode profile of the OM (fiber) structure along the axis directions. The white lines are the physical boundaries of the two structures.

giving rise to an attractive cavity-enhanced optical dipole force (CEODF) between both structures [15]. The intensity distribution along the nanobeam direction is also showed for the case of the lowest-order mode, that is, the one with the highest energy (see Fig. 6.7b). Finite-element method (FEM) simulations using COMSOL Multiphysics [21] have been performed to extract the mechanical band structure of the OM crystal. It shows the localization of certain types of vibrations in the defect region that should present OM coupling with the localized optical modes. These mechanical modes are denominated, from lowest to highest frequency, as the "pinch"

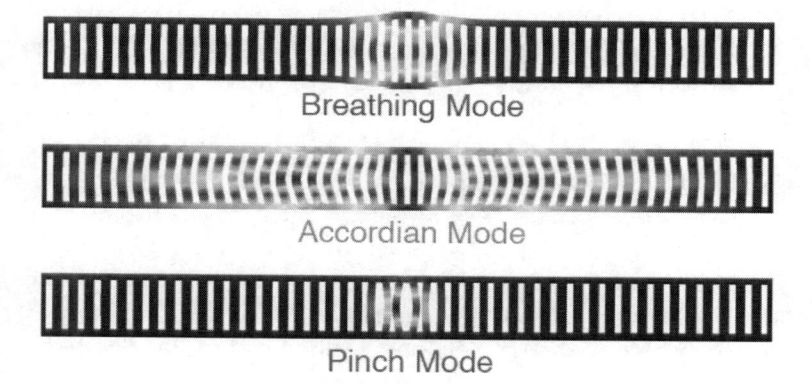

Breathing Mode

Accordian Mode

Pinch Mode

Figure 6.8 From top to bottom, spatial distribution of breathing, accordion, and pinch modes.

(Fig. 6.8, bottom), "accordion" (Fig. 6.8, center), and "breathing" (Fig. 6.8, top) modes. The exact spectral positions of the different mode families depend on the particular design of the OM crystal.

6.4.1 *Optomechanical characterization*

The optical spectrum extracted from the device under study is shown in Fig. 6.9a. The spectral range covered by the tunable laser allows the observation of three localized modes supported by the structure, which are strongly polarized in the plane of the sample (TE polarization). At low input powers, their spectral widths ($\delta\lambda$) are about 0.03 nm, which translates into maximum optical quality factors ($Q_0 = \lambda/\delta\lambda$) of 54,000 (inset of Fig. 6.9a). The oscillations outside the resonant frequencies reflect the presence of whispering gallery modes in the microlooped tapered fiber.

The energy within a given mode (U) can be calculated by using $U = P_c Q_0/\omega$, where P_c is the power coupled into the resonant mode and ω is the optical frequency. Under optimum coupling conditions, the coupled power level can reach up to -1.5 dB of the off-resonance transmitted power. In that case, the energy density ($\rho_u = U/V$, where V is the volume over which U is distributed) in a resonator mode is 6 orders of magnitude higher than that present in the travelling fiber mode. Thus, in terms of the CEODF between both structures, the contribution of the fiber mode field can be neglected with respect to that of the localized mode of the resonator.

As explained in Section 6.3.1, it is possible to analyze the RF spectral components of the transmitted signal. This allows detection of the mechanical modes that are optomechanically coupled to the optical modes. Indeed, when the laser is resonantly probing the cavity, there exists a transduction mechanism in which the thermally activated mechanical motion causes the transmitted field to be phase-modulated around its steady-state value [7]. The coupling strength between optical and mechanical degrees of freedom is directly related to the transduced signal intensity. In addition, the transduced signal intensity also increases with the optical power coupled to the optical mode and with the quality of the optical and mechanical modes [22]. To maximize the transduced signal the laser wavelength should be set at the position of the maximum slope of the resonance. However, as we will show in the following section, it is not straightforward to fulfill this condition, since the resonance experiences a red shift with coupled power.

We have experimentally set the laser wavelength in resonance with the first photonic mode and then obtained the RF spectrum presented in the bottom panels of Fig. 6.9. The OM crystal presents a strong transduced signal for a set of four mechanical modal families

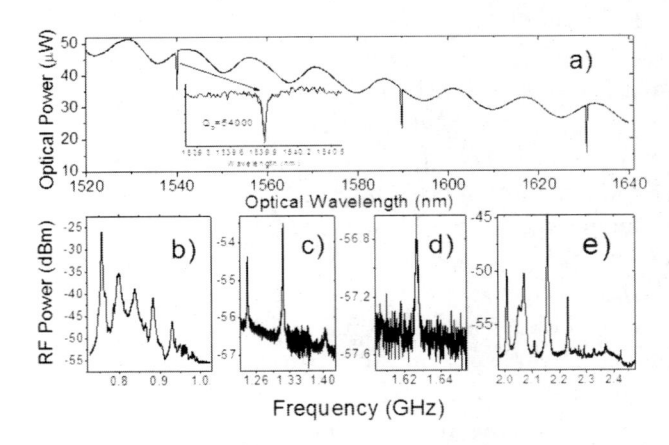

Figure 6.9 Optical (a) and mechanical (bottom panels, log scale in vertical) spectra of the OM crystal.

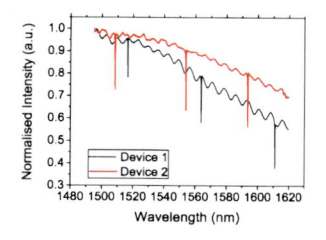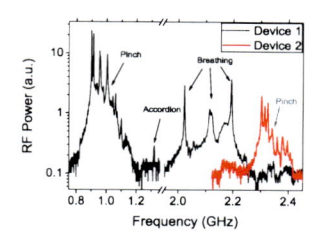

Figure 6.10 Optical (left) and mechanical (right) spectra for Device 1 (dotted black) and Device 2 (continuous red).

placed in different spectral regions. Within these families, there is a rich substructure of different mechanical modes associated to different modal orders. A maximum Q_m of 1050 was measured in the highest-energy region. The first mechanical family corresponds to the pinch modes, the second to the accordion modes, and the fourth to the breathing modes. Similar results were obtained in Ref. [12], although we note that only one mechanical mode was detected around the 1.3 GHz region and the mode at 1.62 GHz was not present.

In Fig. 6.10 we show the comparison between the results obtained on Devices 1 and 2. The spectral range covered by the tunable laser allows the observation of three clear optical resonances for each device, which are strongly polarized in the plane of the sample (TE polarization).

As already said before, Device 1 presents a strong transduced signal for the set of modal families that were expected, that is, pinch, accordion, and breathing modes, each one in a different spectral region. Within the pinch and breathing mode families, a rich substructure of different mechanical modes appears that are associated to different modal orders.

Regarding Device 2, the first transduced components appear above 2.2 GHz and are associated to pinch modes. It is then very interesting to note how the same modal family doubles its eigenfrequency by just carefully scaling the cells. In Fig. 6.11 we illustrate the dependence of the optical and mechanical spectra with the e-beam dose used to fabricate the structures. The results correspond to structures with the nominal design of Device 1. Only

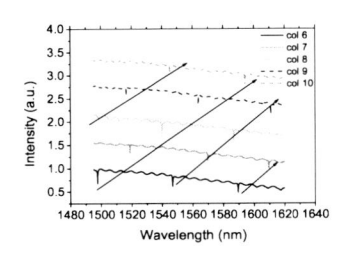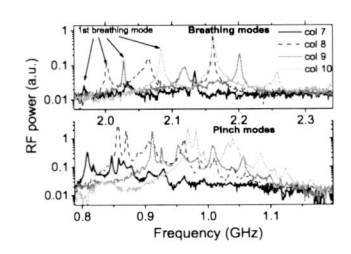

Figure 6.11 Optical (left) and mechanical (right) spectra taken from devices belonging to different columns.

columns 7–10 (e-beam dose $= 1.2 - 1.35 \times 10^2$ µQ/cm^2) showed a nonnegligible transduced signal in the RF spectra. From now on we will only show results concerning the design of Device 1.

Regarding the optical transmission spectra (Fig. 6.11) we observe a clear spectral shift toward longer wavelengths as a function of the column number, that is, by decreasing the e-beam dose. This is an expected result as the effective refractive index of the structure is being increased in this sense. The mechanical modes show instead a spectral shift toward higher frequencies since the effective structure becomes stiffer. The overall shift of the pinch family is evident (bottom panel of Fig. 6.11, right), while in the case of the breathing mode family (top panel of Fig. 6.11, right) it is even possible to follow the behavior of a mode associated to the same order (indicated by arrows for the case of the fundamental mode).

Putting together the whole set of optical transmission spectra corresponding to the different columns we can conclude that at least four different optical eigenmodes are measured. However, it is not possible to label them with the mode orders, since the simulations foresee the presence of five localized modes. Thus, one missing mode (the fundamental or the fifth mode) could be still falling outside (below or above) the spectral range covered by the laser. Indeed, a fundamental mode order could be associated to a certain resonance only by checking that no more modes appear at shorter wavelengths. In any case, it is certain that the optical modes of lower wavelengths observed in the structures of columns 6 to 8 are not the fundamental one. Because the various optical modes have different spatial profiles, each mechanical mode has a different g_0 for each

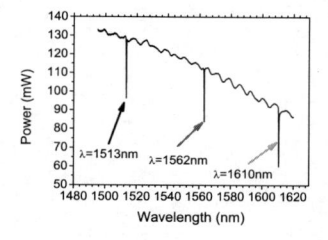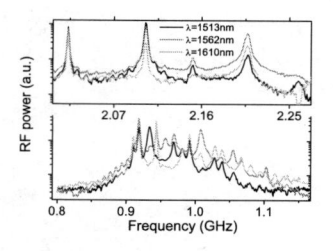

Figure 6.12 Optical (left) and mechanical (right) spectra. The two panels of the right part of the figure cover the spectral regions of the pinch (bottom) and breathing (top) modes. The different curves are obtained by pumping different localized optical modes.

optical mode. This is illustrated in Fig. 6.12, where it is possible to observe that, depending of the optical mode chosen to excite the cavity, it is possible to observe different contributions from the mechanical modes in the pinch and breathing modal families. Indeed, even if the coupled optical powers are comparable, there are clear differences in the relative intensities among modes and even several cases in which mechanical modes seem to be absent. It is worth mentioning that the mechanical amplitude of a mode due to the actuation of the Langevin force depends just on the temperature, so the different transduced intensities are then just associated to the different g_0.

6.4.2 Tuning the Optical and Mechanical Properties of the Optomechanical Crystal with Coupled Optical Power

When the coupling conditions are optimized, a strong red shift of the optical resonance as a function of the power coupled into the OM crystal is observed. Figure 6.13 shows several spectra, associated to different off-resonance transmitted powers through the fiber, taken by sweeping the laser wavelength from short to long wavelengths at a constant speed (40 nm/s). Wavelength drifts up to 70 linewidths are observed and maximum values of 250 linewidths have been reached (see vertical scale of Fig. 6.14a). Furthermore, the change from the symmetric spectral shape at low power (black curve) to the asymmetric spectral shape observed at

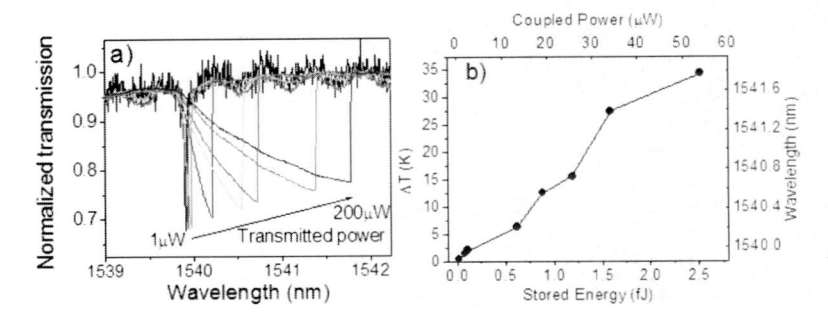

Figure 6.13 (a) Different normalized spectra of the first photonic resonance obtained for different transmitted powers. (b) Average temperature increase (resonant wavelength) as a function of the stored energy within the photonic mode (coupled power).

high powers presents an abrupt transmission jump at the position of the minimum transmitted intensity. Below certain transmitted powers, about 1 μW in the case of Fig. 6.13, the resonances shape does not change anymore. The spectral drift of the resonance to longer wavelengths can be mainly ascribed to thermo-optic effects of the material composing the resonator, that is, expansion of the structure and changes in the material refractive index due to the absorption of part of the high electromagnetic field built up in the resonator. In addition, and as a consequence of the CEODF, the OM crystal studied in this work can also suffer a small deformation that depends on ρ_u within the resonant mode. In the current and next sections we will show that this deformation has a minor impact on the overall spectral drift but may limit the maximum drift achievable. The dynamic evolution of the OM crystal in the high-transmitted-power regime could be explained as follows:

(1) At the start of the measurement, the laser is out of resonance (blue-shifted with respect to the resonance) and there is no optical energy within the OM crystal. The OM oscillator is in a cold configuration.

(2) When the laser wavelength enters the resonance region, the thermo-optic response increases as a function of the coupled power. This results in a dynamic red shift of the resonance

during the spectral scan and while the pump laser is on resonance.

(3) For reasons that will be discussed further in this section, the actual optical resonance becomes blue detuned with respect to the laser wavelength and the system enters an unstable configuration. When the OM crystal relaxes to its original position it gives rise to the abrupt jump in transmission.

For a given value of the coupled power, the stored energy will be distributed according to the optical mode spatial profile. However, even provided with this information, it is an extremely complex problem to infer the exact spatial temperature distribution within the OM crystal. To start with, it will depend on the effective thermal conductivity of the resonator, which is significantly reduced with respect to the bulk values below a thickness of a few micrometers and will be affected by the specific geometric nanostructure of the Si beam as well. Instead, we have extracted an upper limit of the effective temperature increase reached in the region hosting the optical mode. This has been done by assuming that the observed wavelength shift is only associated to an average change in the Si refractive index (n_{Si}). We have also taken the wavelength at the maximum coupled power as the resonant wavelength, which for this purpose is a fair approximation since the actual resonance linewidth, assumed to be independent of the coupled power, is much narrower than the observed spectral shifts. Moreover, there is a linear relation of n_{Si} with temperature (T), $n_{Si} = 3.38 \cdot (1 + 3.9 \cdot 10 - 5 \cdot T(K))$ at $\lambda_0 = 1.55$ μm. FDTD optical simulations also reveal a linear dependence of the optical frequency of the localized mode on n_{Si}. Thus, it is straightforward to establish a relation between the spectral shift and the average temperature increase (ΔT). In Fig. 6.13b we plot the resulting curve that links the energy within the optical mode and ΔT, being about 35 K for the highest power reported in Fig. 6.14a. This energy is distributed over the modal volume ($V = 1 - 2(\lambda_0/n)^3$), resulting in an average ρ_u of about 250 fJ/m³ for this extreme case.

As it has been pointed out, the maximum optical spectral shift increases with ρ_u. We have observed that the spectral shift achieved in higher-order optical modes (presenting similar Q_0 but higher

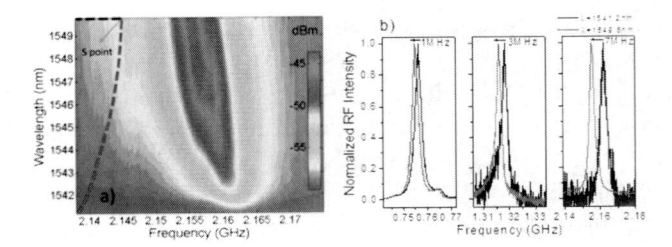

Figure 6.14 (a) Color contour plot of the RF power of a high-frequency mechanical mode as a function of the frequency (x axis) and the optical wavelength (y axis) obtained for a transmitted power (off resonance) of 1.7 mW. The coupled optical power reaches 260 μW for the strongest drift. The optical transmission spectrum at this transmitted power is also showed for clarity (dashed line, b) mechanical modes in different frequency regions at two extreme optical wavelengths.

modal volumes) is lower than in the case showed in Fig. 6.13a, even if the coupled powers are similar.

In our case the resonance contrast becomes smaller with increasing transmitted power. This observation reflects that the absolute minimum of the resonance is not reached. In fact, we have also observed that the maximum coupled power condition red-shifts with the scan velocity (Fig. 6.15), that is, with the rate of change of coupled power. In the steady configuration, in which the RF measurements are performed, the maximum coupling is found at the position reported for the lowest scan velocity (denominated as the S point from now on). The transduced mechanical signal is also experimentally maximized in those conditions because the position of maximum resonance slope has not been achieved. This confirmed to be blue-detuned with respect to the resonance minimum, since the absolute minimum of the resonance would correspond to a minimum in the transduced signal. Thus, from this result we can infer that Q_0 does not degrade meaningfully with the coupled power due to increasing free-carrier absorption losses. Finally, it is worth noting that the S point red-shifts with power in an equivalent way, as that shown in Fig. 6.13a.

The OM crystal state reached with scan velocities higher than 0.5 nm/s is unstable above the S point. However, if only thermo-optic processes were present, one would just expect retardation effects due to a limited temporal response, that is, the thermal adaptation to changes in the coupled power could not be fast enough to keep the resonant condition. In such a case, the behavior with scan velocity should be opposite to that showed Fig. 6.15 and the strongest red shift should occur at the lowest velocity. Moreover, the expansion and refractive index change rates scale with the sweep velocity in the temporal ranges covered by the laser, since the spectral shape before the S point does not vary. Therefore, an independent mechanism must be at play, with a slower temporal response than that of the thermo-optic processes and low impact on the optical drift, which counteracts the OM crystal response to wavelength changes. Indeed, there is an increase of the attractive optical forces among the fiber and the OM crystal leads to a small relaxation of the fiber loading. Moreover, in the following section we will demonstrate that a controlled increase of the fiber loading red-shifts the optical frequency at low coupled powers, when thermo-optic effects are not present. A decrease of the fiber loading would thus lead to a blue shift of the resonance. Since the delay becomes evident at 5 nm/s, the temporal response of the adaptation of the coupled system would be at least in the millisecond range. This long time response is reasonable, given the dimensions of the tapered fiber system.

In addition to the optical resonance tuning, there is also a shift in the frequencies of the mechanical modes when changing ρ_u in the OM crystal, done by changing the optical excitation wavelength at high transmitted powers. In the case illustrated in Fig. 6.14, the off-resonance transmitted power was 1.7 mW and the maximum optical resonance drift was 8 nm, associated to a maximum coupled power of 260 μW, under a steady-state laser configuration. In this extreme condition we have observed a 0.3% frequency downshift (about 7 MHz) in the modes placed around 2 GHz (Fig. 6.14b). Slightly lower relative shifts were observed for the modes present in lower-frequency regions. The reported frequency reduction of the mechanical modes is a consequence of decreasing the Si elastic constants, which is related to the average temperature increase. For the same reasons exposed above, that is, a lack of knowledge of the temperature spatial distribution within the OM crystal, it is not possible to extract the spatial distribution of the elastic

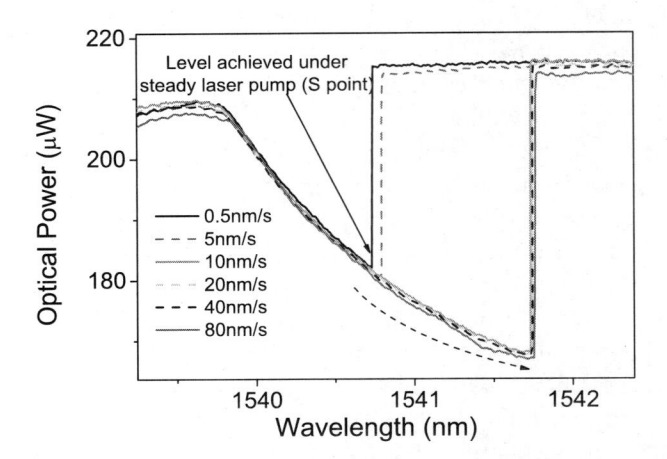

Figure 6.15 Transmission spectrum around the first optical resonance for different sweep velocities.

constants. Nevertheless, by assuming a homogeneous increase of the resonator temperature, associated to the measured coupled powers (see Fig. 6.13b) FEM simulations have provided frequency downshifts compatible with those experimentally reported (0.8% in the extreme condition for the modes around 2 GHz). In summary, the coupled power does not affect the quality factor of the optical and mechanical modes but shifts the optical modes toward longer wavelengths and the mechanical modes to lower frequencies.

6.4.2.1 Effects of fiber loading on optical and mechanical properties

Since the fiber is in contact with the structures, their physical interaction has a significant effect on the spectral position and quality of the optical and mechanical modes of the OM crystal. As mentioned in the experimental part, the tapered fiber is leaning both on the top of the structure and on the surrounding frame. It is then possible to qualitatively control the loading over the OM crystal by changing the relative pressure over the two contact points. The different panels of Fig. 6.16 show the first optical resonance for increasing fiber loading. Each panel shows a solid

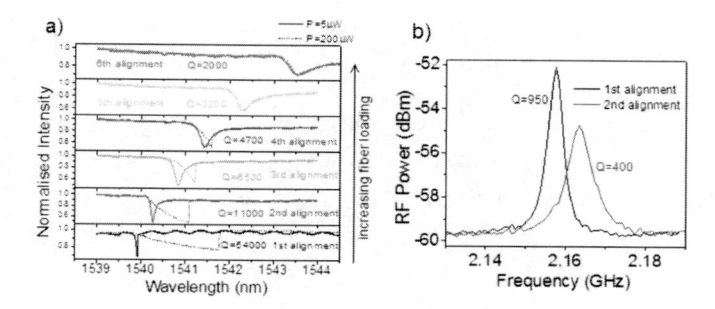

Figure 6.16 Optical (a) and mechanical (b) spectra for different fiber loadings on the OM crystal.

curve corresponding to the low-input-power case and a dashed one corresponding to a transmitted power of about 200 µW off resonance. The coupled fraction in all cases is very similar but there is a remarkable red shift of the resonance position, together with a lowering of the Q_0, which is compatible with a bending of the OM structure. The comparison of the extreme cases at low power reveals a total red shift of 3.6 nm and a 27-fold lowering of the Q_0.

The high-transmitted-power curves show that the maximum achievable red shift, with respect to the resonance observed at low power, decreases with the fiber loading, which is mostly due to the lowering of the Q_0. For the same main reason it is only possible to transduce the mechanical motion in the cases of the two lower loadings. This is illustrated in Fig. 6.16b for one of the highest-frequency mechanical modes. There is a significant shift toward higher frequencies associated to an effective stiffness increase due to the loading. The losses of the supported mechanical modes appear also to be affected and a twofold lowering of the Q_m was observed.

In summary, the physical pressure exerted by the fiber over the structure deforms the latter in a way that the Q-factors strongly degrade, the optical modes shift toward longer wavelengths, and the mechanical modes shift to higher frequencies.

It is worth noting that the optical modes shift in the same sense observed when increasing the coupled power (cf. Fig. 6.13), while the mechanical modes shift in the opposite sense (cf. Fig. 6.16). Moreover, under the lowest fiber loading condition, it is possible to shift the resonance up to more than 8 nm, more than twice the maximum shift reached by increasing the loading, by just increasing the coupled power and yet observe the most intense mechanical modes, without damaging of the mechanical and optical quality factor. Thus, the physical origin behind of the optical and mechanical shifts with coupled powers has to be mainly attributed to thermo-optic effects in the silicon material composing the OM crystal.

6.4.2.2 Experimental observation of dynamical back action

Now we will discuss the dynamical back-action effect when a blue-detuned laser is used. This study has been made on a different structure with respect to what shown until now and shows localized phononic modes around 5.5 GHz.

It is well known that by increasing the ratio Ω_m/κ where κ is the optical mode damping rate) the potential cooling/amplification rates are also increased. Figure 6.17 illustrates this and delimits two regions on the basis of the efficiency of the process: (i) the unresolved band, where $\Omega_m/\kappa < 1$, and (ii) the resolved band, where $\Omega_m/\kappa > 1$.

The mechanical mode subject of this particular study is present as observed at 5.46 GHz and is barely inside the resolved sideband since $\Omega_m/\kappa = 2.02$.

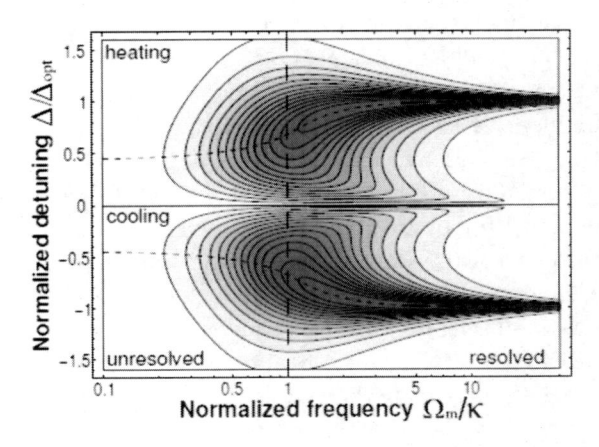

Figure 6.17 The mechanical amplification and cooling rate as a function of detuning and normalized mechanical frequency.

6.4.2.3 Mechanical amplification at 5.46 GHz

We have observed that the mechanical mode present at 5.46 GHz is optically amplified through dynamical back action of the radiation pressure force. The signatures of mechanical amplification are mainly two: (i) a strong power increasing of the transduced signal and (ii) and a linewidth narrowing. Both effects are clearly visible on the main panel of Fig. 6.18 and its inset.

Indeed, we have determined that, above the threshold, the effective mechanical quality factor can reach a value of $Q_m = 24000$, starting from $Q_m = 1600$. There is also a further interesting effect regarding the central frequency of the observed mechanical mode. It seems clear (Fig. 6.18) that in the first curves there is a slight shift toward higher frequencies, while at a certain point the frequency starts to decrease. This can be explained in terms of a combination of optical spring and thermo-optic effects. The former mechanism increases the effective stiffness of the structure and is expected to be present in a situation of blue detuning. The latter process reduces the stiffness as a consequence of a local temperature increasing with the associated reduction of the elastic constants. The observation of mechanical amplification also demonstrates that, in principle,

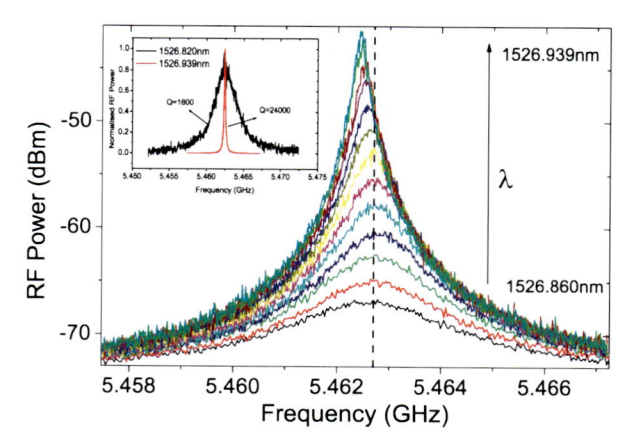

Figure 6.18 Mechanical mode at 5.46 GHz optically amplified through a dynamical back-action process. In the inset, the extreme curves are normalized and represented in linear scale.

it is also possible to cool down this mechanical mode in some extent. However, it is difficult to do that at room temperature since thermo-optic effects are present, which prevents putting the laser wavelength in the cooling side band of the resonance at high powers.

Acknowledgments

The author acknowledges the P2N group at the ICN leaded by Prof. Dr. C. Sotomayor-Torres and, specially, J. Gomis-Bresco with whom most of the research showm in the last part of the chapter was made. The collaboration with the colleagues of the TAILPHOX project is greatly acknowledged. The work was financially supported by the EC project TAILPHOX (ICT-FP7-233883).

References

1. Fuhrmann, D.A., Thon, S.M., Kim, H., Bouwmeester, D., Petroff, P.M., Wixforth, A., and Krenner, H.J. (2011). Dynamic modulation of photonic crystal nanocavities using gigahertz acoustic phonons, *Nat. Photonics*, **5**, pp. 605–609.

2. Chan, J., Alegre, T.P.M., Safavi-Naeini, A.H., Hill, J.T., Krause, A., Gröblacher, S., Aspelmeyer, M., and Painter, O., (2011). Laser cooling of a nanomechanical oscillator into its quantum ground state, *Nature*, **478**, 7367, pp. 89–92.

3. Arcizet, O., Cohadon, P.-F., Briant, T., Pinard, M., and Heidmann, A. (2006). Radiation-pressure cooling and optomechanical instability of a micromirror, *Nature*, **444**, pp. 71–74.

4. Kippenberg, T.J., Rokhsari, H., Carmon, T., Scherer, A., and Vahala, K.J. (2005). Analysis of radiation-pressure induced mechanical oscillation of an optical microcavity, *Phys. Rev. Lett.*, **95**, p. 033901.

5. Winger, M., Blasius, T.D., Mayer Alegre, T.P., Safavi-Naeini, A.H., Meenehan, S., Cohen, J., Stobbe, S., and Painter, O. (2011). A chip-scale integrated cavity-electro-optomechanics platform, *Opt. Express*, **19**, 25, pp. 24905–24921.

6. Bagheri, M., Poot, M., Li, M., Pernice, W.P., and Tang, H.X. (2011). Dynamic manipulation of nanomechanical resonators in the high-amplitude

regime and non-volatile mechanical memory operation, *Nat. Nanotech.*, **6**, 11, pp. 726–732.

7. Kippenberg, T.J., and Vahala, K.J., (2007). Cavity opto-mechanics, *Opt. Express*, **15**, 25, pp. 17172–17205.

8. Aspelmeyer, M., Kippenberg, T.J., and Marquardt, F. (2013). *Cavity Optomechanics*, arXiv preprint arXiv:1303.0733.

9. Hadjar, Y., Cohadon, P.F., Aminoff, C.G., Pinard, M., and Heidmann, A. (1999). High-sensitivity optical measurement of mechanical Brownian motion, *Europhys. Lett.*, **47**, 5, p. 545.

10. Dorsel, A., McCullen, J.D., Meystre, P., Vignes, E., and Walther, H. (1983). Optical bistability and mirror confinement induced by radiation pressure, *Phys. Rev. Lett.*, **51**, 17, pp. 1550–1553.

11. Teufel, J.D., Donner, T., Li, D., Harlow, J.W., Allman, M.S., Cicak, K., Sirois, A.J., Whittaker, J.D., Lehnert, K.W., and Simmonds, R.W. (2011). Sideband cooling of micromechanical motion to the quantum ground state, *Nature*, **475**, 7356, pp. 359–363.

12. Eichenfield, M., Chan, J., Camacho, R., Vahala, K.J., and Painter, O. (2009). Optomechanical crystals, *Nature*, **462**, pp. 78–82.

13. Barclay, P.E., Srinivasan, K., and Painter, O. (2005). Nonlinear response of silicon photonic crystal microresonators excited via an integrated waveguide and fiber taper, *Opt. Express*, **13**, 3, pp. 801–820.

14. Ding, L., Senellart, P., Lemaitre, A., Ducci, S., Leo, G., and Favero, I. (2010). GaAs micro-nanodisks probed by a looped fiber taper for optomechanics applications, *SPIE Photonics Eur.*, pp. 771211–771211.

15. Eichenfield, M., Michael, C.P., Perahia, R., and Painter, O. (2007). Actuation of micro-optomechanical systems via cavity-enhanced optical dipole forces, *Nat. Photonics*, **1**, 7, pp. 416–422.

16. Carmon, T., Yang, L., and Vahala, K.J. (2004). Dynamical thermal behavior and thermal self-stability of microcavities, *Opt. Express*, **12**, pp. 4742–4750.

17. Knight, J.C., Cheung, G., Jacques, F., and Birks, T.A. (1997). Phase-matched excitation of whispering-gallery-mode resonances by a fiber taper, *Opt. Lett.*, **22**, 15, pp. 1129–1131.

18. Cai, M., and Vahala, K. (2000). Highly efficient optical power transfer to whispering-gallery modes by use of a symmetrical dual-coupling configuration, *Opt. Lett.*, **25**, 4, pp. 260–262.

19. Oskooi, A.F., Roundy, D., Ibanescu, M., Bermel, P., Joannopoulos, J.D., and Johnson, S.G. (2010). MEEP: a flexible free-software package for electromagnetic simulations by the FDTD method, *Comput. Phys. Commun.*, **181**, 3, pp. 687–702.

20. Taylor, H.S. (1997). Harmonic inversion of time signals, *J. Chem. Phys.*, **107**, 17, pp. 6756–6769.

21. COMSOL is a multiphysics software package for performing finite-element-method (FEM) simulations, http://www.comsol.com/.

22. Eichenfield, M. (2010). *Cavity Optomechanics in Photonic and Phononic Crystals: Engineering the Interaction of Light and Sound at the Nanoscale*, PhD thesis, California Institute of Technology.

Chapter 7

Quantum Dot–Based Nano-optoelectronics and Photonics

Juan P. Martinez Pastor[a] and Guillermo Muñoz Matutano[b]

[a]*UMDO-FOS Group, Instituto de Ciencia de los Materiales
Universitat de Valencia, P.O. Box 22085, 46071 Valencia, Spain*
[b]*ITEAM Research Institute, Universitat Politecnica de Valencia,
C/ Camino de Vera s/n, 46022 Valencia, Spain*
juan.mtnez.pastor@uv.es

Nanoscience and nanotechnology are progressing along many fronts, often from interdisciplinary and multidisciplinary points of view; the most impressive progress has been made in the area of semiconductor technology. I remember the beginning of the nineties, when I did a posdoc formation on optical properties of semiconductor quantum wells (QWs), the most successful achievement for light-emitting diode (LED) and laser fabrication (one should mention here the achievement of high-power lasers at 808 nm based on GaAs QWs and LEDs—laser diodes based on GaN QWs, the most successful achievement in the field at the time). I particularly remember a one-week course given by Prof. Federico Capasso in the spring of 1992, at the Department of Physics, University of Florence, where I remained impressed by his well-known *bandgap wavefunction engineering* [1] and premonitory *quantum design* concepts [2] to tailor optical and

Nanodevices for Photonics and Electronics: Advances and Applications
Edited by Paolo Bettotti

electronic properties of quantum heterostructures and conceive new electronics, his famous *double-barrier resonant tunneling diode* and optoelectronic devices, and his also well-known *quantum cascade infrared laser* [3].

The applications found for quantum dots (QDs) along the last years extend over many fields, from the most technological, as photonics and optoelectronics, the main topics of this chapter, toward nanomedicine, where colloidal QDs are being used for fluorescent labelling of biomolecules, substituting the well-known fluorophores or organic dyes. The basis for these applications is their electronic (Section 7.3) and optical (Section 7.4) properties, which are analogous to those of atoms and molecules in the sense that they exhibit a discrete electronic/excitonic structure, and for this reason, QDs are often called *artificial atoms*. In Section 7.3 we will also introduce some basic models to calculate confined electronic states in QDs, as the basis for their optical properties. The main aspects and mechanisms regarding exciton recombination will be considered in Section 7.5, together with the corresponding modelling, the master equations of microstates being the most important ones for the study of recombination dynamics in QDs, both colloidal and self-assembled and either a single QD or an ensemble of them.

Photonic based on QDs (Section 7.6), by definition, will take account of the particular interaction of light and QDs, trying to get amplification, modulation, operations, and similar functions usually pursued by using different materials in combination with light propagation on chips (waveguides, couplers, splitters, ...) made on appropriate substrates: silicon, as the most important platform nowadays, but not the only one (glass and flexible substrates) for other applications of photonics, such as sensing and displays.

Of course, the other two important devices to be considered together with the most strict photonics operations are photon sources, LEDs and lasers, and photon detection, that is, optoelectronic devices (Section 7.7), even if we will consider in detail photon sources, given their straightforward combination with photonic structures leading to more efficient devices. Photon-emitting sources are based on electron–hole or exciton recombination in a semiconductor, provided both electron and hole carriers are injected. The combination of QDs and electron/hole injectors is well known for III–V self-assembled QDs, and we will review the advantages and limitations of QD LEDs. An important part of this section will be the very important advances achieved in the case of colloidal QDs to fabricate QD LEDs. This is because the technology is very similar to that of organic light-emitting diodes (OLEDs) but incorporating QDs instead of organic fluorescent molecules. The advantage is clear: An increase in stability and robustness of the device and practically the whole spectrum, from UV to mid-IR can be easily swept by only changing the base material and QD size. Most relevant technological problems arise with the appropriate selection of electron/hole injectors and the QD layer itself in order to build efficient devices.

Semiconductor QDs are at the core of many advanced optical telecommunication technologies like ultralow threshold lasers, amplifiers, and quantum light sources. Some of them are already mature, while others are at different stages of research and development, the seeking of new properties and new applications and device concepts being the goal in the case of quantum light leaving from QDs. This is because, as "artificial atoms" (and two-level systems), they emit single photons and, under certain conditions, indistinguishable and entangled photons. The emission of quantum light on demand has been demonstrated using a single QD by means of incoherent optical pumping, embedding it in microcavities and also in electric charge injection devices. In this way, we can fabricate on-demand single-photon sources (SPSs) and entangled photon sources (EPSs) ready to be used as photon qubits for future quantum cryptography and quantum computing, even if they can be also be used for some photonic operations, as logic gates and optical future quantum photonic devices, and hence paving the road for the manipulation of quantum information and the realization of quantum computing at the most important telecom windows (980, 1300, and 1550 nm) by using integrated quantum photonic devices, as will be discussed in Section 7.8.

7.1 Introduction

7.1.1 *Basics of QD-Based Semiconductors*

The unit cell of III–V semiconductors (e.g., GaAs in Fig. 7.1a) is the atom ensemble representative of the crystal because of its periodical repetition along the three space directions. Setting one corner of the cube as the origin of coordinates the structure is invariant to the point group Td (Zinc-blende group) [4]. The Wigner–Seitz cell associated to the reciprocal lattice of the zinc-blende structure (first Brillouin cell) forms a truncated octahedron (Fig. 7.1b); we can distinguish several symmetry points: The center of this cell is known as the Γ-point, X, L, and K being the larger symmetry points at the Brillouin zone borders.

The electronic wavefunctions can be described by means of periodic functions (Bloch functions) on the basis of the first Brillouin Zone (Fig. 7.1b) and used to obtain the electronic band structure through very different theoretical or pseudo-experimental methods like tight binding or **kp** [4]. The crystalline structures of II–VI and IV–VI semiconductors are different from that of GaAs, which is also reflected by a different band structure (see the case of CdSe and PbSe in Fig. 7.2a); let us observe, for example, the direct nature of the PbSe bandgap at the L-point, whereas it is defined at the Γ-point in CdSe. Most semiconductors present a moderate energy gap (0.5–3.5 eV) between the conduction and valence bands. At 0 K, the valence band of a perfect semiconductor crystal is completely filled by electrons, while the conduction band is completely empty: there are no available carriers in the conduction band and the semiconductor

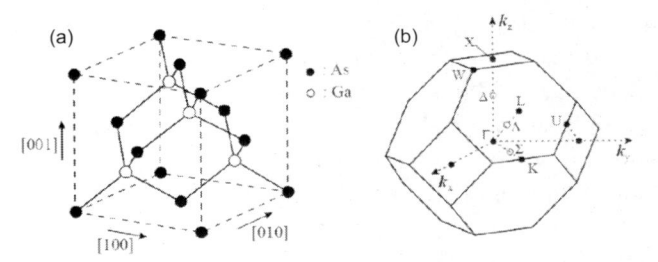

Figure 7.1 (a) Unit cell of the Bravais lattice of a zinc-blende structure, for example, of GaAs, and (b) its corresponding first Brillouin zone.

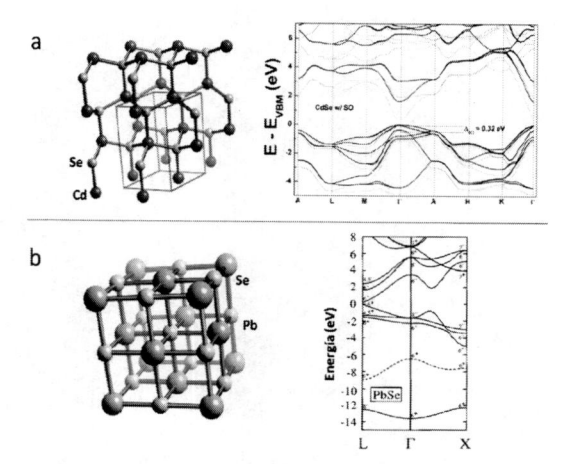

Figure 7.2 (a) Unit cells and energy band structures of CdSe (a) and PbSe (b). The unit cells are taken from cnx.org and band structures from Refs. [5, 6]. Reprinted from Ref. [5], Copyright (2012), with permission from Elsevier. Reprinted figure with permission from Ref. [6]. Copyright (1997) by the American Physical Society.

behaves like an insulator. In contrast, at room temperature, a thermal carrier distribution populates the conduction band (in equal proportion as holes in the valence band) and, consequently, a typical low conductivity is measured. This conductivity can be easily controlled by adding impurities. Acceptor impurities induce hole conduction and donor impurities electron conduction. This ability to change the conduction type (p or n) is the basis for most of the electronic and optoelectronic devices based on p-n homo- and (p-n, p-i-n, or more complex combinations) heterojunctions. Even if the full electronic band structure is rather complex (see right panels in Fig. 7.2), with different energy band dispersions ($E(k)$) at different crystalline directions, most of the semiconductor optical properties can be described, reducing the whole band structure to a simplified $E(k)$ parabolic approximation around the center of the Brillouin zone, the Γ-valley, as shown in Fig. 7.3 for GaAs. The parabolic approximation depicted in Fig. 7.3 for GaAs is the simplest one, describing an electron in the conduction band with an effective mass (m_e), whereas holes in the valence band can be described by heavy and light masses due to the different k-bending of the two

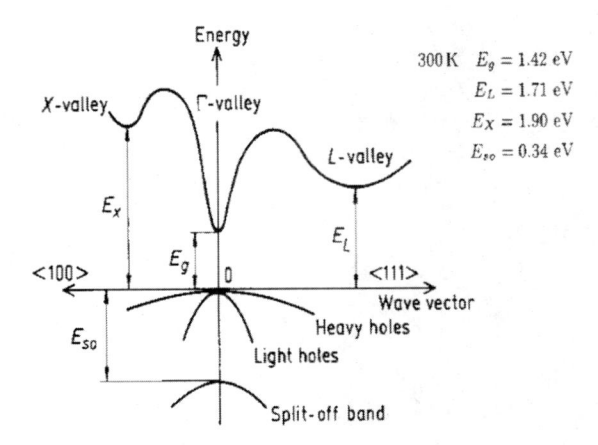

Figure 7.3 Simplified band structure of GaAs at 300 K (from http://www.ioffe.ru/SVA/NSM/Semicond/GaAs/bandstr.html).

bands at the Γ-point. A photon with energy below the bandgap (Eg) cannot be absorbed, while a photon with energy above Eg promotes an electron from the valence band to the conduction band, leaving a hole in the valence band. If electrons and holes remain correlated through Coulomb interaction, such an optical excitation can be described by only one quasiparticle, the exciton (X), instead of the uncorrelated electron–hole pair picture. Once the crystal is excited, it tends to relax that excess of energy until returning to the original crystal ground state (GS), that is, recombination of either an electron–hole pair or an exciton. Depending on the nature of these relaxation channels, we find nonradiative processes (phonon emission, carrier scattering, …) or radiative processes: the electron–hole/excitons recombine by emitting a photon with energy equal to their optical transition energy.

7.2 Synthesis and Growth of Semiconductor Quantum Nanostructures

7.2.1 Colloidal Nanocrystals

Colloidal nanocrystals are obtained by chemical synthesis, where size can be controlled very precisely with high monodispersity. In

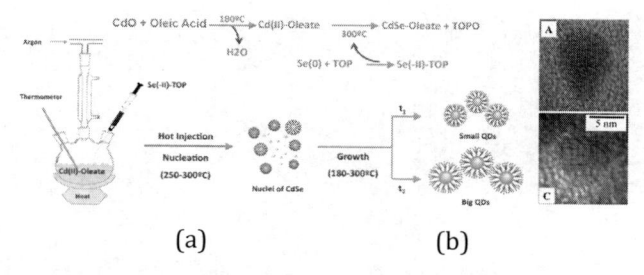

(a) (b)

Figure 7.4 (a) Reaction and reactor for the chemical synthesis of colloidal CdSe QDs; (b) high-resolution TEM image of a core–shell CdS/CdSe QD. Reprinted with permission from Ref. [12a]. Copyright (2003) American Chemical Society.

this way quantum size confinement is achieved, that is, nanocrystals can be considered as quantum dots (QDs). The chemical synthesis is thus considered the easiest and fastest procedure to obtain II–VI colloidal QDs [7, 8]. In the work of Murray et al. [9] was proposed the synthesis of macroscopic quantities of CdX (X = S, Se, Te) QDs in a simple reaction, obtaining QD diameters in the range from 1 to 10 nm, as illustrated in Fig. 7.4. The nucleation of nanocrystals begins immediately after the injection of organometallic reagents into a hot coordinating solvent. When the concentration of reagents reduces from a critical value, nucleation stops and slow growth of the nanocrystals takes place until the desired size value in such a way that aliquots taken from the solution at increasing reaction time will produce colloids with QDs of increasing diameter. If reagents are exhausted during growth bigger nanoparticles can be formed from the material of small nanocrystals, the *Ostwald ripening* growth [10]. Core–shell QDs (e.g., ZnS/CdSe, CdS/CdSe, and other more complex heterostructures) can be also prepared by dropping a chemical precursor of the shell material onto a solution of the core nanocrystals under appropriate reaction conditions [11].

7.2.2 *Self-Assembled Quantum Dots*

Molecular beam epitaxy (MBE) is a commonly used technique for the fabrication of QDs, especially for the III–V compounds (e.g., GaAs, InAs, etc.), whose scheme is shown in Fig. 7.5. The most

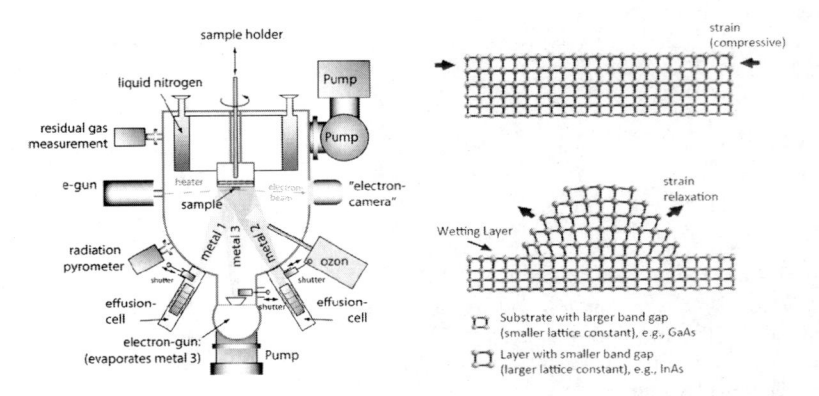

Figure 7.5 (Left) Scheme of the molecular beam epitaxy (MBE) equipment for the growth of semiconductors (from Ref. [12b]); (right) illustration of the Stranski–Krastanov growth mode.

investigated nanostructures are InGaAs self-assembled QDs grown on (100) GaAs and InP substrates by gas source MBE, even if other materials have been also obtained on these substrates, as the case of In alloys based on antimonides. Self-assembled QDs are obtained by the growth of those semiconductor materials that exhibit a mismatch in the lattice parameter with respect to the substrate. This situation enables the growth mode named Stranski–Krastanov, where the stress accumulated during the growth of epitaxial layers with the larger lattice parameter material over the seed matrix relaxes strain by forming isolate islands (the 2D growth mode under these conditions stands up to reaching a critical thickness over which 3D islands appear), standing on a planar and thin layer of the same material, the wetting layer (WL), as illustrated in Fig. 7.5b. Energetically, this layer provides a quantum well (QW)-like energy level structure between the conduction band of the barrier material (usually the matrix) and the confined levels of the QDs. The role of the WL in the overall energetic picture and the dynamics of the system is relevant and will be discussed further later. The density of the QDs can be as high as $10^{11}/cm^2$ and the typical lateral size and height of the dots are in the range of 20–50 nm and 5–10 nm, typically, measured by an atomic force microscope and transversal high-resolution transmission electron microscopy

(TEM) (see Fig. 7.5). See the good review by Stangl et al. on growth and structural properties of self-assembled QDs [13].

7.3 Electronic Structure

Semiconductor nanostructures are low-dimensional systems that result from reducing at least one of the dimensions of a semi-conductor to the nanometric scale. The type I potential profile shown in Fig. 7.6 (positive potential barriers for electrons, $E_{c2} - E_{c1}$, and holes, $E_{v2} - E_{v1}$, for two semiconductors with bandgaps $Eg_2 > Eg_1$) along a given spatial direction is leading to quantum size confinement when the carrier de Broglie length is smaller than the physical dimensions of the nanostructure (L in Fig. 7.6), reducing the available states in k-space and modifying the density of states. For this reason the lowest optical transition energy (ground-state energy, Eg_s) will be higher than Eg_1 and several higher optical transition energies (excited-state energy, E_{Es}) can appear depending on L and the barrier potential energies for electron and holes, $V_e (= E_{c2} - E_{c1})$ and $V_h (= E_{v2} - E_{v1})$, respectively.

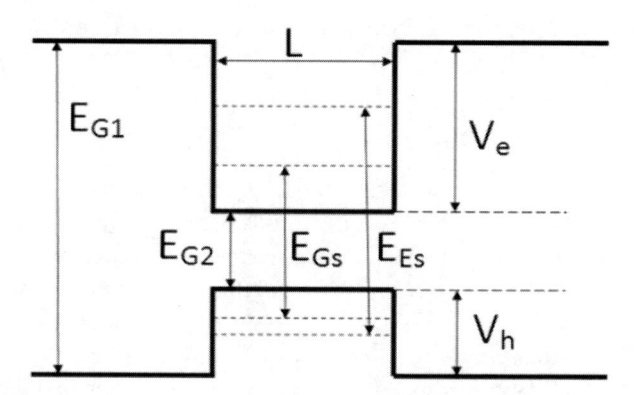

Figure 7.6 1D quantum well potential profile appearing from the type I band alignment between two semiconductors of bandgaps E_{g2} and E_{g1}; V_e and V_e stand for the different band offsets of their conduction and valence bands, respectively, while L is the well length (thickness of the well material).

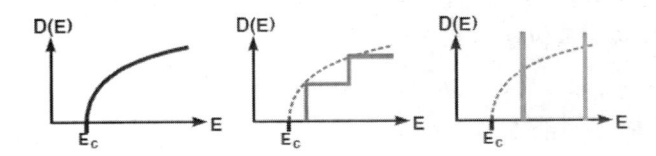

Figure 7.7 Density of states for semiconductor bulk, QW and QD.

There is a striking difference in the density of states if we reduce the dimensions along one, two, or three directions, as plotted in Fig. 7.7. The large density of states concentrated at a given wavelength can be used to increase the performance of laser and optoelectronic devices. The development of QW-based lasers is in a commercial stage, while recent research has actually achieved better performances in QD-based devices and opened the doors to new and exciting applications, like, for example, single-photon sources (SPSs) for quantum cryptography, as will be reviewed later.

QDs exhibit a delta-like density of states:

$$\frac{dN}{dE} \propto \frac{d}{dE} \sum_{\epsilon_1 < E} \Theta(E - \epsilon_i) = \sum_{\epsilon_1 < E} \delta(E - \epsilon_i), \qquad (7.1)$$

which would lead theoretically to threshold-less lasers based on them.

7.3.1 *Colloidal QDs*

The 3D confinement of carriers in a QD results in an atom-like structure. The electronic structure for a real QD can be calculated by taking into account very different aspects: shape, composition profile (intermixing between the QD core and surrounding materials leads to a gradient along the nanostructure), piezoelectric effects, or stress fields that bend the semiconductor band structure.

First of all we can consider a spherical shape, which is the case of colloidal QDs. In this case a first approximation consists in solving a spherical QW with infinite potential barriers for electrons and holes within the effective mass approximation (parabolic bands). The confinement energy of electron and hole levels under this approximation is characterized by the angular momentum quantum number l and can be written as (see Ref. [15])

$$E_{e,h}^{l,n} = \frac{\hbar \phi_{l,n}^2}{2m_{e,h}^* r_0^2}, \tag{7.2}$$

where $m_{e,h}^*$ is the electron and hole effective mass, r_0 is the nanocrystal radius, and $\phi_{l,n}$ is the n-th root of the spherical Bessel function of order l, that is, $j_l(\phi_{l,n}) = 0$ (the four lowest roots are $\phi_{0,0} = \pi$, $\phi_{1,0} = 4.49$, $\phi_{2,0} = 5.76$, and $\phi_{0,1} = 2\pi$). In an early work Kayanuma introduced in the calculation the effect of the electron–hole Coulomb interaction by using a simple variational approach [16]. This term becomes important when the size of the QD is two to three times larger than that of the exciton wavefunction (effective Bohr radius), which is defined as the weak confinement limit. On the opposite, strong confinement means that electron and hole confinement energies are defined by r_0, the Coulomb interaction being a small correction to the ground optical transition energy. The strong confinement limit is the case of most colloidal QDs studied in the literature.

A more convenient model for studying colloidal QDs is the one considering finite potential barriers, where the solution to the radial Schrödinger equation in spherical coordinates, $R(r)$, for the $l = 0$ GS is given in Ref. [17], using spherical Bessel functions of the first and third kind of zeroth order as the wavefunction for an electron (hole) inside and outside the nanocrystal, respectively:

$$R(r) \propto \begin{cases} j_0(k_{in}r) = \dfrac{\sin(k_{in}r)}{k_{in}r}, & r < r_0 \\ h_0^{(1)}(k_{out}r) = -\dfrac{\exp(-k_{out}r)}{k_{out}r}, & r > r_0 \end{cases} \tag{7.3}$$

with

$$k_{in}^2 = \frac{2m^* E}{\hbar^2} \quad \text{and} \quad k_{out}^2 = \frac{2m_0 |E - V_0|}{\hbar^2}, \tag{7.4}$$

m^* being the effective mass of electrons (holes), m_0 the free electron mass, and V_0 the total confining potential:

$$2V_0 = V_{0e} + V_{0h} = E_g(M) - E_g(S), \tag{7.5}$$

where $E_g(M)$ is the surrounding matrix bandgap (formed from the oleate capping agent), $E_g(S)$ the semiconductor bandgap, and V_{0e} and V_{0h} the confining potential for electrons and holes, respectively.

We can choose $V_{0e} = V_{0h} = V_0$, as done in Ref. [17]. By imposing the continuity conditions to the carrier wavefunction and the probability current at the semiconductor–matrix interface and after some mathematical rearrangements, a transcendental equation is obtained, whose solutions are the eigenvalues of the Schrödinger equation:

$$x \cot(x) = 1 - \left(\frac{m^*}{m_0}\right) - \sqrt{\left(\frac{m^*}{m_0}\right)\left(\frac{V_0}{\Delta} - x^2\right)} \qquad (7.6)$$

where $x = K_{in}r_0$ and $\Delta = \hbar^2/2m^* r_0^2$. The carrier effective masses can be taken from Ref. [18] for bulk PbSe (eventually considering nonparabolicity for the carrier effective masses), whereas the value of V_0 is taken as a fitting parameter in the above-described model to reproduce experimental results for PbSe QDs and/or calculated values by using more accurate models, as the case of the tight-binding model used in Ref. [19], parameterized through the equation

$$E_g(D) = E_g(\infty) + \frac{1}{0.0105\,D^2 + 0.2655\,D + 0.0667} \qquad (7.7)$$
$$(E_g \text{ in eV and } D \text{ in nm})$$

where $E_g(D)$ is the effective bandgap energy of a QD of diameter D. Figure 7.8 shows the experimental total carrier size confinement contribution, $E_e + E_h$, as obtained by subtracting the bandgap value for bulk PbSe (taken to be 0.278 eV at 300 K) from the experimentally measured absorption and photoluminescence (PL) exciton peak energies, as a function of the PbSe QD size (the diameters are obtained from statistics on TEM images, as shown in Fig. 7.9), well below the effective Bohr radius of bulk, $a_B^* \approx 43$ nm. It is interesting to note that exciton-binding energy is very low in PbSe, given the small electron and hole effective masses and the large dielectric constant, and hence the optical transitions are determined from electron and hole confinement energies added to the bandgap of bulk PbSe. A rough infinite barrier approach yields confinement energies (dotted line in Fig. 7.8) very far from the experimental ones. Instead, the very simple finite barrier model described above for spherical QDs fits very well the observed blue shift of the QD effective bandgap (solid symbols in Fig. 7.8). At

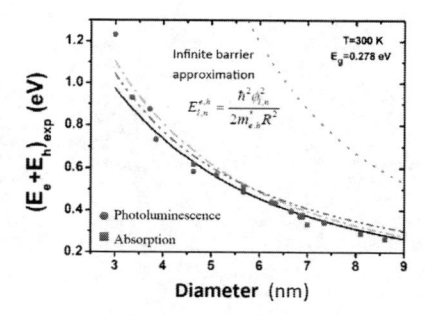

Figure 7.8 Total confinement energy (electron + hole terms) as a function of the PbSe QD diameter: experimental values (solid symbols), calculated through the infinite (dotted line) and finite (dotted-dashed and continuous lines using parabolic and nonparabolic effective masses, respectively) barrier models explained in the text and calculated curve using the expression of Allan and Delerue (see text) [21].

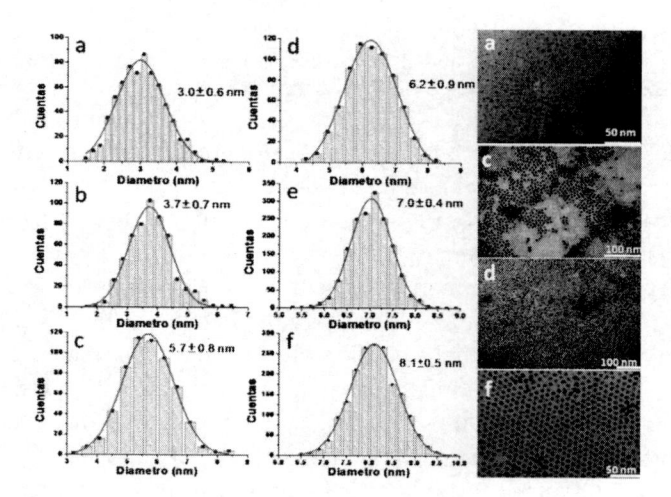

Figure 7.9 (a–f) Size statistics as obtained from TEM images (four right panels).

room temperature the direct application of the model with parabolic effective masses (green dashed-dotted line in Fig. 7.8) yields a reasonable agreement to our experimental values and tight-binding calculations (red dashed-dotted-dotted line in Fig. 7.8) [19] by using a potential barrier value $V_0 = 3$ eV. An additional refinement of this

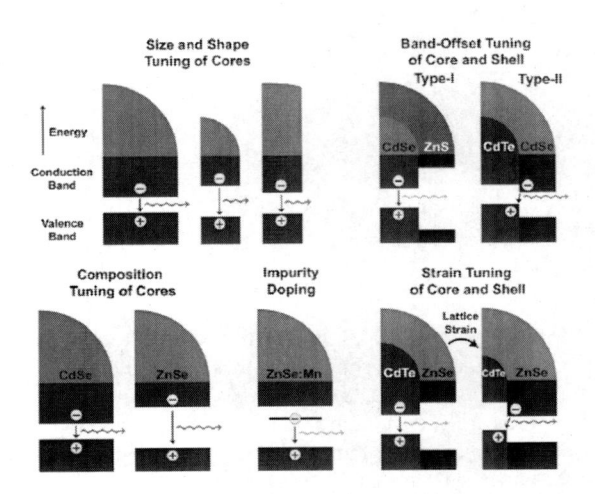

Figure 7.10 Mechanisms of bandgap engineering in semiconductor nanocrystals through size, shape, composition, impurity doping, heterostructure band offset, and lattice strain. Reprinted with permission from Ref. [22]. Copyright (2010) American Chemical Society.

model consists of the introduction of the nonparabolicity effect in the conduction and valence bands of PbSe, as taken from Ref. [20]. The calculated curve of the total carrier confinement energy using nonparabolic effective masses (black solid line in Fig. 7.8) exhibits a smoother variation than in the case of parabolic masses. It is closer to the experimental data in PbSe QDs with diameters above 3 nm and is very similar to the tight-binding prediction (red broken curve in Fig. 7.8). This refined model incorporating nonparabolic masses has been successfully applied to explain the temperature and pressure variation of the exciton transition energies [21] and thus can be of interest to be used for studying the optical properties in bandgap engineering of QD heterostructures, as the cases represented in Fig. 7.10 [22].

7.3.2 Self-Assembled QDs

The shape and composition of self-assembled QDs depend on the growth conditions: substrate temperature, thickness of the deposited layer, atmosphere, ..., and these slightly vary from

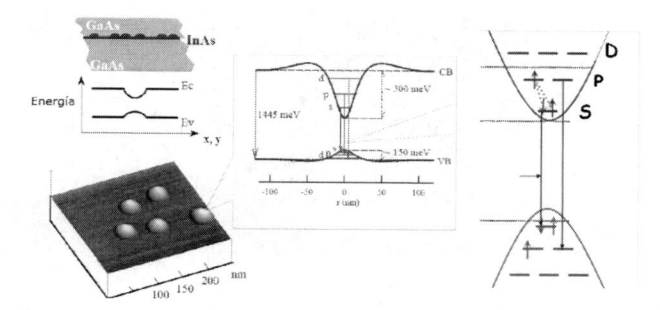

Figure 7.11 Parabolic potential profile of self-assembled QDs along the x-y direction.

QD to QD in the ensemble. Different complex models have been developed under the assumption of particular shapes (pyramidal and lens-like are the most popular ones) and measured or calculated composition gradients and stress fields. The experience, effort, and calculation power needed to get the electronic structure by any of the established methods (8 × 8 **kp**, tight binding, ...) is very high for most of the nanostructures we work with, given the nonstandard shape or the lack of an accurate composition. To combine experimental measurements on QD dynamics with their optical characterization, the energetic positions and degeneracy of the different confined levels are needed. For small QD structures, whose overall size is up to two to three times the effective Bohr radius of the bulk semiconductor free Wannier–Mott exciton, we can consider a strong confinement limit and a simple approach can be applied (in the frame of the effective mass adiabatic model). Typically, the height of a self-assembled QD in the growth direction (z) is really lower than the planar (x-y) dimensions; this allows us to treat the problem in two parts, one regarding the 1D QW confinement problem along the z direction and the approximate 2D infinite parabolic potential along the x-y direction (2D harmonic oscillator problem), as illustrated in Fig. 7.11. The first reference to the use of a parabolic potential for a QD appeared in 1995 [23]. The barrier potential in the z direction will determine the position of the GSs, while the 2D harmonic oscillator in the x-y direction will increase this energy by $\hbar\omega/2$ (zero energy of the quantum oscillator) and it is mainly responsible of the rest of the electronic structure, including the energy difference between excited states (see illustration in Fig. 7.11) that will be constant; $\hbar\omega = \hbar\omega_e + \hbar\omega_h$ and the nature and degeneracy of the confined levels: the ground (first excited) state is s-like (p-like) and doubly (four times) degenerated, if spin is included. See the paper by Ciftja and Kumar for more details on the 2D parabolic potential model [24].

$$\hat{H}(\vec{rho}) = \frac{\hat{p}_p^2}{2m} + \frac{\hat{L}_z^2}{2m\rho^2} + \frac{m}{2}\omega^2\rho^2 \tag{7.8}$$

$$E_{n_p m_z} = \hbar\omega(2n_p + |m_z| + 1) \tag{7.9}$$

$$\Phi_{00}(\rho, \psi) = \frac{\alpha}{\sqrt{\pi}}e^{-\alpha^2\rho^2/2} \tag{7.10}$$

$$\Phi_{n_p m_z}(\rho, \phi) = N_{n_p m_z}\frac{e^{im_z\phi}}{\sqrt{2\pi}}(\alpha\rho)^{|m_z|}e^{-\alpha^2\rho^2/2}L_{n_p}^{|m_z|}(\alpha^2\rho^2), \tag{7.11}$$

where L are the Laguerre polynomials and N the normalization constant:

$$N_{n_p m_z} = \sqrt{\frac{2n_p!\alpha^2}{(n_p + |m_z|)!}}, \ \alpha = \sqrt{\frac{m\omega}{\hbar}} \tag{7.12}$$

7.4 Optical Properties of a Single Quantum Dot

Given that the GS and excited states in a QD can be measured by optical emission spectroscopy and absorption (in QD ensembles but also at the single level), we can deduce some important data from these experiments to validate a given model for the study of optical properties in single QDs. As illustrated in Fig. 7.12, when more than one e^- and/or h^+ populates the QD, we will observe the formation of the neutral (one e^- and one h^+), X^0, and charged excitons, with negative (X^-) and positive (X^+) extra charge over X^0. If two excitons populate the GS (two e^- and two h^+) a biexciton (XX) is formed. Following the same reasoning, if the QD level, or a higher one, can accept more carriers higher charged quasiparticles can be created. This change in the occupancy of the QD levels is leading to slightly different transition energies inside a given state (Fig. 7.12), other than minor effects like screening, strain, and piezoelectric field changes, determining a rich electronic fine structure to the QDs [25].

A typical state involves N electrons in a QD and the approach is to construct an antisymmetric wavefunction for a particular configuration of electrons at the single QD states and then to include the Coulomb interactions with first-order perturbation theory. For the interaction between electrons in states i and j, the matrix elements are of the form

$$E_{ij}^c = \frac{e^2}{4\pi\epsilon_0\epsilon_r} \iint \frac{|\Psi_i^e(\mathbf{r}_1)|^2 |\Psi_j^e(\mathbf{r}_2)|^2}{|\mathbf{r}_1 - \mathbf{r}_2|} d\mathbf{r}_1 d\mathbf{r}_2 \qquad (7.13)$$

for the direct Coulomb interactions and

$$E_{ij}^x = \frac{e^2}{4\pi\epsilon_0\epsilon_r} \iint \frac{\Psi_i^e(\mathbf{r}_1) * \Psi_j^e(\mathbf{r}_2) * \Psi_i^e(\mathbf{r}_2)\Psi_j^e(\mathbf{r}_1)}{|\mathbf{r}_1 - \mathbf{r}_2|} d\mathbf{r}_1 d\mathbf{r}_2 \quad (7.14)$$

for the exchange. These integrals can be evaluated by using the 2D harmonic oscillator wavefunctions. Particularly for S-electrons

$$\Psi_i^e(\mathbf{r}) = \frac{1}{\sqrt{\pi}l_e} \exp(-r^2/2l_e^2), \, l_e = \sqrt{\frac{\hbar}{m^*\omega}}, \, l_h = \sqrt{\frac{\hbar}{m_h^*\omega_h}} \qquad (7.15)$$

that gives

$$E_{ss}^c = \frac{e^2}{4\pi\epsilon_0\epsilon_r} \sqrt{\frac{\pi}{2}} \frac{1}{l_e} \qquad (7.16)$$

for the direct interaction; the other integrals will give a fraction of this quantity both for direct and exchange terms [26]. Similarly, for the interaction between an electron in state i and a hole in state j, which can be interpreted as an exciton-binding energy, the matrix element of the Coulomb interaction is

$$E_{ij}^{eh} = \frac{e^2}{4\pi\epsilon_0\epsilon_r} \iint \frac{|\Psi_i^e(\mathbf{r}_e)|^2 |\Psi_j^h(\mathbf{r}_h)|^2}{|\mathbf{r}_e - \mathbf{r}_h|} d\mathbf{r}_e d\mathbf{r}_h \qquad (7.17)$$

Figure 7.12 shows the energy level splitting by taking into account these energies when charging the QD up to two electrons and two holes (its GS). This configuration is the most common in optical experiments.

Interband transitions with $\Delta m = 0$, typically denoted as $s - s$, $p - p$, and $d - d$, are dominant, even if other transitions can be in the order of 5%–10% of the intensity for $\Delta m = 0$ ones, but not always

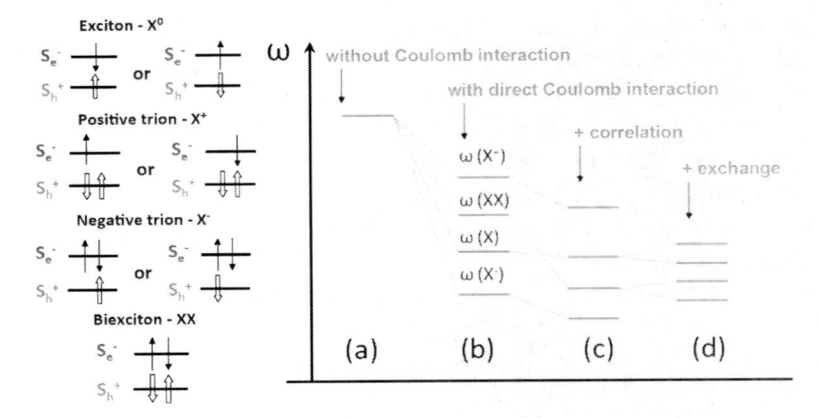

Figure 7.12 Definition of the different exciton species at the ground of the QD (left) and resulting energy level splitting (right-top scheme) due to the different Coulomb and exchange interactions (right-bottom illustration).

negligible, as observed in Fig. 7.11. For neutral QDs, these exciton energies can be written as

$$E_{SS}^X = E_g + E_Z^{eh} + \hbar\omega_e + \hbar\omega_h - E_{SS}^{eh}, \tag{7.18}$$

$$E_{pp}^X = E_g + E_Z^{eh} + 2\hbar\omega_e + 2\hbar\omega_h - E_{pp}^{eh}, \tag{7.19}$$

$$E_{dd}^X = E_g + E_Z^{eh} + 3\hbar\omega_e + 3\hbar\omega_h - E_{dd}^{eh}, \tag{7.20}$$

where E_g and E_Z^{eh} are the material bandgap and the sum of the electron and hole confinement energies along the z direction (QW problem). Figure 7.13 shows the PL and micro-PL spectra of a QD ensemble and a single QD, respectively, under high excitation power in two lens-shape QD families of different sizes (14/54 and 9/36 nm of height/diameter, diameter, from large and small QDs [LQDs] and [SQDs], respectively) present in the same sample [27].

Clearly, when investigating a single QD, its micro-PL spectrum can be representative of any energy interval within the PL spectrum acquired for a QD ensemble, particularly transitions α, β and γ within the Gaussian bands A, B, and C, respectively, in Fig. 7.13a, and i, ii, and iii within the main PL band in Fig. 7.13b. Furthermore, we observe several nominally forbidden transitions (group of β and γ

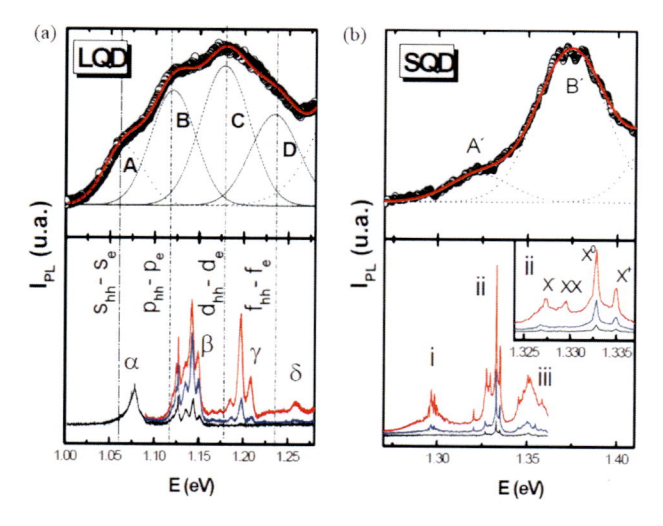

Figure 7.13 PL spectra of two QDs of different sizes: LQD (a) and SQD (b). Top panels correspond to the PL spectrum of a QD ensemble at a sufficiently high excitation power in order to observe the emission from excited states (A, B, and C bands in left panel), if they exist (only for LQD). Bottom panels correspond to micro-PL experiments at three increasing excitation powers.

transitions) at energies around the $p - p$ and $d - d$ shells, as also transition δ above that can only be explained by more complicate models, as the one used by Narvaez and Zunger [28]. The case of transitions i, ii, and iii in Fig. 7.13b cannot be ascribed to excited states but to different single QDs within the only PL band in the SQD family, even if more dedicate experiments are used to identify the observed individual and narrow transitions within those groups of transitions. Typically, micro-PL as a function of excitation power, linear-polarization-resolved micro-PL, and time-resolved micro-PL are used for this purpose, as explained later.

Figure 7.14 shows the basic optical properties of the GS on a well-isolated single InAs QD (from the SQD family presented above) in a low-density sample measured by micro-PL-based experiments at 4 K. Several excitonic emission lines are observed, typically of the order of 50 to 100 μeV in our samples, that should be identified as the neutral exciton or negative or postive trion and biexciton recombination, following the basic model explained above (summarized in Fig. 7.12); in the micro-PL spectrum of Fig. 7.14 the negative trion emission dominates over that of the neutral exciton due to the presence of impurities in the QD surroundings, as reported in Ref. [27]). However, experimental confirmation is needed to support the correct assignation of the different transitions. The integrated intensity as a function of excitation power for all observed peaks can yield a higher slope for the biexciton recombination than that of the negative trion (bottom-left panel in Fig. 7.14). Similarly, micro-PL transients clearly show shorter decay times for biexcitons, typically one-half that of the single-excitonic transitions in the strong confinement limit, as shown in the right panel of Fig. 7.14, even if it can be very close to decay times measured in neutral and charged excitons when the QD size prevents this limit, as occurs in the case of QDs emitting at longer wavelengths (see the discussion on this subject in Ref. [29] and others therein).

The power dependence of the integrated intensity can give us information about the nature of the excitonic transition if this recombination dynamics is conveniently modeled, as explained in the next subsection, but the fine structure of the QD GS, if different from cero (usual case due to anistropy of the QD potential in the plane [30]), can unambiguosly distinguish trions (zero shift) from neutral excitons and biexcitons, whose energy shift takes place in opposite directions, as observed in the central-left panel of Fig. 7.14. A neutral exciton in the QD can be in any one of four spin-state combinations defined by the total angular momentum, $\Delta J = +1, +2$. The degenerate doublet $\Delta J = +1$ (left/right circularly polarized light) is defining the bright exciton because it gives rise to photon emission, while that related to $\Delta J = +2$ is called the dark exciton because it is optically inactive. The degeneracy of the spin doublets of bright and dark excitons is lifted via the anisotropic electron and hole exchange interaction, whose origin is attributed to the nonuniformity of shape and strain in self-assembled InAs/GaAs QDs; for this reason photon emission is linearly polarized, π_x/π_y in Fig. 7.14, revealing the fine structure splitting of the neutral exciton (and biexciton) that can vary from tens to hundreds of μeV and increase from small to big QDs [30]. The QD has two trions energetically degenerate (any energy splitting observed),

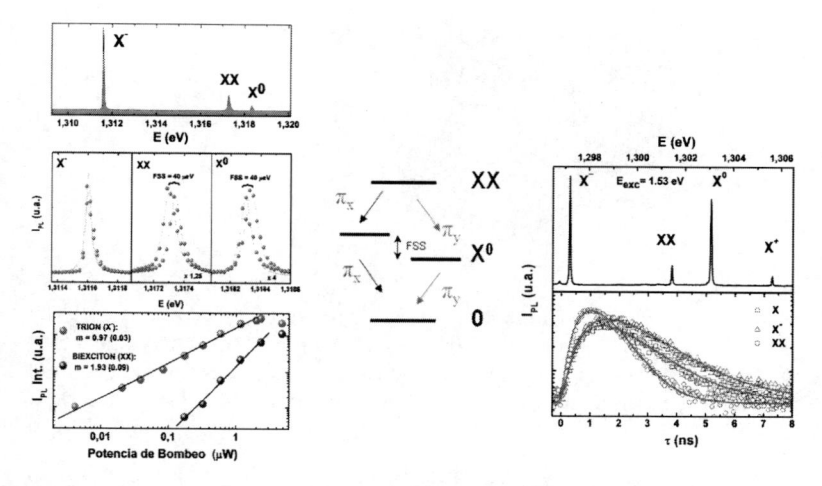

Figure 7.14 (Left) Micro-PL spectrum at medium-excitation-power, polarization-resolved micro-PL and integrated micro-PL intensity vs. power. (Right) Time-resolved micro-PL transient in a similar (but different) single QD whose corresponding continuous-wave spectrum is shown above.

because the electron pair forms a spin singlet state, whereas the hole occupies either spin-up or spin-down. For the biexciton we expect an opposite energy shift to that of the exciton between π_x and π_y polarizations.

Recombination lines from the trion state of the QD can be used for SPSs due to the lack of fine structure splitting (FSS) and the lack of a dark state (DS) [31], whereas the biexciton cascade two-photon emission (biexciton–exciton GS) will give the basis for the development of entangled photon sources (EPSs).

On exciting an electron–hole pair in a dot initially charged by an extra electron we will observe an energy shift to low energies with respect to the neutral exciton due to the domination of the electron–hole attraction over the electron–electron repulsión, as observed experimentally (Fig. 7.15), from measurements in a single quantum ring (QD with a special shape) embedded in a Schottky diode (illustration at the right of Fig. 7.15) [32]. The injection of extra electrons in the InAs nanostructure is giving rise to lower-energy-emitting species, such as the trion, double-charged trion, and other multicharged exciton species, as observed in Fig. 7.15 (left). This

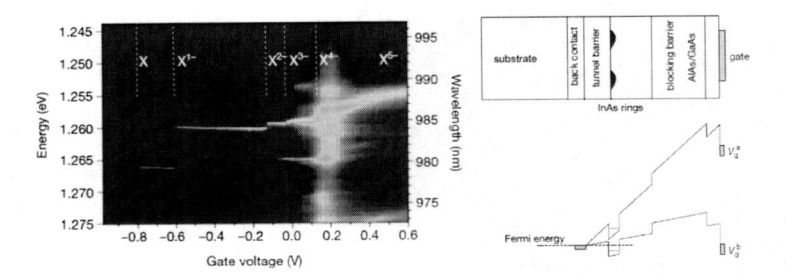

Figure 7.15 Optical transitions in a single InAs quantum ring (left) embedded in a Schotky diode (right) as a function of the gate voltage that tunes the Fermi energy up and down the electron-confined states. Reprinted by permission from Macmillan Publishers Ltd: *Nature* (Ref. [26]), copyright (2000).

kind of device, as a way to initialize the QD with a single electron (or a fixed number of them) and, from here, a spin-defined state by applying an in-plane magnetic field (perpendicular to the growth direction), has been recently used to demonstrate the generation of entanglement between stationary (spin) and propagating (photon) qubits [33] and even quantum teleportation between both qubits when spin and photons are generated on distant QDs [34].

7.5 Recombination Dynamics in Quantum Dots

Here we would briefly introduce two different approaches for the description of the exciton recombination dynamics in QDs—rate equation and master equation of microstates—to study some effects at the core of the use of QDs in optoelectronics and photonics, as in the case of radiative and nonradiative regimes when increasing temperature in QD ensembles (the basis for laser devices), or the interplay between capture of correlated/uncorrelated carriers and the unbalanced escape of electrons/holes out of QDs. In the latter case, these mechanisms will limit the use of these QDs for future devices using quantum light. Even if both models are not formally equivalent, they converge in most of the cases, offering compatible solutions and one can decide to use one or the other, depending on the exciton species and single charges involved in a particular experiment. Finally, we will discuss particularly a very interesting nonradiative recombination mechanism at the core of limitations found in electro-optical devices based on colloidal QDs, the Auger effect.

7.5.1 *Rate Equation Approach: Recombination Dynamics as a Function of Temperature*

Rate equations have been widely used in semiconductor physics for the analysis of carrier recombination in both bulk and QW systems, and they can be extended to the treatment of QD ensembles because they provide a simple description that is analytically solvable in some particular cases. For example, let us examine the application of a rate equation model to explain the recombination dynamics of a QD ensemble as a function of temperature. In principle, the exciton recombination time should be constant with temperature until thermionic emission through the WL states becomes relevant [35], which is not the case if the energy difference between ground excitons and the first excited dark exciton state is not far from KT; in this case the exciton radiative recombination time increases with temperature before thermionic emission appears, as observed experimentally in several kinds of InGaAs nanostructures [36, 37].

A possible recombination kinetic model representative of the above given behavior is depicted in Fig. 7.16 (left). Excitons can be photogenerated in the WL (even if laser excitation is made above the GaAs band edge) at a rate G, where they can recombine with a time constant τ_{WL} (effective decay time accounting for radiative and nonradiative processes at the WL), or they can be captured into QDs with a time constant τ_{QD}. Excitons confined in the QDs can recombine radiatively with the time constant τ_d (associated to the experimental recombination time) or can be activated by thermionic emission to the WL states, a mechanism represented by the time τ_{QD}^*, which includes the Maxwell–Boltzmann thermal factor with an activation energy equal to the difference between the exciton GS and the WL states, as proposed in Ref. [35]. The exchange of carriers between GS and the DS is represented by the time constants τ_0 (from DS to GS state) τ_0^* (from GS to DS state), the latter containing the corresponding thermal factor through the energy difference between GS and DS [36]. Finally, we can include

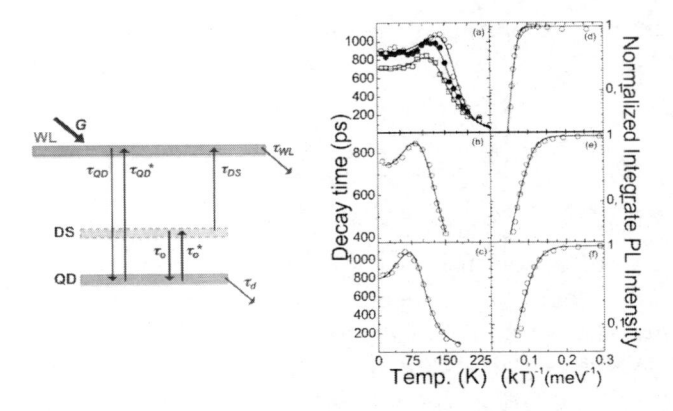

Figure 7.16 Rate equation model to account for GS–DS carrier exchange and thermionic emission through WL states (left) and temperature evolution of the PL-integrated intensity and PL decay time for InGaAs quantum nanostructures with different shapes (right), as reported in Ref. [36]. With kind permission from Springer Science+Business Media: From Ref. [36].

the eventual thermionic emission from the DS to the WL state through the time τ_{DS} [27]. The rate equation system representing this model for the exciton recombination dynamics in most InGaAs quantum nanostructures is

$$\frac{dN_{WL}}{dt} = -\frac{N_{WL}}{\tau_{WL}} - \frac{N_{WL}}{\tau_{WL}^*} + \frac{N_{QD}}{\tau_{QD}^*} + \frac{N_{DS}}{\tau_{DS}} + G \qquad (7.21)$$

$$\frac{N_{DS}}{dt} = -\frac{N_{DS}}{\tau_{DS}} - \frac{N_{DS}}{\tau_0} + \frac{N_{QD}}{\tau_0^*} \qquad (7.22)$$

$$\frac{dN_{QD}}{dt} = -\frac{N_{QD}}{\tau_R} - \frac{N_{QD}}{\tau_{QD}^*} + \frac{N_{QD}}{\tau_0^*} + \frac{N_{DS}}{\tau_0} + \frac{N_{WL}}{\tau_{QD}} \qquad (7.23)$$

Figure 7.16 (right) shows the temperature evolution of the PL-integrated intensity and PL decay time for InGaAs quantum nanostructures with different shapes (obtained by using a kind of overgrowth technique after the QD formation) [36], which are characterized by slightly different energy separation between carrier states. The proposed rate equation model reproduces quite well the evolution of the radiative and nonradiative exciton dynamics in the investigated temperature range. The thermalization between GS and DS dominates the intermediate temperature regime, whereas carrier escape out of the nanostructures becomes important at high temperatures, leading to a strong reduction of the PL decay time and the PL-integrated intensity due to the nonradiative losses through the WL continuum of states. Both regimes are deduced from the above-given rate equation model that yields the following exciton radiative and nonradiative recombination times [36, 37]:

$$\tau_R = \tau_d(13K)[1 + g_2 \exp(-E_{DS-QD}/KT)] \tag{7.24}$$

$$\tau_{NR} = \frac{[1 + g_2 \exp(-E_{DS-QD}/KT)]}{\left[\dfrac{g_1}{\tau_{QD}} \exp(-E_{WL-QD}/KT) + \dfrac{g_2}{\tau_{DS}} \exp(-E_{DS-QD}/KT) \right]} \tag{7.25}$$

7.5.2 Master Equation for Microstates: Carrier Dynamics in a Single QD at the GS

Instead of dealing with average populations, the master equation for microstates (MEM) approach deals with the number of QDs that can be found in a particular electronic configuration (population and level distribution): a microstate. This concept of microstate is based on the idea that carrier capture in QDs is essentially a random process giving rise to the formation of the above-defined exciton quasiparticles (X, X^-, X^+, and XX for GS), and most of the physical mechanisms playing a role in the QD recombination dynamics can be included in this model, as generation+capture and escape of either individual carriers and excitons, as illustrated in Fig. 7.17a,b. The MEM model proposed by Grundmann and Bimberg for a QD ensemble [38] can be extended to incorporate single carrier states and adapted to many physical situations for ensembles of QDs but also for a single QD, such as the examples shown in previous works [39–42]. In this sense, the spectrum of a single QD can be considered as the superposition of temporally different events, that is, generated by a *temporal ensemble* of identical single QDs.

Furthermore, the microstate cannot capture an additional charge/exciton when completely filled because of the *Pauli blockade* (e.g., the biexciton is formed), that is, we approximate the QD by having only S-states.

Other important assumptions in the model are (a) no initial charge in the QD GS and (b) exciton and uncorrelated electron–

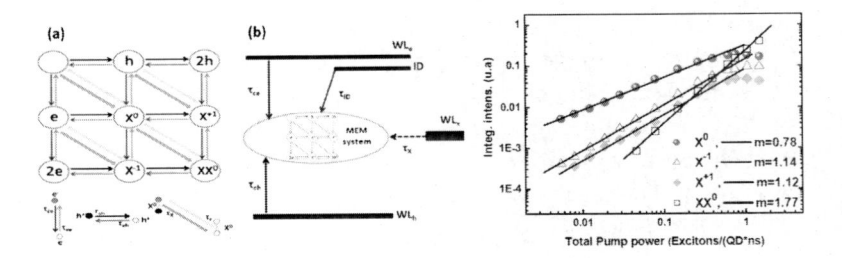

Figure 7.17 (a) Radiative and nonradiative transitions between different microstates defining the model equations. (b) Interaction between the QD and its environment (capture of excitons and uncorrelated carriers from WL states, capture of single electrons from impurities). (c) Power dependence of the micro-PL-integrated intensity measured in a single InAs self-assembled QD emitting at 905 nm (neutral exciton) under laser excitation at 790 nm (see similar variations and micro-PL spectra in Refs. [40–42]).

hole capture processes taking place from WL states (Fig. 7.17b). In addition, we considered for the special case of the studied sample in Refs. [40–42] an extra reservoir of only electrons (Fig. 7.17b) to simulate the case of resonant excitation to the impurity levels (extra electron injection mechanism). Figure 7.17a represents the QD population dynamics scheme. Arrows refer to mechanisms changing the charge inside the QD, where τ_{ce}, τ_{ch}, and τ_X are the capture times of electrons, holes, and excitons, respectively, taken from the literature; τ_r the radiative lifetime for the different exciton species, taken from the experiment (time-resolved micro-PL); and τ_{ee} and τ_{eh} the electron and hole escape times that are output parameters of the model. The carrier dynamics of the GS in a single QD following the MEM model can be written as

$$\frac{d\eta_{ij}}{dt} = R_{ij}^X WL_X + R_{ij}^{ce} WL_e + R_{ij}^{ch} WL_h + R_{ij}^{ee} + R_{ij}^{eh} + R_{ij}^{ID} ID + R_{ij}^r \tag{7.26}$$

$$\frac{dWL_X}{dt} = -\sum_{ij} N_{ij} R_{ij}^X WL_X - \frac{WL_X}{\tau_W} + G_X \tag{7.27}$$

$$\frac{dWL_e}{dt} = -\sum_{ij} i R_{ij}^{ce} WL_X - \frac{WL_e}{\tau_W} + G_{eh} \tag{7.28}$$

$$\frac{dWL_h}{dt} = -\sum_{ij} j R_{ij}^{ch} WL_h - \frac{WL_h}{\tau_W} + G_{eh} \qquad (7.29)$$

$$\frac{dID}{dt} = -\sum_{ij} i R_{ij}^{ID} WL_e + G_e, \qquad (7.30)$$

where $\eta_{ij}(i, j = 0, 1, 2)$ are the microstate population probabilities, G_X and G_{eh} are the generation rates for correlated (X^0) and uncorrelated electrons/holes at the WL states (WL_X, WL_e and W_h), respectively, and G_e is the generation rate for extra electrons from ID levels, if this mechanism is activated by selective optical pumping of impurity levels (residual doping present in the sample) [27, 41]. These generation rates are also output parameters of the MEM model to fit experimental results. Finally, the R_{ij} term represents the capture and escape processes for correlated and uncorrelated carriers (inverse of the above-defined times), following the definitions and compactness of the equations from Ref. [41]. As an example, Fig. 7.17c shows the experimental power evolution of the micro-PL-integrated intensity for all exciton species in the investigated single QD under laser excitation at 790 nm (no extra charge capture from impurity levels). These results are nicely reproduced by using the MEM model with the following output parameters: τ_{ee} and τ_{eh} equal to 5.5 and 3.5 ns and G_X and G_{eh} equal to 0.014 and 0.012 carriers/ns. It is worth noting that the difference in the escape time between electrons and holes is responsible of the appearance and relative importance of X^- over that of X^+; these optical transitions are 1 order of magnitude smaller than the neutral exciton without any other extra charge injection. Furthermore, the MEM model is also able to reproduce the slope observed for the different optical transitions, theoretically 1 and 2 for X^0 and XX^0, but experimentally always below these values. The origin is the two correlated and uncorrelated carrier-feeding mechanisms and the unbalanced escape of electrons and holes out of the QD.

7.5.3 Nonradiative Processes in Colloidal QDs: Blinking and Auger Recombination

Single-nanocrystal QDs exhibit an intermittent/telegraphic light emission or blinking effect that has been attributed to the Auger nonradiative recombination path [43]. In this way, the emitting on state corresponds to a neutral QD, while the off state takes place when an extra charge (electron or hole) populates the QD. In this case the Auger recombination can occur when the photogenerated exciton energy is transferred to that extra carrier, instead of emitting a photon. The extra charge in the QD charging may result from intrinsic processes (Auger-assisted photoionization, Auger-assisted tunneling) but also from traps on the QD surface or in its surroundings (ligands, matrix, substrate). Auger recombination in standard core QDs is very efficient because of the strong confinement potential that enhances exciton–exciton Coulomb interactions [44] and the important spatial overlap between electron and hole wavefunctions [45]. More recently, two sorts of blinking manifestations have been individuated in nanocrystal QDs, the on/off one accompanied by important changes in their corresponding PL decay times (above-given mechanism related to charge/discharge of the QD core) and a second type where large changes in the emission intensity are observed without significant changes in emission dynamics, which is attributed to charge fluctuations in the electron acceptor traps at the surface (this mechanism can be controlled by allowing the occupation of these traps, leading to this blinking-type suppression) [46]. See the excellent review by Hollingsworth for further reading on this subject [47].

Auger recombination quenches emission of trions and other multiple-excited QDs (under high excitation powers), limiting the utility of colloidal QDs to form emitting layers in laser devices or extract multiple excitons generated in solar cells. The insight into the nature of the blinking effect is suggesting that the interfaces (charge traps on them) and shape of the confinement potential (reducing the overlap between the initial and the final state) might play an important role in Auger recombination. In this sense, the most used approaches to reduce Auger decay rates are consisting of the engineering of core–shell QDs: thick shell, type II alignment between core and shell, multiple-shell structure, and alloyed shells.

7.6 Quantum Dot Photonics

The term "photonics" was coined in 1967 by Pierre Aigrain, and since then it is accepted as the science of generating, controlling, and detecting photons, and therefore nanophotonics should mean the same but at the nanometer scale [48]. Given that QDs are nanoscale materials, their use in standard photonic structures can be also considered within the definition of nanophotonics. In this section we center on structures and devices regarding the use of QDs as an active medium able to manage and make operations with light. For the generation and detection of light, electronic phenomena are included, other than photons, and hence this is commonly grouped in the field of optoelectronic devices. Of course, all these parts of the whole are needed to increase the performance of a given device and/or build the heart of a different apparatus to store, transmit, and receive information related to the real world (cameras, televisions, tablets, smartphones, computers, information interfaces, ...). Silicon photonics [49] is the technology based on silicon chips and hence silicon (e.g., SOI technology) can be the best platform to incorporate other materials, especially active media such as QDs, to define new functions and eventually increase the functionalities of the chip and reduce its final footprint. A natural active material for silicon technology would be Si QDs, but it is difficult to conserve their intrinsic properties under ambient conditions (oxidation). Their light emission is limited to the range of 400–800 nm, approximately, due to recombination of confined carriers at the Γ-point, approximately. Moreover, very small QDs are needed to reach quantum confinement, given the small 3D excitonic Bohr radius of Si, and very good size control of produced QDs by a given technique is needed. This is the reason for the use of other QD materials in Si photonics. However, other substrates are very useful for some important applications (sensing devices and displays, for example), constituting a growing research field, organic photonics being one of the most important subfields [50]. In this case, photonic elements such as waveguides, couplers, ring resonators, and interferometers, typically fabricated using SOI technology, can be also fabricated by using polymers (such as poly(methyl methacrylate) [PMMA] and the

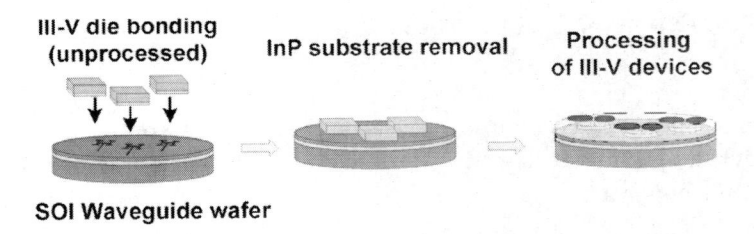

III-V die bonding (unprocessed)   **InP substrate removal**   **Processing of III-V devices**

SOI Waveguide wafer

Figure 7.18 Illustration of the die-to-wafer bonding process: III–V structures are bonded to the SOI waveguide, substrate removal, and processing of III–V devices. Reprinted from Ref. [52], Copyright (2007), with permission from Elsevier.

SU8 resist, for example), perfectly compatible with SOI, but also on glass and flexible substrates. It is worth mentioning that most of the nanocrystal QDs can be incorporated on polymers, as mentioned below.

Photodetectors (as photon receivers) are needed in Si-based photonic chips. Evidently, Si is nowadays the material for fabricating the best photodetectors because its indirect bandgap nature is not a drawback here, as in the case for emitters, but again, no detection is possible for wavelengths larger than 1100 nm, which is the reason to combine with Ge; particularly, photodetectors for visible and near-infrared (NIR) light based on conducting layers of QDs are an attractive alternative [51], as will be summarized in the next section.

The first question is then straightforward: how should QDs be integrated with silicon photonic chips? If we are thinking in III–V materials (self-assembled QDs), the technology was solved during the last years by using wafer-bonding technology in such a way that InP/InGaAs QW lasers and III–V photonic structures in general, for example, can be easily integrated on Si photonic chips (see the review made by pioneering groups of this technique [52]). This technique consists of die bonding (using an adhesive polymer or van der Waals bonding between SiO2-on-Si and III–V face) of the III–V active structure, its substrate removal, and final processing of devices, as illustrated in Fig. 7.18. In this way, photonic/optoelectronic devices and structures based on QDs could be entirely implemented on a III–V substrate (without lattice

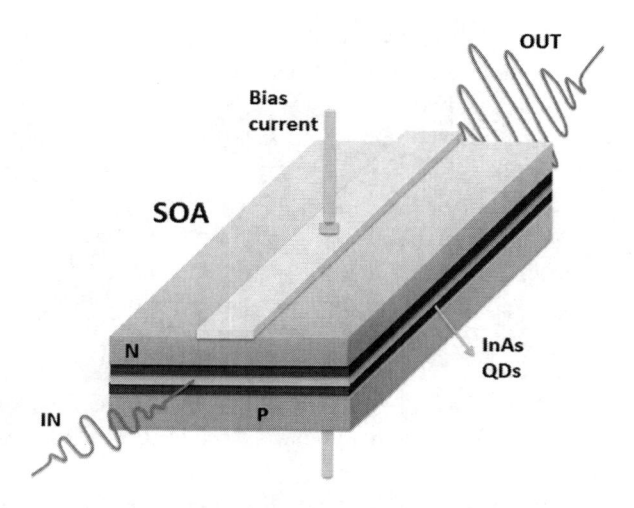

Figure 7.19   Illustration of an InAs QD-based SOA.

mismatch or the small lattice mismatch needed to grow self-assembled QDs) and transferred into a Si photonic chip. Now we will examine some (nano)photonic devices and structures based on III–V self-assembled and colloidal nanocrystal QDs.

Among these devices, QD-based semiconductor optical amplifiers (SOAs) (Fig. 7.19), are better than any other optical amplifier (based on QWs) [53] because QD SOAs offer an improved linewidth enhancement factor, low-temperature operation sensitiveness, and low chirp. For Ethernet networks working above 10 Gbit/s, high modulation bandwidth is needed for these SOAs while emitting in the spectral windows for telecommunications [54], the latter still being a challenge. For a high modulation bandwidth a fast gain recovery is needed, which is mainly limited by carrier capture and relaxation in QDs [55]. The intrinsic limitations of the QD SOA system can be overcome by cold carrier tunnel injection [56] and p-modulated doping [58]. This latter strategy is widely used to improve temperature operation insensitiveness, because QDs can be populated by holes that enhances carrier capture and relaxation. Impressive results have been lately obtained in undoped QD-based devices [59, 60], and better performance is expected from new processed p-doped samples.

The colloidal QDs allow solution-processed techniques (spin coating, spray, inkjet printing, microplotting, …) for deposition on practically any substrate. While these QDs became one of the most attractive photonic nanomaterials, they have still not found widespread application in practical photonic devices. This is in part due to the difficulty in incorporating these QDs into photonic structures using controlled densities and exhibiting sufficient stability/photostability in air. In spite of this, enhancement of luminescence and lasing action has been reported in colloidal QDs incorporated in different light-confining structures [60–63]. Practical ultrafast all-optical switches, modulators, flexible emitters, and even room-temperature SPSs using these QDs can be realized if they could be integrated in a given material matrix and patterned into desired photonic structures.

A promising technique to implement alternative SOAs and other photonic structures (waveguides, interferometers, microdisks/microrings, phase shifters for microwave photonics, …) consists of the use of patternable polymers embedding colloidal QDs (see Refs. [64–68] and others therein). Furthermore, if one bears in mind the possibility to combine with semiconductor photonic technology, the possibilities for applications of SOI technology become vast in the integrated photonics field. The amplification in an SOA based on a QD-doped polymer by using continuous- wave, either optical or electrical, pumping is still a challenge due to the drawback related to the Auger nonradiative recombination mechanism on most colloidal QD materials, as explained in the third part of Section 7.5. Very recently, white-light waveguiding was demonstrated on planar structures based on PMMA incorporating CdS, CdSe, and CdTe QDs [66], as shown in Fig. 7.20. The fabrication of 2D waveguides was also shown by using the SU8 resist incorporating visible- and NIR-emitting QDs, even if for this resist a ligand exchange is necessary to dissolve the QD colloidal solution into that of the resist; furthermore, SU8 has a background fluorescence band very intense under UV excitation.

One of the limitations in the use of QD-doped polymers is the reabsorption effect of emitted light by QDs if their concentration in the nanocomposite is very high [64]. A possibility to increase the concentration and reduce the undesired reabsorption effect is the

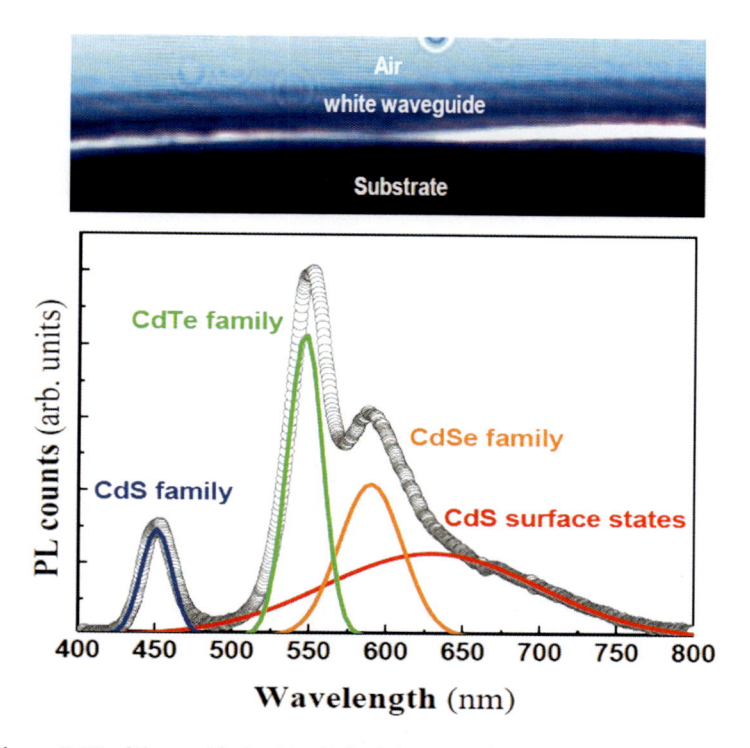

Figure 7.20 Waveguided white light (photograph on top of the spectrum) from PL bands of three different QD materials (CdS, CdTe, and CdSe, as indicated in the plot) in appropriate proportions, homogeneously dispersed in a PMMA thin film deposited on a Si–SiO$_2$ substrate.

use of bilayer structures where the active nanocomposite (QDs in PMMA) is deposited on top of the Si–SiO$_2$ substrate; a film of the SU8 resist is then spin-coated and 2D ridges patterned on it (see illustration in Fig. 7.21a) that is able to waveguide the QD-emitted light by an evanescent field (see Fig. 7.21b) [67]. The reabsorption can be also controlled by using engineered core–shell QDs because most of the absorbed light can take place at the shell (ZnS, CdS, for example), well above the PL peak energy at the core (CdSe and PbS, for visible and NIR-emitting materials, for example). Moreover, the use of appropriate core–shell QDs and the core material itself can reduce the Auger nonradiate mechanism, as discussed in Section

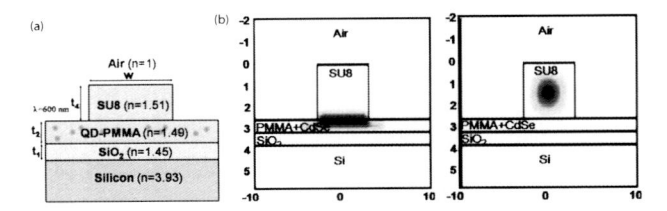

Figure 7.21 (a) Bilayer waveguide based on a patterned SU8 resist on top of the active QD-PMMA layer and (b) transversal section of the laser pumping and PL signal beams at different distances during propagation (in microns), simulated by means of a spontaneous emission model programmed into an active beam propagation method.

7.5, and hence multiply the chances to use this simple technology to fabricate SOAs and lasers.

Single-layer QD-PMMA waveguides have been recently used to implement a phase shifter at telecom wavelengths for microwave photonics [68]. When they are pumped at wavelengths where PbS QDs have efficient absorption (980 or 1310 nm), a phase shift in a signal carried at 1550 nm is induced. In a proof-of-concept experiment, a continuous phase shift up to 35° at 25 GHz was achieved. The use of other QD materials and patterned 2D waveguides can help to increase this shift significantly.

Linear microcavities can be made by using an appropriate couple of dielectric materials (eventually compatible with Si technology) with sufficiently high refractive index contrast, as in the case of SiO_2/TiO_2 that can be deposited by sputtering. Microcavities were fabricated in this way and used recently to obtain lasing at visible wavelengths by optical pumping using core–shell CdSe/ZnCdSe colloidal QDs [69], as shown in Fig. 7.22. Instead of linear microcavities, other resonators have been used in the literature, such as 2D photonic crystal cavities [60, 61] and pillar microcavities [62]. Mostly PL enhancement is observed in these cases, possibly due to the use of standard core QDs instead of core–shell QDs, as used in Ref. [59]. The major advantage of colloidal QDs will be for developing certain telecom devices using Si-based integrated photonics.

The electrical pumping of colloidal QD-based SOA structures will be also possible in the future, possibly by using conducting colloidal

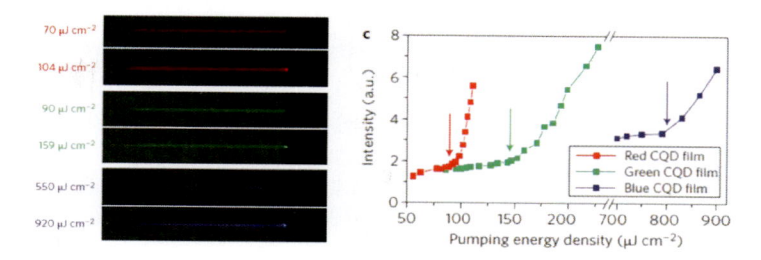

Figure 7.22 (a) Plane-view photographs of emission from excited stripes of QD films. (b) Edge emission as a function of pump energy density for QD films, with arrows indicating amplified spontaneous emission thresholds of 90 (red), 145 (green), and 800 mJ/cm^2 (blue), respectively. Reprinted by permission from Macmillan Publishers Ltd: Nature Nanotechnology (Ref. [69]), copyright (2012).

QD layers (layers of QDs with short ligands maintaining the 3D confinement but allowing carrier transport) or QDs embedded in conductive polymers, as will be presented in the next section on QD–light-emitting diodes (LEDs) and solar cells.

7.7 QD-Based Optoelectronic Devices

As stated in the last section QD-based LEDs and laser diodes using III–V substrates have not experienced the foreseen market success, given that high densities of QDs in multilayer stacks are technologically complex and expensive as compared to a single QW. The most important argument for using QDs as an active medium for semiconductor electroluminescent and laser diodes is that they exhibit a zero-dimensional density of states and unique optical properties, as summarized in Sections 7.3 and 7.4.

First of all, one should note that the number of III–V materials giving rise to self-assembled QDs is smaller than that for QW heterostructures, given that an important (but not too high) mismatch (3% and 7% for the case of InAs on InP and GaAs, respectively) is necessary to obtain the Stransky–Krastanov growth regime. In this sense, Ge/Si QDs can be grown on Si substrates, InGaN/GaN on different anisotropic substrates like CSi, InP/GaAs and InP/InGaP on GaAs and GaP, InGaAs/GaAs on GaAs and InP, and

InGaSb/GaAs on GaAs substrates, as the most known examples of QD material/barrier giving rise to QDs on listed substrates. These combinations give restricted regions for emission (green-yellow, red, and NIR mainly), which makes more appropriate to design lasers for the telecommunication market.

On the other hand, QD LEDs and laser diodes based on III–V self-assembled QDs are not typically offering sufficiently high light intensity as compared to QW-based devices, given that achieved sheet densities are around 10^{11} cm^{-2} and multilayer stacks are necessary. This also implies the relaxation of the strain field along the structure and the consequent change of size and the increase of structural defects. In a single layer of QDs the size dispersion is leading to very wide PL bands and hence further reduction in the light intensity at a given wavelength.

7.7.1 Quantum Dot LEDs

Given the above issues regarding the extensive use of III–V and IV self-assembled QDs for the fabrication of QD LEDs, a relatively small number of studies is found in the literature. The most important case is the use of InGaN QDs capped by GaN and Ge QDs capped with Si. In the first case, they are proposed to circumvent the green gap in visible LEDs based on InGaN/GaN multi-QWs due to their low efficiency as compared to that emitted in the green-blue region by the development of dislocations and other structural defects [71]. This failure of multi-QW LEDs for yellow-green wavelengths leaves a place for the success of LEDs based on InGaN/GaN multi-QD layers [72]. In the meantime rare-earth oxides are used commercially as down-shift wavelength converters to obtain white light from blue QW LEDs.

However, the cost of light of these semiconductor QW LEDs is around 2 orders of magnitude more expensive than incandescent lightbulbs, other than the limits in the device area (1–2 mm) due to structural defects and postprocessing difficulties. These drawbacks pushed the research interests in OLED technology. Red and green OLEDs based on phosphorescent iridium complexes showed external quantum efficiencies (EQEs) above 20% [73] and 63% by using optimized structures for light out-coupling [74],

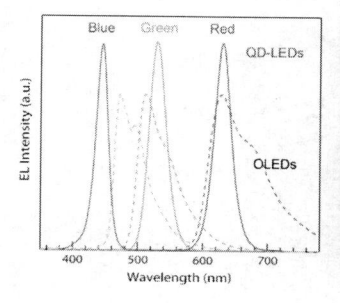

Figure 7.23 (a) Comparison of color-rendering quality of electroluminescence in RGB QD LEDs (solid lines) and OLEDs emitting at similar wavelengths (dashed lines). From Ref. [78], reproduced with permission. (b) Commercial LED bulb using a QD down-converter (QD Vision Inc.).

whereas blue OLEDs exhibit EQEs around 14% [75]. Of course, any technology is exempt of problems, the most important in this case being the poor cost-efficiency due to the high number of processing steps, the chemical stability/photostability, and color purity (linewidth of the emission band greater than 40 nm, as observed in Fig. 7.23). As the reader can easily guess, a solution for these problems could be the use of colloidal QDs [76–79]. Nowadays the wet-chemistry synthesis of QDs is able to achieve monodisperse size QD colloids and hence high color purity, as observed in Fig. 7.23 (continuous lines) in comparison to dyes, other than improvements in stability and durability over organic molecules. As also occurs for these materials, colloidal QDs can be deposited in the form of thin films on practically any substrate through inexpensive solution-processing techniques (spin coating, deep coating, doctor blading, inkjet printing, spray, …). These thin films can be very dense by a simple deep-coating or drop-casting deposition of the QD colloid on a substrate (glass, plastics), given the relatively broad selection of organic solvents to favor a self-assembling process of the nanocrystals that produce hexagonal close-packed layers whose lattice parameter is determined by the QD size and the QD–ligand length. This is of particular interest because conductive layers of QDs can be easily fabricated by using the same solution-processing technique and the appropriate exchange ligand, more popular for the use of QD films as absorber

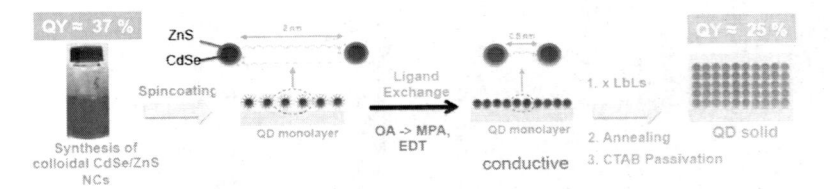

Figure 7.24 Layer-by-layer method to deposit a conducting QD layer (to form a QD solid) on top of a given substrate.

layers in photodetectors and solar cells, usually known as the layer-by-layer (LBL) deposition method. The length of these ligands should define appropriate short distances between QDs in order to make possible ballistic or tunneling transfer of carriers and hence allow the carrier transport in the QD solid. Natural ligands of QDs (oleate) are exchanged by shorter ones (3-mercaptopropionic acid [MPA], ethanedithiol [EDT], . . .), other than enabling the matching with the other layers of an LED or solar cell structure. The use of core–shell QDs will increase the emission quantum yield (QY) and maintain the passivation of the surface, increasing photostability against oxygen and other ambient agents (and other advantages referred above). The LbL method is illustrated in Fig. 7.24 using CdSe-ZnS core–shell QDs whose QY passes from 37% in the initial colloidal solution to 25% after the ligand exchange process. Another successful technology is to disperse in a plastic matrix (as in the case of waveguides discussed in Section 7.6) to be used as down-shift wavelength converters, as in the case of rare-earth oxides, but this time in the form of a cover of the LED bulb (Fig. 7.23). In this case, QDs are excited optically by absorbing blue radiation from an InGaN/GaN multi-QW LED (emitting at 405 or, better, at 450 nm), before re-emitting green or red light (see a simple example consisting in a GaN LED encapsulated by a plastic cover made of QDs dispersed in a plastic and resulting spectra of the system in Fig. 7.25) [80–83], even if color conversion is also observed by using a single layer of QDs spin-coated on top of the GaN structure due to nonradiated energy transfer [83, 84].

QD LEDs are quickly approaching industry standards for lighting and displays, as probed by the interest of the most important companies in the field, such as Sony Corporation, Samsung, LG,

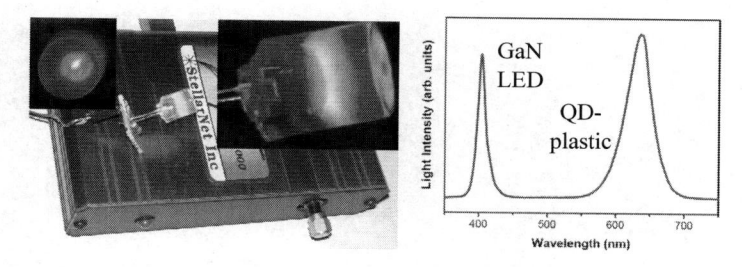

Figure 7.25 Wavelength down-shift conversion of an UV LED by a plastic containing red-emitting core–shell QDs: photograph of the QD plastic on an LED and its PL spectrum.

etc. In fact, the first QD television is now a reality, the FHD and UHD Triluminous models from Sony. These TVs use QDs from the US company QD Vision Inc. and their color quality is about that offered by the advanced AMOLED technology but at a fraction of the cost. In the last few years, LEDs based on colloidal QDs have been successfully fabricated under several approaches [85–87] and their efficiency is now very close to that of OLEDs, as summarized in Fig. 7.26. The EQE of a QD LED is defined as the ratio of the number of photons emitted by the LED to that of injected electrons and can be formulated as $EQE = \eta_r \eta_{PL} \eta_{oc}$, where η_r is the fraction of injected charges that form excitons in the QDs, η_{PL} is the emission QY associated to the exciton transitions, and η_{oc} is the fraction of emitted photons that are coupled out of the device. The internal quantum efficiency (IQE) is the efficiency of the charge recombination process, $IQE = EQE/\eta_{oc}$.

The QD LED architecture has evolved from polymer–QD bilayer structures (type II) to devices employing direct charge injection from finely tuned electron and hole transport layers (type III–IV). Type IV QD LED hybrid organic–inorganic architecture is giving the best performance at the moment, as shown in Fig. 7.26 [76, 77], and consists of inorganic/organic electron/hole transporting layers to inject both carriers on the QD layer (Fig. 7.26, right) [87], other than the electrodes (ITO/PEDOT and Ag or higher work function metal). The working mechanism of these type IV QD LEDs is possibly very similar to that of OLEDs. For example, Qian et al. reported red, green, and blue (RGB) solution-processed QD LEDs whose EQEs

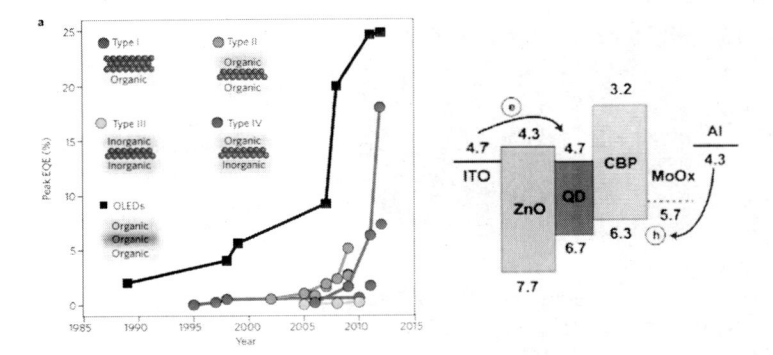

Figure 7.26 Evolution of EQE in different types of QD LEDs as compared to that of OLEDs (left). Band scheme of a type IV using an organic hole injector (4,4′-bis(*N*-carbazolyl)-1,1′-biphenyl [CBP]) layer and inorganic electron injector (ITO/ZnO) layers. Reprinted by permission from Macmillan Publishers Ltd: *Nature* (Ref. [76]), copyright (2013).

were 1.7%, 1.8%, and 0.22% and maximum brightness values were 31.000, 68.000, and 4.200 cd/m^2 for red, green, and blue devices, respectively, values among the highest reported for QD LEDs [88]. Parallel engineering efforts with similar architecture have led to full-color active-matrix-driven QD displays that were fabricated by microcontact printing in Ref. [89]. Recently, QD Vision reported a QD LED having EQE = 18% using a refinement of this hybrid type IV structure [90], which greatly surpassed previous efficiencies and is near the theoretical maximum of 20%. Nevertheless, such a high EQE was maintained for only less than one hour, which can be attributed to both chemical (photochemical) degradation of the QDs and photophysical effects. The presence of excess charges in the QDs might lead to both reversible degradation of the LED efficiency due to Auger nonradiative recombination as well as irreversible changes due to photoactivation of chemical reactions at the QD surface and/or ligands. Shirasaki et al. have studied and elucidated in QD LEDs the efficiency roll-off (limitation of electroluminescence at high driving currents) to be due to the quantum-confined Stark effect (QCSE) associated to the internal electric field in the QD layer [91]; the QCSE produces a spatial separation between electrons and holes and an consequent increase of the exciton radiative lifetime. In any case, the explanation of the roll-off in this paper is satisfacotory but still phenomenological because the recombination times are mainly determined by nonradiative effects at the QD layer and small radiative changes are not appreciated in the PL transients. In InGaAs self-assembled nanostructures, where the recombination times are mainly radiative at low temperatures one is able to clearly measure the increase of the exciton lifetime by QCSE (low-intermediate electric field region) and further reduction (high electric fields, greater than 100–150 kV/cm) by carrier tunneling out of the QD, even if this mechanism depends on the interplay of QD size (higher-energy GS) and potential barrier height [92]. The realization of high-efficiency LEDs will require QD materials that exhibit high QY in the single-exciton and multicarrier regimes (high driving currents). For this reason, the Auger recombination path should be suppressed by appropriate core–shell engineering (Sections 7.2 and 7.5); other electrodes used to lower the operational voltage (i.e., the built-in electric field within the QD layer) can be also useful in order to reduce the roll-off effect.

7.7.2 Laser Diodes

The operation of semiconductor laser diodes should be characterized by a low threshold current, a large temperature parameter of the threshold current (T_0, from the dependence $J_{th} = J_{th0} \exp(-T/T_0)$), a large direct modulation bandwidth (related to the frequency response of a laser), and a small linewidth enhancement factor (α-parameter, related to the ratio between refractive index and gain changes with carrier density) and chirp (proportional to α-parameter). Due to the 3D quantum confinement and δ-like density of states, self-organized QDs lasers would yield a very small J_{th}, a large T_0 values, a large differential gain, and a large modulation bandwidth.

QD lasers was firstly discussed by Arakawa and Sakaki [93] from the point of view of the dimensionality reduction in the density of states, but the first real injection laser based on self-organized InGaAs QDs was fabricated in 1994 [94], demonstrating a low threshold current (120 and 950 A/cm^2 at 77 K and room temperature, respectively) and a temperature-insensitive threshold current ($T_0 = 350$ K below 120 K, much higher than the expected

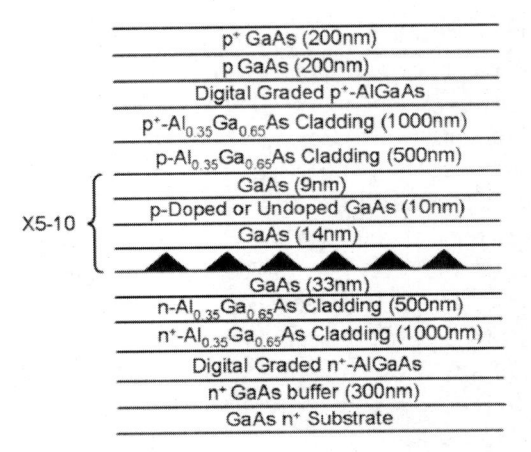

Figure 7.27 Scheme of a laser at 1.3 μm based on a modulated p-doped heterostructure containing InGaAs QDs (right). Reprinted with permission from Ref. [97]. Copyright 2004, AIP Publishing LLC.

limit for QWs, 285 K). The same group demonstrated room-temperature continuous-wave lasing in QD stacks (several layers of QDs) grown by metal-organic chemical vapor deposition (MOCVD) [95], which was a great advance due to the difference in the final cost of devices as compared to MBE. These lasers exhibited a threshold current density of 12.7 and 181 A/cm^2 at 100 and 300 K (under pulsed operation), respectively, but their lasing emission was occurring at the high-energy side of the QD GS PL peak; in the case of a triple QD layer stack the laser emission was observed on the QDs' GS and allowed for continuous-wave operation at room temperature. Further advances in QD lasers included the achievement of an ultralow threshold current as low as $J_{th} = 13$ A/cm^2 [96] (thanks to the introduction of QDs in a QW that allows also a shift of the lasing wavelength toward 1300 nm); temperature-invariant operation, that is, $T_0 = \infty$ (due to the role played by Auger nonradiative recombination above room temperature in the laser structure depicted in Fig. 7.27) [97]; a record output power of 3.9 W (diode laser emitting at 940 nm at room temperature based on an active layer consisting of 10 submonolayer InAs QD layers) [98]; a large modulation bandwidth up to 25 GHz (achieved by the reduction of hot carrier effects in undoped and p-doped tunnel injection lasers emitting at 1.1 and 1.3 µm at room temperature, [99]; and near-zero chirp and α-parameter [89, 100]. As mentioned earlier, laser diodes with an emission wavelength up to 1.3 µm can be achieved by using InGaAs QDs grown on GaAs, but the reader can guess that higher technological interest lies in the extension toward 1.55 µm. For this reason, the pseudomorphic and metamorphic growth and optical properties of InAs QDs on GaAs emitting at long wavelength have been intensively investigated by many groups [37, 101, 102] after the first metamorphic QD-based laser fabricated on GaAs emitting at 1.5 µm by Ledentsov et al. [103]. In this approach, the buffer layer grown on GaAs substrates consists of an InGaAs alloy that implies a smaller biaxial strain in the subsequent growth of InAs or InGaAs materials forming the QDs, and hence bigger dots and longer emission wavelengths [92]. Vertical cavity surface emitting lasers (VCSELs) with QD-active regions emitting at room temperature were fabricated only some years after the first edge-emitting laser [104]. These lasers exhibited relatively low threshold currents ($J_{th} = 500$ A/cm^2) under pulsed mode operation at an emission wavelength of 960 nm. The first 1.3 µm QD-VCSEL laser on GaAs was fabricated several years later due to the relatively low gain associated to the exciton GS transition in these QDs, which are characterized by $J_{th} = 2800$ A/cm^2 [105], a value rather high attributed to the large QD size inhomogeneity leading to very broad emission bands. Several years later continuous-wave operation of a QD-VCSEL at 1.3 µm was achieved by incorporating multistack layers of InAs QDs between fully n- and p-doped distributed Bragg reflectors (DBRs) [106], as illustrated in Fig.7.28.

Nowadays, optically pumped semiconductor disk lasers (SDLs) are known to produce multiwatt output powers with excellent beam quality [107]; these devices were initially based on QWs, but the first device using QDs was demonstrated very recently [108]. The operation of this SDL was based on the use of an intracavity diamond heat spreader or flip-chip design (see Fig. 7.29). An output power of 2 W at 1200 nm was reached in a laser gain structure comprising 35 QD layers grouped and placed at antinodes of the optical field. The flip-chip design is a promising technology for the construction of high-power QD SDLs operating in the 1200–1300 nm spectral range, as demonstrated by its rapid transfer to the commercial stage

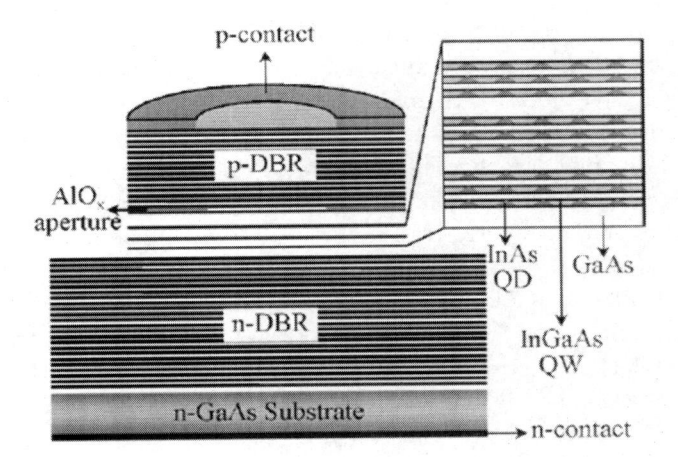

Figure 7.28 VCSEL structure containing multistack layers of InAs QDs between fully n- and p-doped distributed Bragg reflectors (DBRs). Reprinted with permission from Ref. [4]. Copyright 1996, AIP Publishing LLC.

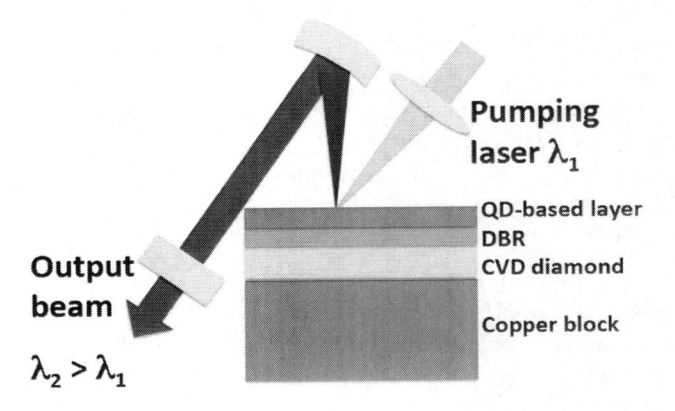

Figure 7.29 Cavity scheme of an optically pumped QD SDL.

(Innolume GmbH). This approach can be very interesting to develop similar lasers based on colloidal QDs embedded between DBRs constituted by oxides (SiO_2 and TiO_2, for example) [69], directly deposited on the diamond layer. Finally, another potential laser system containing QDs is based on photonic crystal microcavities, given their strong confinement factor and small modal volume.

Ellis et al. demonstrated an electrically pumped photonic crystal nanocavity QD laser that exhibits ultralow thresholds of 181 and 287 nA at 50 and 150 K, respectively, 3 orders of magnitude lower than using QWs in the microcavity [109], with a consumption power of only 208 and 296 nW at 50 and 150 K, respectively, lower than to date. More recently, Khajavikhan et al. demonstrated threshold-less lasing at 4 K [110].

7.7.3 Solar Cells and Photodetectors Based on Colloidal QDs

In recent years, colloidal QDs have been found to be promising for next-generation solar cells because of the size confinement effect increasing their effective bandgap [111]. QDs have been explored due to their size and compositional-dependent absorption [112, 113]. Previously unachievable with bulk semiconductors, the QDs allow energy level matching between desired donor and acceptor materials, which are crucial in designing efficient photovoltaic devices.

QDs can be solution-processed into films and hence they could be an alternative to commonly employed sensitizer molecules [114]. During the last decade, colloidal QDs have been integrated in different types of solar cells such as Schottky solar cells [115, 116] (Fig. 7.30a), depleted heterojunction solar cells [117] (Fig. 7.30b), and inorganic–organic heterojunction solar cells [118]. The QD solar cell technology has the following promising characteristics: (1) It is a solution-processed deposition technology, (2) QDs can be fabricated at low temperatures ($<200°C$), and (3) most of the sun's emission spectrum can be absorbed because of the effective bandgap tuning of the QDs by the quantum size effect, empowering a single material strategy to produce stacked multijunction solar cells. Schottky junction QD solar cells based on PbS have attracted intense attention in the past few years [115, 116], but they have several limitations: While large short-circuit currents (J_{sc}) of over 20 mA/cm^2 are achieved, the open-circuit voltage (V_{oc}) is low compared to the bandgap energy (E_g) and dependent on the QD size [115]. As a result, low conversion efficiencies (around 2%) were reached. With an expectation to improve the efficiencies of Schottky

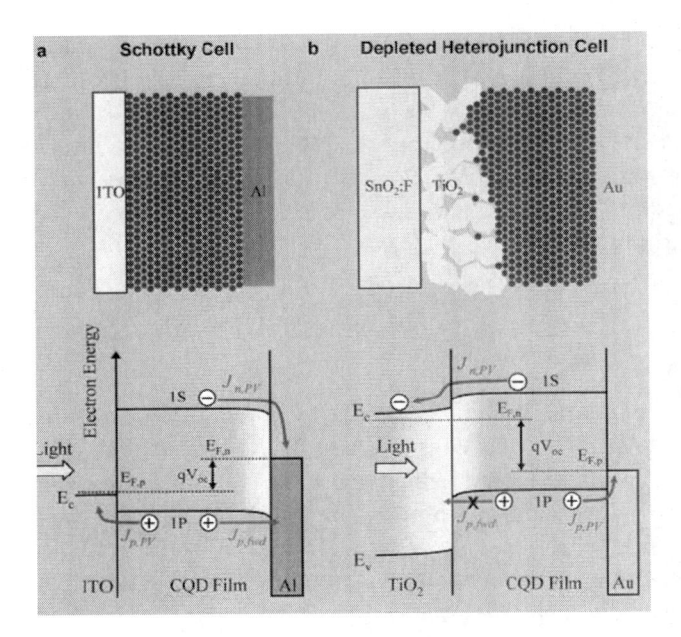

Figure 7.30 Schottky (a) vs. depleted heterojunction (b) solar cells. Reprinted with permission from Ref. [117]. Copyright (2010) American Chemical Society.

solar cells ternary PbS_xSe_{1-x} QDs have been used as well [119]. Most of the reported investigations on depleted heterojunction solar cells are concentrated on lead chalcogenide QDs, PbS and PbSe. The use of PbS QDs in a depleted heterojunction device architecture (Fig. 7.30b), exhibiting the band structure depicted in the same Fig. 7.30b, is able to offer power conversion efficiencies in the range of 5%–6% (100 mW/cm², AM1.5) for PbS QDs characterized by a band edge around 1000 nm, as reported in Ref. [117] The photovoltaic cell architecture in general consists of a QD layer sandwiched between an electron-transporting layer (TiO₂) and a metal electrode (Fig. 7.30a). In such kind of structure, electrons flow toward the TiO₂ layer rather than the evaporated metal contact (Fig. 7.30b), thus creating an inverted polarity [120]. Moreover, hole transfer from TiO₂ to QDs is prohibited, which allows efficient carrier separation. The QD depleted heterojunction design overcomes the limitations of prior work on QD–Schottky devices in three principal ways. First, the electron-accepting TiO_2 contact is transparent and placed on the illumination side, thus benefiting more efficiently from minority carrier separation. Second, whereas the V_{oc} of the QD–Schottky device is limited by Fermi-level pinning due to defect states at the metal–semiconductor interface, the interface between the electrode and the QD film benefits from passivation during the solution-phase deposition of the QDs. Third is a large discontinuity in the valence band of the device, combined with the minimization of back carrier recombination. From energy diagram perspectives nanocrystalline PbS is an ideal light-harvesting material in the NIR region since it can be used as an electron donor for wide-bandgap materials (TiO_2 or ZnO) in heterojunction solar cells. In this sense, Luther et al. [121] demonstrated that p-type PbS QDs (with an effective bandgap around 1.3 eV) in contact with n-type ZnO nanoparticles form a p-n heterojunction, achieving $J_{sc} = 9$ mA/cm^2, $V_{oc} = 0.44$ V, $FF = 0.56$, and yield around 3%.

At this point it is interesting to note that only a few groups are reporting conversion efficiencies in the ranges given above (2%–5%). This is surely because film-processing conditions and (optical) quality of used nanocrystals are limiting the technology success of the photovoltaic device. The chemistry of QDs and QD layers (addressed in Section 7.7.3) is evidently a key point in the solar cell technology. It is clear that QD processing can be done under globe-box conditions to minimize the incorporation of oxygen as a carrier trap at the QD surface, like the work developed in Ref. [115], even if air-stable layers seem to be obtained by other authors (Sargent's group), as reviewed in Refs. [122–124], also because PbS QDs show higher air stability than other alternative nanocystals (i.e., PbSe nanocrystals) [125]. This group has reported the highest conversion efficiencies, now around 7.3% (photocurrent 22.5 mA/cm^2), by using a hollow TiO_2 nanowire array instead of a planar layer in the depleted heterojunction architecture, and 7.4% (photocurrent 21.8 mA/cm^2), by introducing halide anions during the end stages of the synthesis process for passivating trap sites not accessible by the standard ligands (mercaptopropionic acid, MPA) [126].

Other lead-free materials can be used as alternative materials (and may be less sensible to oxygen traps), for example, substituting Pb by Sn or other compound semiconductors like Bi_2S_3 [127, 128],

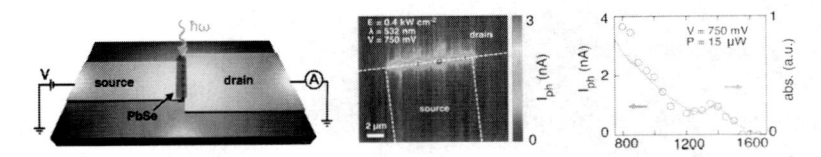

Figure 7.31 Nanogap photodetector with 10 μm wide electrodes separated 4 nm where a monolayer of PbS QDs are placed (left); map of the photocurrent (center) and wavelength dependence of the photocurrent of the device as compared to the optical absorption spectrum (right). Reprinted with permission from Ref. [129]. Copyright (2012) American Chemical Society.

for which a p-n (p-PbS QDs and n-Bi_2S_3 nanocrystals) heterojunction is formed.

From the above given summary one can guess that semiconducting QD layers can be also the basis for the fabrication of efficient photoconductors/photodetectors at the NIR, with the advantage to be integrated in silicon, glass, or flexible substrates. Reported responsivities in the recent literature are in the range of 0.1–3.9 A/W for a PbS QD monolayer in a nanogap photoconductor (see this interesting concept in Fig. 7.31) [129], larger than 100 A/W in a PbS layer thicker than 200 nm deposited on a standard interdigitated photoconductor electrode [130], and more than 10^6 A/W in an metal-oxide semiconductor (MOS) structure based on a PbS layer (60–80 nm thick) on graphene [131]. The QD–solid approximation is considered one of the latest advanced concepts for photodetection, given the high absorption of QDs, the low cost of the solution-processing technique used to deposit the material, and its integration in Si technology.

Graphene, a single layer of carbon atoms bonded into a 2D hexagonal crystal structure, has attracted a great deal of attention due to its extraordinary physical and electrical properties [132]. Further to the many applications proposed and corroborated for graphene layers, graphene could also serve as a transparent conducting electrode (TCE) in solar cells [133–135], where high optical transparency (97%–98%) and low sheet resistance are essential. The latter is possible by the small addition of alkaline atoms, water, or other molecules, which allows tuning the graphene work function (around 4.5 eV) by ±0.5 eV [136]. The concept of using graphene as a TCE in solar cells emerges as a very practicable option for future devices, although solar cell performance is far from being optimized. The outstanding thermal and chemical stability of graphene, as well as its atomic flatness, make graphene suitable for ultrathin solar cells (organic, for example) [137]. Recently, graphene has been also proposed as an electrode for QD- [138] and silicon-based [139] photovoltaics.

7.7.4 Other QD Optical Devices

The use of QDs as memory elements has been discussed for a number of years. In a conventional flash memory, the charge is stored in a polysilicon floating gate, which is isolated by two $SiO2$ barriers that must be overcome to write or erase the data. This is a slow process and eventually can damage the device. The first advantage of a QD memory is that electrons are captured by the QD potential well, which is an intrinsically fast process, whereas the erasing operation is slow, but controllable up to a certain extent. The second advantage of the QD memory is that the height of the QD-confining potential barrier can be tuned by either an applied electric field (see typical modulation-doped field-effect transistor structure in Fig. 7.32a [140]) or changing materials, so different memory types can be fabricated by using the same concept, from nonvolatile to volatile memories, by decreasing the QD potential barriers, that also will affect the speed of writing and refresh operations. The QD plays the role of the floating gate in the flash memory and is charged and discharged via the channel by applying a voltage to the control gate. The basic principle is to create carriers optically and then sweep out due to the strongly differing tunneling rates of electrons and holes under an applied gate voltage, as illustrated in Fig. 7.32b–d [140]. The remaining carrier, typically the hole, has a long residence time in the QD. The use of InAs-based dots limits operating temperatures below 140 K due to the relatively small energy barriers for carrier thermal excitation and the loss of carriers to the WL or continuum, but other materials are being used to over-come this issue, as GaSb nanostructures (further reading in Refs. [140, 141]).

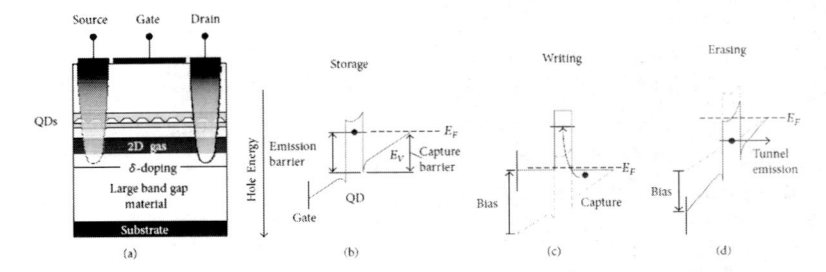

Figure 7.32 (a) Generic structure of a QD-based memory device consisting of a layer of self-organized QDs embedded into a modulation-doped field-effect transistor made of a large-bandgap material. (b) Storage of information requires an emission and a capture barrier. (c) Writing operation. (d) Erasing operation. From Ref. [140] (http://creativecommons.org/licenses/by/3.0/).

7.8 Nanophotonics Based on Single QDs

The applications of a single QD in nanophotonics will be based mainly on the nature of the light emitted by this nanoscale semiconductor. A QD, as an atomic two-level system for excitons (electrons and holes correlated by Coulomb attraction, as described in Section 7.4), emits a single photon with a probability of 1 (0 for less/more than 1 photon) in response to an external trigger. That is, the QD can be considered as a single-photon or quantum light source and hence new concepts, devices, and detection schemes can be introduced in future nanophotonics for quantum information technologies.

Figure 7.33 shows three different photon statistics in the Fock-state basis: Bose–Eisntein (Fig. 7.33a), Poissonian (Fig. 7.33b), and single photon (Fig. 7.33c), the latter being that characterizing quantum light. It is worth noting that any of the classical light sources will always yield some probability of obtaining photon numbers different from 1, as calculated in Fig. 7.33a,b, assuming an average number of photons equal to 1. The QD, as an atom-like system, can emit a single photon through either optical excitation or electrical injection. In the case of optical excitation, the incoming laser light is in a coherent state (Poissonian statistics in Fig. 7.33b). The QD converts this light into an antibunched single-photon

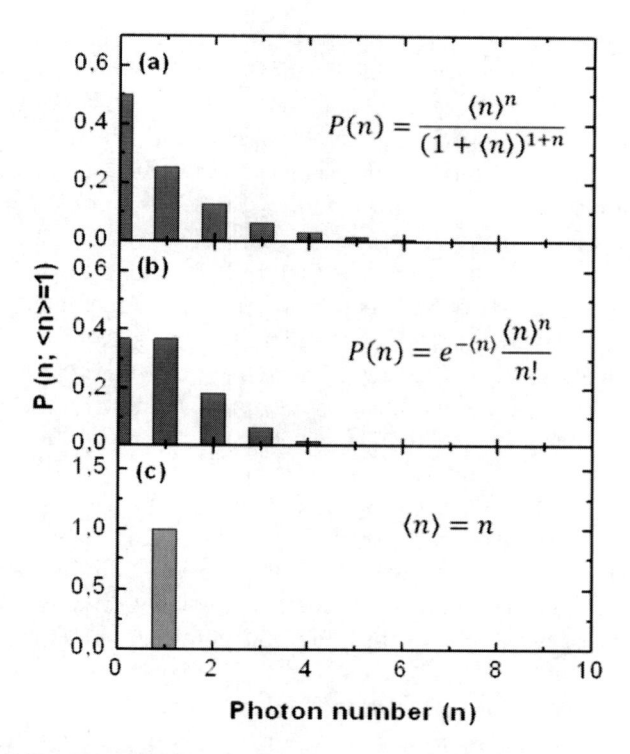

Figure 7.33 Photon statistics. (a) Bose–Einstein, (b) Poissonian, and (c) single photon.

stream. This regime consists of the following process: The QD in the excited state (an exciton quasiparticle) emits a single photon via spontaneous emission to decay into the zero-exciton state, but it can no longer re-emit a photon until it is excited again.

In Section 7.4 we described most of the important optical properties of single QDs, particularly those of the GS, where very narrow emission lines were observed at precise wavelengths, each of them having a characteristic emission lifetime that will determine the speed of the SPS. These lines exhibit phonon-induced linewidth broadening and hence their single-photon characters are maintained only up to relatively low temperatures, typically below 77 K. At lower temperatures it is also necessary that emission lines be narrow enough to be lifetime limited and ensure the indistinguishability of the emitted photons. The broadening of emission lines in QDs over this limit can arise from fluctuations of the optical resonance frequency or spectral diffusion by, for example, an internal electric field from impurities or other defects. Moreover, time jittering in the single-photon emission can be present due to the relaxation time of carriers from higher states if excitation/injection is not performed resonantly with excitons at the QD GS. A last optical property of a QD emitter that will determine the efficiency and/or degradation of SPSs is the polarization of the emission; as we discussed in Section 7.4 the existence of a measurable FSS under orthogonally polarized light will introduce particular effects if the emission occurs preferentially into the optical mode of a microcavity mode, which in general is strongly polarized; on the opposite, if FSS = 0 entangled photons can be emitted by the QD during the biexciton–exciton cascade recombination.

Self-assembled QDs are excellent optical systems, given that they join their unique optical properties and the feasibility of technical processing of semiconductors, which is important to integrate QDs into devices for electrical injection and photonic structures, as is the case of optical microcavities [142]. Michler et al. reported the first demonstration of a self-assembled semiconductor QD as an SPS [143]. Colloidal QDs constitute a very interesting system to develop quantum photonic structures and devices integrated in silicon substrates, as the most interesting near-future applications. This is because colloidal QDs can be deposited by using very simple and low-cost techniques (see Sections 7.6 and 7.7) and are very efficient nanomaterials emitting at room temperature. The weak optical properties of colloidal QDs restricting their use for production of quantum light are long recombination times, broad emission lines due to dephasing and spectral diffusion effects, and Auger nonradiative recombination. In spite of these negative features, the emission of single photons has been demonstrated at room temperature in specially engineered core–/graded-shell QDs [144].

The demonstration that a source is emitting single photons needs the measurement of the second-order photon intensity autocorrelation function, $g^2(\tau)$; $g^2(0) < 5$ evidences the antibunched emission of photons and $g^2(0) = 0$ (<0.5 as an extended range) certifies the emission of single photons. The first-order correlation function is the light intensity (proportional to the number of photons) measured by one photodetector, whereas $g^2(\tau)$ distinguishes between the different statistics of emitted light. In the general case of light correlation between optical transitions i and j

$$g_{ij}^2(\tau) = \frac{< I_i(t)I_j(t+\tau) >}{< I_i(t) >< I_j(t) >},\qquad (7.31)$$

where I_i denotes the intensity measured with a detector for transition i and τ the time delay imposed by the interferometer arm between the two detectors, as firstly proposed by Hanbury-Brown and Twiss (HBT) [145] (Fig. 7.34). The time differences τ between detection events from the two light signals are registered by time-correlated single-photon counting (TCSPC) electronics. Intensity interferometry allows also the determination of quantum coherence of sources, that is, the HBT experimental technique opened the field of quantum optics. In this sense, the HBT technique can

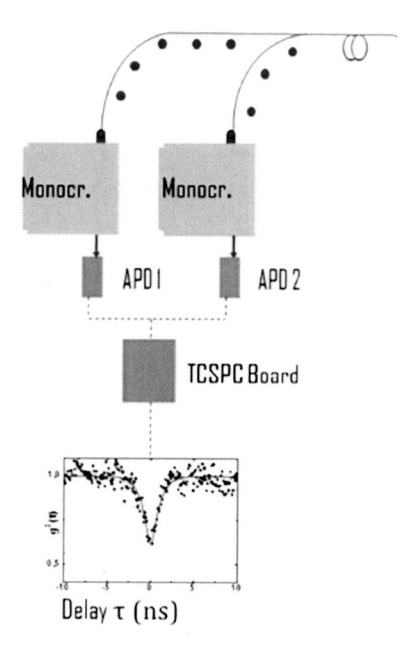

Figure 7.34 Hanbury–Brown and Twing setup using two monochromators and two avalanche photodiode detectors (APDs).

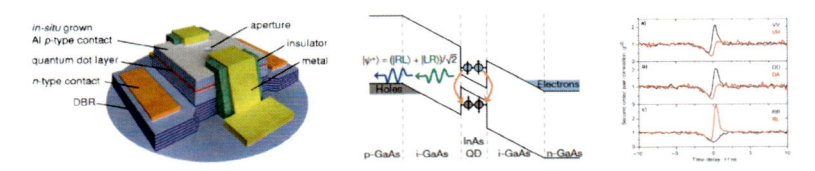

Figure 7.35 Schematic of the LED structure formed by a single quantum dot (QD) in a p-i-n junction. Carrier recombination (orange arrows) produces the emission of a polarization entangled photon pair (blue and green arrows). Second-order correlation function under carrier electrical injection measured in the rectilinear (a), diagonal (b), and circular (c) bases. Reprinted by permission from Macmillan Publishers Ltd: *Nature* (Ref. [148]), copyright (2010).

be considered as a tool for a quantitative analysis of quantum light signals in the incoming quantum information technologies. QD ensembles have been presented in precedent sections as potential active media to develop many optical devices, but quantum information management will need the use of single QDs. Entangled photon emitters [146, 147] and diodes [148] (see Fig. 7.35) were proposed as interesting devices for gating processes. Quantum logic information is provided by the electronic coupling between QD states [149], the exciton–biexciton Rabi rotation [150], and even the spin–orbit interaction [151], and quantum versions of logic gate operations have been shown [152]. All these efforts suggest near-future possibilities to implement all-in-one quantum optical devices to obtain efficient single-photon logic operations. In a recent work we showed how to use bistable single-photon emission from trion and neutral exciton states of single InAs QDs as a logic transference function using a two-color excitation scheme (see Fig. 7.36) [42]. In the experiments, different photon energies are switched on and off by the possibility to create those excitonic states, depending on the laser excitation color (one excitonic state per pumping laser), as illustrated in Fig. 7.36; the generation of only trions is possible by the impurity surroundings of the QD, as was explained in Section 7.4.

Photons from X^0 and X^- generated under the two-color laser excitation scheme exhibit an antibunching pattern registered by cross-correlation interferometry, given that they proceed from exciton recombination at the same fundamental QD GS. In the single

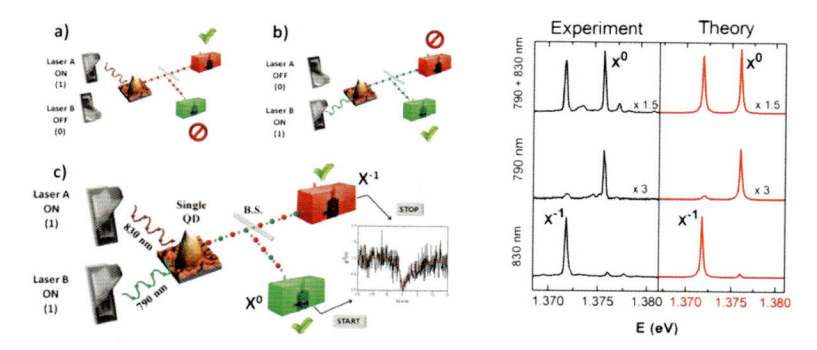

Figure 7.36 Schematics of (a) and (b), single-color excitation (the single QD generates single-photon emission at a specific wavelength), where laser A generates red photons (from X states) and laser B green photons (from X^0 states), respectively. (c) Two-color excitation scheme: generation of two-color single-photon emission (red + green photons) if the QD is excited by using both lasers. In the right panel the corresponding micro-PL spectra (measured and simulated by the master equations of microstates) are shown. Reprinted with permission from Ref. [42]. Copyright (2014) American Chemical Society.

QD literature, the two-color excitation scheme has been also used for gating the resonant response of the QD [153]. Further advances in such photonic gates using quantum light and quantum logic gates can be obtained in the near future by using external fiber optic circuitry and devices. In this sense, we developed for the first time the coupling of emitted light at around 1.3 µm by a single QD into a tunable fiber Bragg grating that filters this emission and hence is able to select photons coming from different excitonic transitions [29]. Since the first demonstration of a QD as an SPS a lot of work has been published on the topic, both on single QDs and QDs inside optical microcavities that can be fabricated around them by using the nowadays well-developed III–V semiconductor processing technology. In fact, demonstration of strong coupling between excitons confined in QDs and optical modes in micropillars [154], microdisks [155], and photonic crystal [156] (see Fig. 7.37) microcavities was achieved some years ago. Additionally, a control of either the emission wavelength of these emitters or that of the cavity modes is needed to allow strong coupling between them, as

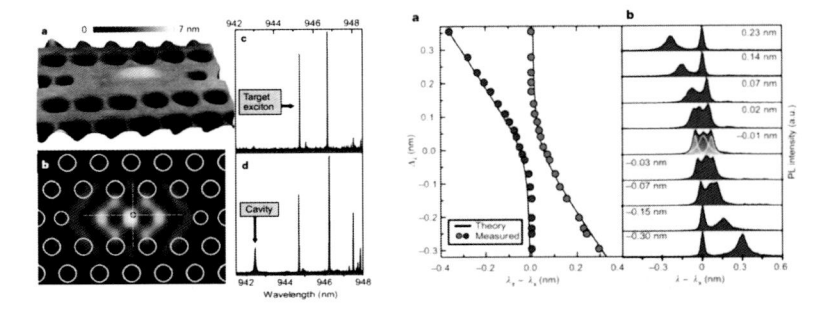

Figure 7.37 (Left) (a) AFM topography of a photonic crystal nanocavity aligned to a QD, (b) electric field intensity of the photonic crystal cavity mode with the buried QD overlapping the field maximum, (c, d) micro-PL spectrum of the target single QD before and after cavity fabrication. (Right) (a) Wavelength of the polaritons for various cavity–QD detunings and (b) spectra of the two anticrossing polariton states near zero detuning. Reprinted by permission from Macmillan Publishers Ltd: *Nature* (Ref. [156]), copyright (2007).

observed in Fig. 7.37; an anticrossing effect is observed in this case whose energy difference at resonance is proportional to the QD–cavity coupling (Rabi splitting). This is one of the bottlenecks to be overcome for allowing future SPSs and EPSs based on this approach. In a single QD deterministically positioned into a node of an optical microcavity the spontaneous emission lifetime is reduced by a factor of around 1 order of magnitude (depending on the confinement of the electromagnetic field in a particular microcavity) by the Purcell effect [157]. This is very desirable for a number of reasons because of higher repetition rates, high quantum efficiencies, and increased indistinguishability of emitted photons (see the very nice review by Buckley et al. [158]). Efficient electrical injection into photonic crystal cavities containing single QDs is not easy and is hence the subject of intense research nowadays and a reality very soon.

Acknowledgments

I would like to thank all members of my group, former and present, because most of the work referred to in this chapter has been developed during their PhD or posdoc formation. Among them I would like to mention Benito Alén, Jordi Gomis, Josep Canet, David Rivas, Isaac Suárez, Rafael Abargues, and Pedro. J. Rodríguez, whose work has been mostly centered on the optical properties of self-assembled QDs and synthesis-layers-applications of colloidal QDs.

References

1. Capasso, F. (1987). Band-gap engineering: from physics and materials to new semiconductor devices, *Science*, **235**, p. 172.

2. Capasso, F., Faist, J., and Sartori, C. (1996). Mesoscopic phenomena in semiconductor nanostructures by quantum design, *J. Math. Phys.*, **37**, p. 4775.

3. Faist, J., Capasso, F., Sivco, D.L., Sirtori, C., Hutchinson, A.L., and Cho, A.Y. (1994). Quantum cascade laser, *Science*, **264**, p. 553.

4. Yu, P.Y., and Cardona, M. (2005). *Fundamentals of Semiconductors: Physics and Materials Properties*, 3rd. ed., Springer.

5. Ribeiro Jr., M., Ferreira, L.G., Fonseca, L.R.C., and Ramprasad, R. (2012). CdSe/CdTe interface band gaps and band offsets calculated using spin-orbit and selfenergy corrections, *Mater. Sci. Eng. B: Adv. Funct. Solid-State Mater.*, **177**, p. 1460.

6. Wei, S.-H., and Zunger, A. (1997). Electronic and structural anomalies in lead chalcogenides, *Phys. Rev. B*, **55**, p. 13605.

7. Yu, W.W., Falkner, J.C., Shih, B., and Colvin, V.L. (2004). Preparation and characterization of monodisperse PbSe semiconductor nanocrystals in a noncoordinating solvent, *Chem. Mater.*, **16**, pp. 3318–3322.

8. Trindade, T., O'Brien, P., and Pickett, N.L. (2001). Nanocrystalline semiconductors: synthesis, properties, and perspectives, *Chem. Mater.*, **13**, pp. 3843–3858.

9. Murray, C.B., Norris, D.J., and Bawendi, M.G. (1993). Synthesis and characterization of nearly monodisperse CdE (E = sulfur, selenium, tellurium) semiconductor nanocrystallites, *J. Am. Chem. Soc.*, **115**, pp. 8706–8715.

10. Ostwald, W. Z. (1901). Blocking of Ostwald ripening allowing long-term stabilization, *Phys. Chem.*, **37**, p. 385.

11. Reiss, P., Protiére, M., and Li, L. (2009). Core/shell semiconductor nanocrystals, *Small*, **5**, p. 154.

12. (a) Li, J.J., Wang, Y.A., Guo, W., Keay, J.C., Mishima, T.D., Johnson, M.B., and Peng, X. (2003). Large-scale synthesis of nearly monodisperse CdSe/CdS core/shell nanocrystals using air-stable reagents via successive ion layer adsorption and reaction, *J. Am. Chem. Soc.*, **125**, p. 12567. (b) Baiutti, F., Christiani, G., and Logvenov, G. (2014). Towards precise defect control in layered oxide structures by using oxide molecular beam epitaxy, *Beilstein J. Nanotechnol.*, **5**, pp. 596–602.

13. Stangl, J., Holyý, V., and Bauer, G. (2004). Structural properties of self-organized semiconductor nanostructures, *Rev. Mod. Phys.*, **76**, p. 725.

14. Liu, H.Y., Tey, C.M., Sellers, I.R., Badcock, T.J., Mowbray, D.J., Skolnick, M.S., Beanland, R., Hopkinson, M., and Cullis, A.G. (2005). Mechanism for improvements of optical properties of 1.3-μm InAs/GaAs quantum dots by a combined InAlAs-InGaAs, *J. Appl. Phys.*, **98**, 8, p. 083516.

15. Wikipedia. (2014). *Particle in a Spherically Symmetric Potential*, http:// en.wikipedia.org/wiki/Particle_in_a_spherically_symmetric_potential.

16. Kayanuma, Y. (1986). Wannier exciton in microcrystals, *Solid State Commun.*, **59**, p. 405.

17. Pellegrini, G., Mattei, G., and Mazzoldi, P. (2005). Finite depth square well model: applicability and limitations, *J. Appl. Phys.*, **97**, p. 073706.

18. Preier, H. (1979). Recent advances in lead-chalcogenide diode lasers, *Appl. Phys.*, **20**, p. 189.

19. Allan G., and Delerue, C. (2004). Confinement effects in PbSe quantum wells and nanocrystals, *Phys. Rev. B*, **70**, p. 245321.

20. Wu, H., Dai, N., and Mc Cann, P.J. (2002). Experimental determination of deformation potentials and band nonparabolicity parameters for PbSe, *Phys. Rev. B*, **66**, p. 045303.

21. Pedrueza, E., Segura, A., Abargues, R., Bosch, J., and Martínez-Pastor, J.P. (2013). Effect of quantum size confinement on the optical properties of PbSe nanocrystals as a function of temperature and hydrostatic pressure, *Nanotechnology*, **24**, p. 205701.

22. Smith, A.M., and Nie, S. (2010). Semiconductor nanocrystals: structure, properties, and band gap engineering, *Acc. Chem. Res.*, **43**, p. 190.

23. Ahopelto, J., H. Lipsanen, H., Sopanen, M., Koljonen, T., and Niemi, H.E.-M. (1995). Selective growth of InGaAs on nanoscale InP islands, *Appl. Phys. Lett.*, **65**, p. 1662.

24. Ciftja, O., and Kumar, A.A. (2004). Ground state of two-dimensional quantum-dot helium in zero magnetic field: perturbation, diagonalization, and variational theory, *Phys. Rev. B*, **70**, p. 205326.

25. Schliwa, A., and Winkelnkemper, M. (2008). Theory of excitons in InGaAs/GaAs quantum dots. In Bimberg, D. (Ed.), *Semiconductor Nanostructures*, pp. 139–169, Springer, Berlin.

26. Warburton, R.J., Miller, B.T., Dürr, C.S., Bödefeld, C., Karrai, K., Kotthaus, J.P., Medeiros-Ribeiro, G., Petroff, P.M., and Huant, S. (1998). Coulomb interactions in small charge-tunable quantum dots: a simple model, *Phys. Rev. B*, **58**, p. 16221.

27. Muñoz-Matutano, G., Alén, B., Martínez-Pastor, J., Seravalli, L., Frigeri, P., and Franchi, S. (2008). Selective optical pumping of charged excitons in unintentionally doped InAs quantum dots, *Nanotechnology*, **19**, p. 145711.

28. Narvaez, G.A., and Zunger, A. (2006). Nominally forbidden transitions in the interband optical spectrum of quantum dots, *Phys. Rev. B*, **74**, p. 045316.

29. Muñoz-Matutano, G., Rivas, D., Ricchiuti, A., Barrera, D., Fernández-Pousa, C.R., Martínez-Pastor, J., Seravalli, L., Trevisi, G., Frigeri, P., and Sales, S. (2014). Time resolved emission at 1.3 μm of a single InAs quantum dot by using a tunable fibre Bragg grating, *Nanotechnology*, **25**, p. 035204.

30. Seguin, R., Schliwa, A., Rodt, S., Pötschke, K., Pohl, U.W., and Bimberg, D. (2005). Size-dependent fine-structure splitting in self-organized InAs/GaAs quantum dots, *Phys. Rev. Lett.*, **95**, p. 257401.

31. Strauf, S., Stoltz, N.G., Rakher, M.T., Coldren, L.A., Petroff, P.M., and Bouwmeester, D. (2007). High-frequency single-photon source with polarization control, *Nat. Photonics*, **1**, p. 704.

32. Warburton, R.J., Schaflein, C., Haft, D., Bickel, F., Lorke, A., Karrai, K., Garcia, J.M., Schoenfeld, W., and Petroff, P.M. (2000). Optical emission from a charge-tunable quantum ring, *Nature*, **405**, p. 926.

33. Gao, W.B., Fallahi, P., Togan, E., Miguel-Sanchez, J., and Imamoğlu, A. (2012). Observation of entanglement between a quantum dot spin and a single photon, *Nature*, **491**, pp. 426–430.

34. Gao, W.B., Fallahi, P., Togan, E., Delteil, A., Chin, Y.S., Miguel-Sanchez, J., and Imamoğlu, A. (2013). Quantum teleportation from a propagating photon to a solid-state spin qubit, *Nat. Commun.*, **4**, p. 2744.

35. Sanguinetti, S., Henini, M., Grassi Alessi, M., Capizzi, M., Frigeri, P., and Franchi, S. (1999). Carrier thermal escape and retrapping in self-assembled quantum dots, *Phys. Rev. B*, **60**, p. 8276.

36. Gomis, J., Martínez-Pastor, J., Alén, B., Granados, D., García, J.M., and Roussignol, P. (2006). Shape dependent electronic structure and

exciton dynamics in small In(Ga)As quantum dots, *Eur. Phys. J. B*, **54**, p. 471.

37. Trevisi, G., Suárez, I., Seravalli, L., Rivas, D., Frigeri, P., Muñoz, G., Grillo, V., Nasi, L., and Martínez Pastor, J.P. (2013). The effect of high-In content capping layers on low-density bimodal-sized InAs quantum dots, *J. Appl. Phys.*, **113**, p. 194306.

38. Grundmann, M., and Bimberg, D. (1997). Theory of random population for quantum dots, *Phys. Rev. B*, **55**, p. 940.

39. Baier, M.H., Malko, A., Pelucchi, E., Oberli, D.Y., and Kapon, E. (2006). Quantum-dot exciton dynamics probed by photon-correlation spectroscopy, *Phys. Rev. B*, **73**, p. 205321.

40. Muñoz-Matutano, G., Gomis, J., Alén, B., Martínez-Pastor, J., Seravalli, L., Frigeri P., and Franchi, S. (2008). Exciton, biexciton and trion recombination dynamics in a single quantum dot under selective optical pumping, *Physica E*, **40**, pp. 2100–2103.

41. Gomis-Bresco, J., Muñoz-Matutano, G., Martínez-Pastor, J., Alén, B., Seravalli, L., Frigeri, P., Trevisi, G., and Franchi, S. (2011). Random population model to explain the recombination dynamics in single InAs/GaAs quantum dots under selective optical pumping, *New J. Phys.*, **13**, p. 023022.

42. Rivas, D., Muñoz-Matutano, G., Canet-Ferrer, J., García-Calzada, R., Trevisi, G., Seravalli, L., Frigeri, P., Martínez-Pastor, J. (2014). Two-color single-photon emission of InAs quantum dots: toward logic information management using quantum light, *Nano Lett.*, **14**, p. 456.

43. Nirmal, M., Dabbousi, B.O., Bawendi, M.G., Macklin, J.J., Trautman, J.K., Harris, T.D., and Brus, L.E. (1996). Fluorescence intermittency in single cadmium selenide nanocrystals, *Nature*, **383**, p. 802.

44. Klimov, V.I., Mikhailovsky, A.A., McBranch, D.W., Leatherdale, C.A., and Bawendi, M.G. (2000). Quantization of multiparticle auger rates in semiconductor quantum dots, *Science*, **287**, p. 1011.

45. Wang, L.-W., Califano, M., Zunger, A., and Franceschetti, A. (2003). Pseudopotential theory of auger processes in CdSe quantum dots, *Phys. Rev. Lett.*, **91**, p. 056404.

46. Galland, C., Ghosh, Y., Steinbrück, A., Sykora, M., Hollingsworth, J.A., Klimov, V.I., and Htoon, H. (2011). Two types of luminescence blinking revealed by spectroelectrochemistry of single quantum dots, *Nature*, **479**, p. 203.

47. Hollingsworth, J.A. (2013). Heterostructuring nanocrystal quantum dots toward intentional suppression of blinking and auger recombination, *Chem. Mater.*, **25**, p. 1318.

48. Gaponenko, S.V. (2010). *Introduction to Nanophotonics*, Cambridge University Press.

49. Lockwood, D.J., and Pavesi, L. (Eds.). (2011). *Silicon Photonics II: Components and Integration, Topics in Applied Physics 119*, Springer, Berlin.

50. Clark, J., and Lanzani, G. (2010). Organic photonics for communications, *Nat. Photonics*, **4**, pp. 438–446.

51. Konstantatos, G., and Sargent, E.H. (2010). Nanostructured materials for photodetection, *Nat. Nanotech.*, **5**, p. 391.

52. Roelkens, G., Van Campenhout, J., Brouckaert, J., Van Thourhout, D., Baets, R., Rojo Romeo, P., Regreny, P., Kazmierczak, A., Seassal, C., Letartre, X., Hollinger, G., Fedeli, J.M., Di Cioccio, L., and Lagahe-Blanchard, C. (2007). III-V/Si photonics by die-to-wafer bonding, *Mater. Today*, **10**, p. 36.

53. Akiyama, T., Sugawara, M., and Arakawa, Y. (2007). Quantum-dot semiconductor optical amplifiers, *Proc. IEEE*, **95**, p. 1757.

54. Sugawara, M., Hatori, N., Ishida, M., Ebe, H., Arakawa, Y., Akiyama, T., Otsubo, K., Yamamoto, T., and Nakata, Y. (2005). Recent progress of self-assembled quantum dot optical devices for optical telecommunication: temperature-insensitive 10 Gb/s directly modulated lasers and 40 Gb/s signal-regenerative amplifiers, *J. Phys. D*, **38**, p. 2126.

55. Chow, W.W., and Koch, S.W. (2005). Theory of semiconductor quantum-dot laser dynamics, *IEEE J. Quant. Electron.*, **41**, p. 495.

56. Mi, Z., Bhattacharya, P., and Fathpour, S. (2005). High-speed 1.3 μm tunnel injection quantum-dot lasers, *Appl. Phys. Lett.*, **86**, p. 153109.

57. Otsubo, K., Hatori, N., Ishida, M., Okumura, S., Akiyama, T., Nakata, Y., Ebe, H., Sugawara, M., and Arakawa, Y. (2004). Temperature-insensitive eye-opening under 10-Gb/s modulation of 1.3 μm p-doped quantum-dot lasers without current adjustments, *Jpn. J. Appl. Phys.*, **43**, p. L1124.

58. Lämmlin, M., Fiol, G., Meuer, C., Kuntz, M., Hopfer, F., Kovsh, A.R., Ledentsov, N.N., and Bimberg, D. (2006). Distortion-free optical amplification of 20-80 GHz modelocked laser pulses at 1.3 μm using quantum dots, *Electron. Lett.*, **42**, pp. 697–699.

59. Dommers, S., Temnov, V.V., Woggon, U., Gomis, J., Martínez-Pastor, J., Lämmlin, M., and Bimberg, D. (2007). Complete ground state gain recovery after ultrashort double pulses in quantum dot based semiconductor optical amplifier, *Appl. Phys. Lett.*, **90**, p. 033508.

60. Lodahl, P., van Driel, A.F., Nikolaev, I.S., Irman, A., Overgaag, K., Vanmaekelbergh, D., and Vos, W.L. (2004). Controlling the dynamics

of spontaneous emission from quantum dots by photonic crystals, *Nature*, **430**, p. 654.

61. Fushman, I., Englund, D., and Vučković, J. (2005). Coupling of PbS quantum dots to photonic crystal cavities at room temperatura, *Appl. Phys. Lett.*, **87**, p. 241102.

62. Kahl, M., Thomay, T., Kohnle, V., Beha, K., Merlein, J., Hagner, M., Halm, A., Ziegler, J., Nann, T., Fedutik, Y., Woggon, U., Artemyev, M., Pérez-Willard, F., Leitenstorfer, A., and Bratschitsch, R. (2008). Colloidal quantum dots in all-dielectric high-Q pillar microcavities, *Nano Lett.*, **7**, p. 2897.

63. Menon, V.M., Luberto, M., Valappil, N., and Chatterjee, S. (2008). Lasing from quantum dots in a spin-coated flexible microcavity, *Opt. Express*, **16**, p. 19535.

64. Suárez, I., Gordillo, H., Abargues, R., Albert, S., and Martínez-Pastor, J. (2011). Photoluminiscence wave-guiding in CdSe and CdTe QDs-PMMA nanocomposite films, *Nanotechnology*, **22**, p. 435202.

65. Gordillo, H., Suárez, I., Abargues, R., Rodríguez-Cantó, P., and Martínez-Pastor, J.P. (2013). Color tuning and white light by dispersing CdSe, CdTe and CdS in PMMA nanocomposite waveguides, *IEEE Photonics J.*, **5**, p. 2201412.

66. Gordillo, H., Suárez, I., Abargues, R., Rodríguez-Cantó, P., Albert, S., and Martínez-Pastor, J.P. (2012). Polymer/QDs nanocomposites for wave-guiding applications, *J. Nanomater.*, 960201, pp. 1–9.

67. Gordillo, H., Suárez, I., Abargues, R., Rodríguez-Cantó, P., Almuneau, G., and Martínez-Pastor, J.P. (2013). Quantum-dot double layer polymer waveguides by evanescent light coupling, *J. Lightwave Technol.*, **31**, pp. 2515–2525.

68. Ricchiuti, A.L., Suárez, I., Barrera, D., Rodríguez-Cantó, P.J., Fernández-Pousa, C.R., Abargues, R., Sales, S., Martínez-Pastor, J.P., and Capmany, J. (2014). Colloidal quantum dots-PMMA waveguides as integrable microwave photonic phase shifters, *Photonics Technol. Lett.*, **26**, pp. 402–404.

69. Dang, C., Lee, J., Breen, C., Steckel, J.S., Coe-Sullivan, S., and Nurmikko, A. (2012). Red, green and blue lasing enabled by single-exciton gain in colloidal quantum dot films, *Nat. Nanotech.*, **7**, p. 335.

70. Feezell, D.F., Schmidt, M.C., Den Baars, S.P., and Nakamura, S. (2009). Development of nonpolar and semipolar InGaN/GaN visible light-emitting diodes, *MRS Bull.*, **34**, p. 318.

71. Cho, H.K., Lee, J.Y., Kim, C.S., and Yan, G.M. (2002). Influence of strain relaxation on structural and optical characteristics of InGaN/GaN multiple quantum wells with high indium composition, *J. Appl. Phys.*, **91**, p. 1166.

72. Lv, W., Wang, L., Wang, J., Hao, Z., and Luo, Y. (2012). InGaN/GaN multilayer quantum dots yellow-green light-emitting diode with optimized GaN barriers, *Nanoscale Res. Lett.*, **7**, p. 617.

73. Tanaka, D., Sasabe, H., Li, Y., Su, S., Takeda, T., and Kido, J. (2007). Ultra high efficiency green organic light-emitting devices, *Jpn. J. Appl. Phys.*, **46**, p. 10.

74. Wang, Z.B., Helander, M.G., Qiu, J., Puzzo, D.P., Greiner, M.T., Hudson, Z.M., Wang, S., Liu, Z.W., and Lu, Z.H. (2011). Unlocking the full potential of organic light-emitting diodes on flexible plastic, *Nat. Photonics*, **5**, 12, pp. 753–757.

75. Yeh, S.J., Wu, M.F., Chen, C.T., Song, Y.H., Chi, Y., Ho, M.H., Hsu, S.F., and Chen, C.H. (2005). New dopant and host materials for blue-light-emitting phosphorescent organic electroluminescent devices, *Adv. Mater.*, **17**, 3, pp. 285–289.

76. Shirasaki, Y., Supran, G.J., Bawendi, M.G., and Bulović, V. (2013). Emergence of colloidal quantum-dot light-emitting technologies, *Nat. Photonics*, **7**, p. 13.

77. Supran, G.J., Shirasaki, Y., Song, K.W., Caruge, J.-M., Kazlas, P.T., Coe-Sullivan, S., Andrew, T.L., Bawendi, M.G., and Bulović, V. (2013). QLEDs for displays and solid-state lighting, *MRS Bull.*, **38**, p. 703.

78. Bae, W.K., Brovelli, S., and Klimov, V.I. (2013). Spectroscopic insights into the performance of quantum dot light-emitting diodes, *MRS Bull.*, **38**, p. 721.

79. Talapin, D.V., Lee, J.-S., Kovalenko, M.V., and Shevchenko, E.V. (2010). Prospects of colloidal nanocrystals for electronic and optoelectronic applications, *Chem. Rev.*, **110**, p. 389.

80. Lee, J., Sundar, V.C., Heine, J.R., Bawendi, M.G., and Jensen, K.F. (2000). Full color emission from II–VI semiconductor quantum dot–polymer composites, *Adv. Mater.*, **12**, 15, pp. 1102–1105.

81. Jang, E., Jun, S., Jang, H., Lim, J., Kim, B., and Kim, Y. (2010). You have full text access to this content white-light-emitting diodes with quantum dot color converters for display backlights, *Adv. Mater.*, **22**, 28, pp. 3076–3080.

82. Woo, H., Lim, J., Lee, Y., Sung, J., Shin, H., Oh, J.M., Choi, M., Yoon, H., Bae, W.K., and Char, K. (2013). Robust, processable, and bright quantum

dot/organosilicate hybrid films with uniform QD distribution based on thiol-containing organosilicate ligands, *J. Mater. Chem.*, **C1**, 10, pp. 1983–1989.

83. Achermann, M., Petruska, M.A., Kos, S., Smith, D.L., Koleske, D.D., and Klimov, V.I. (2004). Energy-transfer pumping of semiconductor nanocrystals using an epitaxial quantum well, *Nature*, **429**, 6992, pp. 642–646.

84. Achermann, M., Petruska, M.A., Koleske, D.D., Crawford, M.H., and Klimov, V.I. (2006). Nanocrystal-based light-emitting diodes utilizing high-efficiency nonradiative energy transfer for color conversion, *Nano Lett.*, **6**, 7, pp. 1396–1400.

85. Caruge, J.M., Halpert, J.E., Wood, V., Bulovic, V., and Bawendi, M.G. (2008). Colloidal quantum-dot light-emitting diodes with metal-oxide charge transport layers, *Nat. Photonics*, **2**, 4, pp. 247–250.

86. Cho, K.-S., Lee, E.K., Joo, W.-J., Jang, E., Kim, T.-H., Lee, S.J., Kwon, S.-J., Han, J.Y., Kim, B.-K., Choi, B.L., and Kim, J.M. (2009). High-performance crosslinked colloidal quantum-dot light-emitting diodes, *Nat. Photonics*, **3**, 6, pp. 341–345.

87. Kwak, J., Bae, W.K., Lee, D., Park, I., Lim, J., Park, M., Cho, H., Woo, H., Yoon, D.Y., Char, K., Lee, S., and Lee, C. (2012). Bright and efficient full-color colloidal quantum dot light-emitting diodes using an inverted device structure, *Nano Lett.*, **12**, 5, pp. 2362–2366.

88. Qian, L., Zheng, Y., Xue, J., and Holloway, P.H. (2011). Stable and efficient quantum-dot light-emitting diodes based on solution-processed multilayer structures, *Nat. Photonics*, **5**, pp. 543–548.

89. Kim, L., Anikeeva, P.O., Coe-Sullivan, S.A., Steckel, J.S., Bawendi, M.G., and Bulovic, V. (2008). Contact printing of quantum dot light-emitting devices, *Nano Lett.*, **8**, p. 4513.

90. Mashford, B.S., Stevenson, M., Popovic, Z., Hamilton, C., Zhou, C., Breen, C., Steckel, J., Bulovic, V., Bawendi, M., Coe-Sullivan, S., and Kazlas, P.T. (2013). High-efficiency quantum-dot light-emitting devices with enhanced charge injection, *Nat. Photonics*, **7**, 5, pp. 407–412.

91. Shirasaki, Y., Supran, G.J., Tisdale, W.A., and Bulovic, V. (2013). Origin of efficiency roll-off in colloidal quantum-dot light-emitting diodes, *Phys. Rev. Lett.*, **110**, p. 217403.

92. Alén, B., Bosch, J., Granados, D., Martínez-Pastor, J., García, J.M., and González, L. (2007). Oscillator strength reduction induced by external electric fields in self-assembled quantum dots and rings, *Phys. Rev. B*, **75**, p. 045319.

93. Arakawa, Y., and Sakaki, H. (1982). Multidimensional quantum well laser and temperature dependence of its threshold current, *Appl. Phys. Lett.*, **40**, p. 939.

94. Kirstaedter, N., Ledentsov, N.N., Grundmann, M., Bimberg, D., Ustinov, V.M., Ruvimov, S.S., Maximov, M.V., Kopév, P.S., Alferov, Z., Richter, U., Werner, P., Gosele U., and Heydenreich, J. (1994). Low threshold, large T_0 injection laser emission from (InGa)As quantum dots, *Electron. Lett.*, **30**, 17, pp. 1416–1417.

95. Heinrichsdorff, F., Mao, M.-H., Kirstaedter, N., Krost, A., Bimberg, D., Kosogov, A.O., and Werner, P. (1997). Room-temperature continuous-wave lasing from stacked InAs/GaAs quantum dots grown by metalorganic chemical vapor deposition, *Appl. Phys. Lett.*, **71**, p. 22.

96. Eliseev, P.G., Li, H., Liu, T., Newell, T.C., Lester, L.F., and Malloy, K.J. (2001). Ground-state emission and gain in ultralow-threshold InAs-InGaAs quantum-dot lasers, *IEEE J. Sel. Top. Quant. Electron.*, **7**, p. 135.

97. Fathpour, S., Mi, Z., Bhattacharya, P., Kovsh, A.R., Mikhrin, S.S., Krestnikov, I.L., Kozhukhov, A.V., and Ledentsov, N.N. (2004). The role of Auger recombination in the temperature-dependent output characteristics ($T_0 = \infty$) of p-doped 1.3 mm quantum dot lasers, *Appl. Phys. Lett.*, **85**, p. 5164.

98. Zhukov, A.E., Kovsh, A.R., Mikhrin, S.S., Maleev, N.A., Ustinov, V.M., Livshits, D.A., Tarasov, I.S., Bedarev, D.A., Maximov, V.M., Tsatsulnikov, A.F., Soshnikov, I.P., Kopev, P.S., Alferov, Z.I., Ledentsov, N.N., and Bimberg, D. (1999). 3.9 W cw power from sub-monolayer quantum dot diode laser, *Electron. Lett.*, **35**, p. 1845.

99. Fathpour, S., Mi, Z., and Bhattacharya, P. (2005). High-speed quantum dot lasers, *J. Phys. D: Appl. Phys.*, **38**, p. 2103.

100. Kondratko, P.K., Chuang, S.L., Walter, G., Chung, T., and Holonyak, N. (2003). Observation of near-zero linewidth enhancement factor in a quantum-well coupled quantum-dot laser, *Appl. Phys. Lett.*, **83**, p. 4818.

101. Le Ru, E.C., Howe, P., Jones, T.S., and Murray, R. (2003). Strain-engineered InAs/GaAs quantum dots for long-wavelength emission, *Phys. Rev. B*, **67**, p. 165303.

102. Seravalli, L., Frigeri, P., Nasi, L., Trevisi, G., and Bocchi, C. (2010). Metamorphic quantum dots: quite different nanostructures, *J. Appl. Phys.*, **108**, p. 064324.

103. Ledentsov, N.N., Kovsh, A.R., Zhukov, A.E., Maleev, N.A., Mikhrin, S.S., Vasil'ev, A.P., Semenova, E.S., Maximov, M.V., Shernyakov, Y.M.,

Kryzhanovskaya, N.V., Ustinov, M., and Bimberg, D. (2003). High performance quantum dot lasers on GaAs substrates operating in 1.5 μm range, *Electron. Lett.*, **39**, p. 1126.

104. Saito, H., Nishi, K., Ogura, I., Sugou, S., and Sugimoto, Y. (1996). Room-temperature lasing operation of a quantum-dot vertical-cavity surface-emitting laser, *Appl. Phys. Lett.*, **69**, p. 3140.

105. Ustinov, V.M., Zhukov, A.E., Maleev, N.A., Kovsh, A.R., Mikhrin, S.S., Volovik, B.V., Musikhin, Y.G., Shernyakov, Y.M., Maximov, M.V., Tsatsul'nikov, A.F., Ledentsov, N.N., Alferov, Z.I., Lott, J.A., and Bimberg, D. (2001). 1.3 μm InAs/GaAs quantum dot lasers and VCSELs grown by molecular beam epitaxy, *J. Cryst. Growth*, **227–228**, p. 1155.

106. Yu, H.C., Wang, J.S., Su, Y.K., Chang, S.J., Lai, F.I., Chang, Y.H., Kuo, H.C., Sung, C.P., Yang, H.P.D., Lin, K.F., Wang, J.M., Chi, J.Y., Hsiao, R.S., and Mikhrin, S. (2006). 1.3 μm InAs-InGaAs quantum-dot vertical-cavity surface-emitting laser with fully doped DBRs grown by MBE, *IEEE Photonics Technol. Lett.*, **18**, p. 418.

107. Kuznetsov, M., Hakimi, F., Sprague, R., and Mooradian, A. (1999). Design and characteristics of high-power (>0.5−W CW) diode-pumped vertical external-cavity surface-emitting semiconductor lasers with circular TEM00 beams, *IEEE J. Sel. Top. Quant. Electron.*, **5**, p. 561.

108. Rantamäki, A., Rautiainen, J., Toikkanen, L., Krestnikov, I., Butkus, M., Rafailov, E.U., and Okhotnikov, O. (2012). Flip chip quantum-dot semiconductor disk laser at 1200 nm, *IEEE Photonics Technol. Lett.*, **24**, p. 1292.

109. Ellis, B., Mayer, M.A., Shambat, G., Sarmiento, T., Harris, J., Haller, E.E., and Vuckovic, J. (2011). Ultralow-threshold electrically pumped quantum dot photonic-crystal nanocavity laser, *Nat. Photonics*, **5**, p. 297.

110. Khajavikhan, M., Simic, A., Katz, M., Lee, J.H., Slutsky, B., Mizrahi, A., Lomakin, V., and Fainman, Y. (2012). Thresholdless nanoscale coaxial lasers, *Nature*, **482**, p. 204.

111. Kamat, P.V. (2007). Meeting the clean energy demand: nanostructure architectures for solar energy conversion, *J. Phys. Chem. C*, **111**, p. 2834.

112. Emin, S., Loukanov, A., Wakasa, M., Nakabayashi, S., and Kaneko, Y. (2010). Photostability of water-dispersible CdTe quantum dots: capping ligands and oxygen, *Chem. Lett.*, **39**, p. 654.

113. Bear, D., Qian, L., Tseng, T.-K., and Holloway, P.H. (2010). Quantum dots and their multimodal applications: a review, *Materials*, **3**, pp. 2260–2345.

114. Bang, J.H., and Kamat, P.V. (2009). Quantum dot sensitized solar cells. A tale of two semiconductor nanocrystals: CdSe and CdTe, *ACS Nano*, **3**, p. 1467.

115. Luther, J.M., Law, M., Song, Q., Reese, M.O., Beard, M.C., Ellingson, R.C., and Nozik, A.J. (2008). Schottky solar cells based on colloidal nanocrystal films, *Nano Lett.*, **8**, p. 3488.

116. Tang, J., Wang, X., Brzozowski, L., Barkhouse, D.A.R., Debnath, R., Levina, L., and Sargent, E.H. (2010). Schottky quantum dot solar cells stable in air under solar illumination, *Adv. Mater.*, **22**, p. 1398.

117. Pattantyus-Abraham, A.G., Kramer, I.J., Barkhouse, A.R., Wang, X., Konstantatos, G., Debnath, R., Levina, L., Raabe, I., Nazeeruddin, M.K., Grätzel, M. and Sargent, E.H. (2010). Depleted-heterojunction colloidal quantum dot solar cells, *ACS Nano*, **4**, p. 3374.

118. Chang, J.A., Rhee, J.H., Im, S.H., Lee, Y.H., Kim, H.-J., Seok II, S., Nazeeruddin, M.K., and Grätzel, M. (2010). High-performance nanostructured inorganic–organic heterojunction solar cells, *Nano Lett.*, **10**, p. 2609.

119. Ma, W., Luther, J.M., Zheng, H., Wu, Y., Alivisatos, A.P. (2009). Photovoltaic devices employing ternary PbSxSe1-x nanocrystals, *Nano Lett.*, **9**, p. 1699.

120. Debnath, R., Greiner, M.T., Kramer, I.J., Fischer, A., Tang, J., Barkhouse, D.A.R., Wang, X., Levina, L., Lu, Z.-H., and Sargent, E.H. (2010). Depleted-heterojunction colloidal quantum dot photovoltaics employing low-cost electrical contacts, *Appl. Phys. Lett.*, **97**, p. 023109.

121. Luther, J.M., Gao, J., Lloyd, M.T., Semonin, O.E., Beard, M.C., and Nozik, A.J. (2010). Stability assessment on a 3% bilayer PbS/ZnO quantum dot heterojunction solar cell, *Adv. Mater.*, **22**, p. 3704.

122. Tang, J., and Sargent, E.H. (2011). Infrared colloidal quantum dots for photovoltaics: fundamentals and recent progress, *Adv. Mater.*, **23**, pp. 12–29.

123. Kim, J., Voznyy, O., Zhitomirsky, D., Sargent, E.H. (2013). 25th Anniversary article: colloidal quantum dot materials and devices; a quarter-century of advances, *Adv. Mater.*, **25**, p. 4986.

124. Kramer, I.J., and Sargent, E.H. (2013). The architecture of colloidal quantum dot solar cells: materials to devices, *Chem. Rev.*, **114**, p. 863.

125. Moreels, I., Justo, Y., De Geyter, B., Haustraete, K., Martins, J.C., and Hens, Z. (2011). Size-tunable, bright, and stable PbS quantum dots: a surface chemistry study, *ACS Nano*, **5**, p. 2004.

126. Ip, A.H., Thon, S.M., Hoogland, S., Voznyy, O., Zhitomirsky, D., Debnath, R., Levina, L., Rollny, L.R., Carey, G.H., Fischer, A., Kemp, K.W., Kramer,

I.J., Ning, Z., Labelle, A.J., Chou, K.W., Amassian, A., and Sargent, E.H. (2012). Hybrid passivated colloidal quantum dot solids, *Nat. Nanotech.*, **7**, p. 577.

127. Rath, A.K., Bernechea, M., Martinez, L., and Konstantatos, G. (2011). Solution-processed heterojunction solar cells based on p-type PbS quantum dots and n-type Bi_2S_3 nanocrystals, *Adv. Mater.*, **23**, p. 3712.

128. Rath, A.K., Bernechea, M., Martinez, L., Garcia de Arquer, F.P., Osmond, J., and Konstantatos, G. (2012). Solution-processed inorganic bulk nano-heterojunctions and their application to solar cells, *Nat. Photonics*, **6**, p. 529.

129. Prins, F., Buscema, M., Seldenthuis, J.S., Etaki, S., Buchs, G., Barkelid, M., Zwiller, V., Gao, Y., Houtepen, A.J., Siebbeles, L.D.A., and van der Zant, H.S.J. (2012). Fast and efficient photodetection in nanoscale quantum-dot junctions, *Nano Lett.*, **12**, p. 5740.

130. Konstantatos, G., and Sargent, E.H. (2011). Colloidal quantum dot photodetectors, infrared physics & technology, *Infrared Phys. Technol.*, **54**, p. 278.

131. Konstantatos, G., Badioli, M., Gaudreau, L., Osmond, J., Bernechea, M., Garcia de Arquer, F.P., Gatti, F., Koppens, F.H.L. (2012). Hybrid graphene–quantum dot phototransistors with ultrahigh gain, *Nat. Nanotech.*, **7**, p. 363.

132. Novoselov, K.S., Geim, A.K., Morozov, S.V., Jiang, D., Zhang, Y., Dubonos, S.V., Grigorieva, I.V., and Firsov, A.A. (2004). Electric field effect in atomically thin carbon films, *Science*, **306**, p. 666.

133. Wang, X., Zhi, L., Tsao, N., Tomovic, Z., Li, J., and Müllen, K. (2008). Transparent carbon films as electrodes in organic cells, *Angew. Chem.*, **120**, p. 3032.

134. Kim, K.S., Zhao, Y., Jang, H., Lee, S.Y., Kim, J.M., Kim, K.S., Ahn, J.-H., Kim, P., Choi, J.-Y., and Hong, B.H. (2009). Large- scale pattern growth of graphene films for stretchable transparent electrodes, *Nature*, **457**, p. 706.

135. Ihm, K.,Lim, J.T., Lee, K.-J., Kwon, J.W., Kang, T.-H., Chung, S., Bae, S., Kim, J.H., Hong, B.H., and Yeom, G.Y. (2010). Number of graphene layers as a modulator of the open-circuit voltage of graphene-based solar cell, *Appl. Phys. Lett.*, **97**, p. 032113.

136. Kasry, A., Kuroda, M.A., Martyna, G.J., Tulevski, G.S., and Bol, A.A. (2010). Chemical doping of large-area stacked graphene films for use as transparent, conducting electrodes, *ACS Nano*, **4**, p. 3839.

137. Liu, Z., Liu, Q., Huang, Y., Ma, Y., Yin, S., Zhang, X., Sun, W., and Chen, Y. (2008). Organic photovoltaic devices based on a novel acceptor material: graphene, *Adv. Mater.*, **20**, p. 3924.

138. Guo, C.X., Yang, H.B., Sheng, Z.M., Lu, Z.S., Song, Q.L., and Li, C.M. (2010). Layered graphene/quantum dots for photovoltaic devices, *Angew. Chem.*, **49**, p. 3014.

139. Li, X., Zhu, H., Wang, K., Cao, A., Wei, J., Li, C., Jia, Y., Li, Z., Li, X., and Wu, D. (2010). Graphene-on-silicon Schottky junction solar cells, *Adv. Mater.*, **22**, p. 2743.

140. Nowozin, T., Bimberg, D., Daqrouq, K., Ajour, M.N., and Awedh, M. (2013). Materials for future quantum dot-based memories, *J. Nanomater.*, 2013, Article ID 215613, 6 pages (http://www.hindawi.com/journals/jnm/2013/215613/).

141. Hayne, M., Young, R.J., Smakman, E.P., Nowozin, T., Hodgson, P., Garleff, J.K., Rambabu, P., Koenraad, P.M., Marent, A., Bonato, L., Schliwa, A., and Bimberg, D. (2013). The structural, electronic and optical properties of GaSb/GaAs nanostructures for charge-based memory, *J. Phys. D: Appl. Phys.*, **46**, p. 264001.

142. Vahala, K.J. (2003). Optical microcavities, *Nature*, **424**, p. 839.

143. Michler, P., Kiraz, A., Becher, C., Schoenfeld, W.V., Petroff, P.M., Zhang, L.D., Hu, E., and Imamoglu, A. (2000). A quantum dot single-photon turnstile device, *Science*, **290**, p. 2282.

144. Galland, C., Brovelli, S., Bae, W.K., Padilha, L.A., Meinardi, F., and Klimov, V.I. (2013). Dynamic hole blockade yields two-color quantum and classical light from dot-in-bulk nanocrystals, *Nano Lett.*, **13**, p. 321.

145. Hanbury Brown, R., and Twiss, R.Q. (1954). A new type of interferometer for use in radio astronomy, *Philos. Mag.*, **7**, p. 663.

146. Dousse, A., Suffczynski, J., Beveratos, A., Krebs, O., Lemaitre, A., Sagnes, I., Bloch, J., Voisin, P., and Senellart, P. (2010). Ultrabright source of entangled photon pairs, *Nature*, **466**, p. 217.

147. Stevenson, R., Young, R., Atkinson, P., Cooper, K., Ritchie, D., and Shields, A. (2006). A semiconductor source of triggered entangled photon pairs, *Nature*, **439**, p. 179.

148. Salter, C., Stevenson, R., Farrer, I., Nicoll, C., Ritchie, D., and Shields, A. (2010). An entangled-light-emitting diode, *Nature*, **465**, p. 594.

149. Robledo, L., Elzerman, J., Jundt, G., Atatüre, M., Högele, A., Fält, S., and Imamoglu, A. (2008). Conditional dynamics of interacting quantum dots, *Science*, **320**, p. 772.

150. Li, X., Wu, Y., Steel, D., Gammon, D., Stievater, T.H., Katzer, D.S., Park, D., Piermarocchi, C., and Sham, L.J. (2003). An all-optical quantum gate in a semiconductor quantum dot, *Science*, **301**, p. 809.

151. Szumniak, P., Bednarek, S., Partoens, B., and Peeters, F.M. (2012). Spin-orbit-mediated manipulation of heavy-hole spin qubits in gated semiconductor nanodevices, *Phys. Rev. Lett.*, **109**, p. 107201.

152. Qureshi, M.S., Sen, P., Andrews, J.T., and Sen, K. (2009). All optical quantum CNOT gate in semiconductor quantum dots, *IEEE J. Quant. Electron.*, **45**, p. 59.

153. Nguyen, H., Sallen, G., Voisin, P., Roussignol, P., Diederichs, C., and Cassabois, G. (2012). Optically gated resonant emission of single quantum dots, *Phys. Rev. Lett.*, **108**, p. 057401.

154. Reithmaier, J.P., Sek, G., Löffler, A., Hofmann, C., Kuhn, S., Reitzenstein, S., Keldysh, L.V., Kulakovskii, V.D., Reinecke, T.L., and Forchel, A. (2004). Strong coupling in a single quantum dot–semiconductor microcavity system, *Nature*, **432**, p. 197.

155. Peter, E., Senellart, P., Martrou, D., Lemaître, A., Hours, J., Gérard, J.M., and Bloch, J. (2005). Exciton-photon strong-coupling regime for a single quantum dot embedded in a microcavity, *Phys. Rev. Lett.*, **95**, p. 067401.

156. Hennessy, K., Badolato, A., Winger, M., Gerace, D., Atatuüre, M., Gulde, S., Fält, S., Hu, E.L., and Imamoglu, A. (2007). Quantum nature of a strongly coupled single quantum dot-cavity system, *Nature*, **445**, p. 896.

157. Purcell, E.M. (1946). Spontaneous emission probabilities at radio frequencies, *Phys. Rev.*, **69**, p. 681.

158. Buckley, S., Rivoire, K., and Vuckovic, J. (2012). Engineered quantum dot single-photon sources, *Rep. Prog. Phys.*, **75**, p. 126503.

Chapter 8

Group III–V on Silicon: A Brand-New Optoelectronics

Badhise Ben Bakir, Corrado Sciancalepore, Antoine Descos, Héléne Duprez, Damien Bordel, and Sylvie Menezo

CEA-Leti, Minatec, Departement of Optoelectronics,
17 rue des Martyrs, 38054 Grenoble, France
badhise.ben-bakir@cea.fr

Silicon, so far the mainstay of the electronics industry, is fast becoming the triggering material of a photonics-based computational era. This chapter is intended to provide readers with the recent developments in the field of silicon photonics, specifically concerning the heterogeneous integration of group III–V materials on silicon. This approach—unfolding into several basic building blocks such as lasers, photodetectors, and modulators, as well as novel proofs of concepts—is paving the way toward a brand-new optoelectronics showing a considerable integration potential with scalable, cost-effective complementary metal-oxide semiconductor (CMOS) technology and silicon-based photonic integrated circuits (PICs).

Nanodevices for Photonics and Electronics: Advances and Applications
Edited by Paolo Bettotti
Copyright © 2016 Pan Stanford Publishing Pte. Ltd.
ISBN 978-981-4613-74-3 (Hardcover), 978-981-4613-75-0 (eBook)
www.panstanford.com

8.1 Introduction

We are living in the exponential times of an increasingly data-driven, densely interconnected, global society. This mirrors the relentless growth of data traffic in today's telecommunication infrastructures, driving the demand for increasing transmission rates and computing capabilities. Global IP traffic will grow at a compound annual growth rate (CAGR) of 29% from 2011 to 2016 and will surpass the zettabyte (1021 bytes or 1 trillion gigabytes) threshold by the end of 2016 [1]. Such an ever-growing data stream is challenging the intrinsic limit of copper-based, short-reach interconnects and microelectronic circuits in server architectures and data centers to offer enough modulation bandwidth at reasonable power dissipation. Against this backdrop, a dramatic bottleneck is constituted by the difficulties in moving digital information, at every level: from worldwide links to chip-to-chip and intrachip interconnections. By contrast, optics-based telecommunications has the potential of light to carry information over large distances at very high data rates with minor power dissipation. However, while photonics can be already found at the heart of today's communication networks, providing enormous performance levels to core, metro, and access systems, at shorter distances, the challenges implied by signal speeds, power consumption, miniaturization, and overall costs are still only partially addressed. Pure electronics and hybrid optoelectronic integration is not looking effective anymore whenever a drastic level of miniaturization has to be achieved, whenever performance levels at hundreds of gigabits/s are to be achieved, and whenever the costs of implementation and the energy consumption are to be cut significantly. Within this mind-set, the functional integration of photonic devices in electronics and complementary metal-oxide semiconductor (CMOS) technology nowadays constitutes an answer of strategic importance to solve the copper dead-end. CMOS technology has been used for more than 30 years to manufacture complex electronic chips: it has very advanced process capability and huge commercial fabrication capacity. The idea of applying a high-volume, low-cost technique from the electronics industry to manufacture photonic integrated circuits (PICs) has led to silicon photonics technology—that is, integrating photonic functions with electronic circuits—at the wafer-scale manufacturing level. By developing a small number of generic integration technologies with different levels of functionality and by making such technologies accessible via foundries, silicon photonics can address a broad range of advanced applications.

Silicon lies at the heart of the aforementioned heterogeneous integration, constituting the common bridging platform between photonics opportunities—on the one side—and the micro-nano-electronics technological maturity on the other [2]. In this sense, we can claim that the functional merge of photonics on electronic chips and cards is triggering the beginning of a photonics-based computational era.

This chapter aims at introducing the reader to the main recent achievements concerning the heterogeneous integration of group III–V on silicon, thus focusing on key active building blocks constituting the silicon photonics toolbox, such as lasers, modulators, and innovative proofs of concepts.

Here follows the remainder of this chapter, which unfolds the inherent potential of heterointegrating group III–V materials on silicon. Reasons behind the push for bridging laser technology to CMOS fab maturity are exposed and discussed in Section 8.2. Group III–V-on-silicon distributed Bragg reflectors (DBRs) as well as distributed feedback (DFB) using high-index-contrast silicon-/semiconductor-on-insulator (SOI) gratings are presented in Section 8.3, along with current performances, in both static and dynamic operation regimes. Within the same section a brief overview is given over III–V on Si electroabsorption modulators (EAMs), which take advantage from a common material, design, and fabrication approach of the laser building blocks. Section 8.4 will instead focus on the innovative proofs of concepts. In particular, high-contrast-grating vertical-cavity surface-emitting lasers (HCG-VCSELs) using III–V on Si heterointegration recently came to the fore, owing to their inherent compactness, low power consumption, high-speed direct modulation potential, and spectral purity. Section 8.5 summarizes and concludes the chapter, while shedding light on forthcoming research and developments.

8.2 Bridging Laser Technology to Silicon Maturity

Such staggering growth in the volume of data traffic running through our telecommunication infrastructures is approaching fast the performance bottleneck of copper-based interconnects of standard electronic circuits and systems in present computing nodes and data centers. In this sense, the onset of the social networking phenomenon, as well as the exponential increase in richer multimedia content exchange over the Net, is contributing to calling into question the capability of existent datacom backbones to withstand such a relentlessly growing stream of data. Moreover, continuing shrinking microelectronic nodes as a long-term solution puts at risk its technological as well as economic viability, as the pin/interconnect density is limited by the shrinking available space over rack and board areas, while heat management proves harder and costs for cooling systems are ballooning in present data center architectures.

Within such a scenario, the emerging solution of adopting power-efficient broadband optical interconnects capable to face the mounting demand for data transmission bandwidth as well as aiming at increasing computing capabilities has gained momentum in the last 10 years within both research and industrial environments. In detail, the intrinsic capability of light to transport information over large distances at very high data rates/low latency with minor power dissipation can be thus scaled from rack-level optical links down to card-to-card to chip-to-chip interconnects. This translates into solid perspectives for *revolutionizing* the way we nowadays design and conceive telecommunication system and server architectures [3] as well as high-performance computing [4].

As well as that, the second strategic pillar of this paradigm shift in datacom architectures consists of using silicon as a common bridging material platform between existing standard CMOS integrated circuits (ICs) and novel photonics-enhanced nodes. This would "leverage" CMOS technology maturity, reliability, and cost-effectiveness into silicon photonic chips, resulting in a new generation of transceivers featuring highly integrated functions and scalable speeds as well as promoting a faster research-to-market transfer on more solid perspectives.

The effect of leveraging silicon technological maturity into next-generation photonics-based optical links is concrete and topical. First proofs of concepts and demonstrators are nowadays gradually exiting research and development pipelines and entering into main semiconductor industrial roadmaps, ready to access product maturity, commercial service launch, and, ultimately, technology open sourcing [5, 6].

Silicon has represented the mainstay of electronics and IC manufacturing over the last decades, mainly owing to low production costs and its good electronic properties. Concerning photonics and optoelectronics, however, its indirect bandgap material proves very inefficient for light amplification, thus seeming unfit for the on-chip integration of lasers and sources meant for the generation of optical carriers.

Considerable efforts have been spent in the last 10 years to transform silicon into an active optical material, even reaching remarkable turning points in laser and photonics technology. For instance, the first Raman-scattering-based continuous-wave silicon lasers called for hopes in the photonics community [7, 8]. However, this solution needs very high pumping powers and looks unfit to be scaled onto a CMOS-compatible platform at affordable costs for mass-scale production. As well as that, light emission from silicon nanocrystals [9] and Er^{3+}-doped silicon nanoclusters [10] proves to be quite inefficient, thus setting a far horizon over the concrete exploitation of this approach for industrial demonstrators and, ultimately, commercial products. Although other approaches to avoid this intrinsic limitation of silicon materials such as nanoporous silicon [11] and rare-earth-doped silica glasses [12] have been investigated, the common critical missing element, however, is electrical pumping, since an external light source is necessary to pump the devices and achieve laser emission. A last notable example to cope with this issue is represented by the use of strained, heavily doped Germanium (Ge) as a gain-enabling material for silicon photonics. Nevertheless, though the first electrically pumped CMOS-compatible Ge-on-Si laser was realized in 2012, more development is necessary to increase laser reliability, power efficiency, and process stability [13].

In the absence of practically efficient lasers achievable directly in silicon or other group IV materials, Si photonic transmitter sources must be made by hybrid integration with III–V gain materials. One commercial solution makes use of external bulk-InP-processed laser dies. The laser light is coupled into the PIC by means of a lens, which is followed by an optical isolator and a mirror for directing the light to a surface grating coupler in the Si PIC [14]. As we will see in the next section, other approaches consist of butt-coupling a III–V semiconductor reflective semiconductor optical amplifier (SOA) to the Si PIC waveguide that comprises a Bragg mirror for defining the laser cavity [15]. Alternatively, the heterogeneous integration of group III–V direct-bandgap materials on SOI platforms is increasingly becoming a more promising approach for a cost-effective on-chip laser technology in silicon photonics [16, 17]. An InP substrate having the laser III–V gain layers grown on top is bonded with loose alignment requirements ($\sim$50 μm), the III–V gain layers facing down to the bottom silicon wafer on which silicon-based components and passive circuitry are preprocessed. In a more economical route, the InP substrate having the laser III–V gain layers on top is first diced, and the dies are collectively bonded, where needed. Then the InP substrate is removed, and the laser process is continued on the III–V bonded dies in a regular wafer-level process flow. Putting the expensive III–V gain material only where needed saves on cost. In addition to lower cost, photonic integration promises improved reliability and performance as well as a reduced footprint over discrete components systems.

While high-gain III–V layers provide efficient light amplification, the high-index-contrast SOI architecture allows for the low-footprint implementation of optical functions constituting the whole silicon photonics toolbox such as optical resonators and filters [18], input/ouput (I/O) couplers [19, 20], high-speed modulators [21, 22], Si-Ge photodiodes [23], and wavelength (de)multiplexers [24]. This results in a *win-win* approach, where mature high-throughput CMOS technology is combined with the excellent optoelectronic behavior of III–V materials. The heterogeneous integration of group III–V semiconductor alloys on silicon offers then an approach where efficient light amplification is achieved within the III–V direct-bandgap layers, while optical functions are implemented in the passive silicon architectures. Within this mind-set, laser sources can be fabricated on silicon and optical carriers can be generated for data transfer at very high bit rates [25, 26].

In the following sections, the reader is swiftly introduced to the chapter core concerning the III–V heterogeneous integration onto micro-nano-patterned silicon via molecular bonding for the realization of power-efficient high-performance *III–V-on-silicon* lasers. The different device architectures made through this common fabrication strategy are investigated and discussed.

8.3 III–V-on-Silicon Lasers

A laser source is the combination of a light emitter medium—called gain medium—and a light-confining structure retaining photons in proximity to the gain medium in order to promote stimulated photon emission. The essential challenge common to the entire world of micro-nano-photonics, lies in achieving light confinement within the smallest possible space during the longest time duration to allow the production of highly compact energy-efficient photonic devices such as microlasers. This is indeed very challenging due to the natural propensity of light to escape freely because of its dual nature of tiny massless photon particles and electromagnetic waves at optical frequencies. However, as electron transport in solids is governed by the atomic scale architecture, in a similar way a periodic patterning of the optical medium at light-wave scale can be used for molding the light flow.

Silicon-based compounds are commonly used in commercial optical waveguide devices for applications such as passive optical interconnects and biomedical sensors. However, the integration of lasers together with Si integrated photonic-electronic circuits has proved to be much more challenging. As we have seen in the previous section, the complexity lies in the fact that silicon is a poor light-emitting material due to its indirect energy bandgap. In addition, the direct growth of standard III–V materials on Si substrates is still a major obstacle because of the mismatch in lattice constants and in thermal expansion coefficients [27].

Although light emission from silicon is not straightforward, the development of an efficient electrically pumped laser is essential to make silicon the material of choice for full optoelectronic integration.

A different way consists of coupling laser beams emerging from III–V heterostructures to silicon waveguides. This so-called hybrid integration can be done using different techniques like flip-chip bonding [28] or self-assembly [29]. Both approaches present the disadvantage of requiring submicron precision alignment to enable efficient coupling between lasers and silicon waveguides. Even if the cost of a silicon photonic circuit is generally low, aligning precisely a laser chip to a planar photonic circuit is quite expensive, time consuming, and unsuitable for high-volume fabrication.

A particularly promising approach instead is based on molecular bonding of III–V materials on top of a patterned SOI substrate [30]. and can be performed at the die or te wafer level. Then, hybrid Si/III–V lasers are realized following a collective fabrication procedure, enabling complex photonic integrated systems onto the silicon platform [31]. Using this technology, Fabry–Pérot [32, 33], racetrack [34], and DFB lasers [35] were demonstrated. Very often, the bonded structure is designed to support a common optical mode— that we will later define as *supermode*—whose electromagnetic field is distributed between the III–V structure and the underlying Si waveguide. In fact, the major part of the field is located in the silicon waveguide and only a few percent, that is, the tail of the optical mode, overlap the multiple quantum wells of the III–V active region. By doing that the laser mode is then mainly concentrated in the passive silicon waveguide to the detriment of the modal gain. Another difficulty of such an approach is the tight control of the low-index bonding layer, whose thickness must be kept as thin as possible, typically lower than 10 nm. These technological constraints constitute an important limiting factor at the design stage as well as during the fabrication process, impacting laser reproducibility and performance stability. Therefore, although these innovative, evanescently coupled laser structures show great integration potential, such solutions do not use III–V materials as their foremost efficiency. This ultimately results in the inevitable trade-off between the modal gain—the supermode confinement factor in the III–V waveguide—and the output coupling efficiency in the silicon passive circuitry, impinging on overall output power available for on-chip and out-of-chip light routing to I/O components to be fed at the system level.

More recently, electrically pumped hybrid laser prototypes such as DBR- and DFB-based emitters have been demonstrated by INTEL Corp., UC Santa Barbara [36], and Aurrion Inc. [37], respectively. In all these approaches, the cavity optical mode is again mainly confined within the passive silicon waveguides, being just evanescently coupled in the III–V mesa for light amplification. This restrains the modal gain available to optical modes cycling in the resonators, resulting in the trade-off between coupling efficiency of III–V to the Si waveguide and power output levels.

Only in 2013, Kotura Inc. (now Mellanox Technologies Ltd.) and Oracle Labs have proposed a different solution by butt-coupling a III–V reflective SOA chip to a passive 3 μm thick Si waveguide containing a DBR [15]. Even if this demonstrator shows excellent performance in terms of modal selection (SMRS > 45dB), wall-plug efficiency (9.5%), low threshold (20 mA), and output power (6 mW), the technological processing of butt coupling suffers from stiff and time-consuming alignment requirements, showing weak wafer-scale integration potential as well as impinging on cost-effectiveness.

As mentioned earlier in the present chapter, the heterogeneous integration of III–V materials on silicon and SOI platforms draws a strong advantage from both the silicon technological maturity issuing from the CMOS fab environment and the excellent light amplification properties of III–V compounds. Along with the well-known inherent processing stability, repeatability, and good etch and surface quality, CMOS-compatible materials such as silicon, silicon dioxide (SiO_2), and silicon nitride (SiN_x, Si_3N_4) bring in the notable key factor of exhibiting a high-index contrast in their effective indices in all main datacom and telecom wavelength bands. Primarily, the use of high-index-material systems such as Si/SiO_2 permits the design of waveguides and I/O components where the optical mode is tightly confined within submicrometer silicon cores. This allows shrinking down of the silicon photonic circuit's footprint, therefore enabling the increase in the density of optical functions, including modulation, wavelength (de)multiplexing, optical signal filtering, and detection, which are available on-chip.

As a second point, the well-controllable micro-nano-patterning of silicon endows device designers with the capability to fashion at will the optical medium, thus contributing to achieve a fully deterministic mold of light at the wavelength scale. This fact highlights once again the robustness of the strategic pillar holding up the paradigm shift brought about silicon photonics. It is in fact not just a question of reproducing the silicon maturity on optoelectronics materials and devices. It is instead mainly about *siliconizing photonics*, thus providing photonic devices with the whole potential offered by silicon maturity. Therefore, low-cost, high-throughput fabrication of silicon photonic chips featuring complex highly dense functionality using power-efficient components can be integrated into small-footprint systems and modules.

The approach developed in our group at CEA-Leti exploits the adiabatic tapering of SOI [38] as well as III–V [39] waveguides in order to maximize the optical mode amplification in the III–V waveguide, while maintaining a high coupling efficiency of photons to the silicon wires underneath, yielding significantly higher output power available for on-chip signal processing at low current threshold levels.

In the following sections, we report on the front-to-back-end fabrication roadmap of the III–V-on-Si heterogeneous integration as well as on experimental demonstrations of electrically driven hybrid Si/III–V lasers, including DBR and DFB emitters in the C-band, which are based on the supermode control of a two-coupled waveguide system [40–42]. The proposed architectures definitely overcome the aforementioned trade-off inherent to the Si evanescent lasers previously reported. The novelty consists of fashioning of the optical supermode along the cavity length to obtain a strong overlap with the gain region (rather than the evanescent tail), ending up with a larger gain available for amplification and maintaining a high coupling efficiency with the bottom silicon waveguide, thus preserving the integration potential as well as thermal and power consumption budgets.

8.3.1 *III–V-on-Si Integration for Hybrid Lasers*

As previously stated, the heterointegration of III–V materials on silicon and SOI platforms draws an advantage from CMOS fab maturity. In this sense, the electronics/photonics convergence can be played along different strategic roadmaps such as flip-chip bonding [28] or self-assembly [29]. However, the limited integration potential of these approaches highlighted the need for a cost-effective very-large-scale integration (VLSI) with micro-nano-electronic technology nowadays without impinging on the quality of the III-/V-based optoelectronic components.

Alternatively, 3D photonic integration is inherently *integration-friendly* as III–V/Si photonics layers are vertically distributed, where needed, onto SOI electronics layers, providing high-throughput fabrication, cost-effective yields, and greater functionality.

Basically, our approach stems from the use of low-temperature molecular wafer bonding of III–V epilayers to micro-nano-patterned silicon and it is composed of three main processing phases. In brief, as illustrated in Fig. 8.1, III–V-on-silicon lasers processing starts from the patterning of a 200 mm SOI wafer by means of standard CMOS-compatible tools, including 193 nm deep ultraviolet (DUV) lithography and HBr-based dry reactive ion etching (RIE) for silicon patterning. This first phase of the fabrication aims at the realization of the whole silicon passive circuitry, including optical resonators, waveguides, and I/O components such as wafer-to-fiber couplers and wavelength (de)multiplexers. Subsequently, processed SOI wafers are planarized via a chemical mechanical polishing (CMP) step prior to the III–V epilayers to SOI molecular wafer bonding, which is carried out at relatively low temperatures (200°C). Bonding yields well above 90% over a full 2-inch-wide III–V layer surface have been achieved, as shown in Fig. 8.1d. In detail, strict control over the bonded surface particle and carbon contaminations is essential to achieve such high III–V on silicon bonding yields. Moreover, a III–V low epitaxial defect density ($<10/in^2$) and substrate bow (<8 μm over 2-inch wafers) are necessary as well. The InP substrate of the III–V epilayer is then removed by H_2O/HCl-based wet etching, thus ending the second phase of the process flow.

Heterogeneous integration

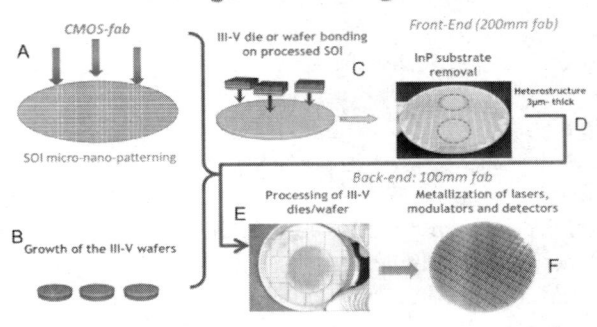

Figure 8.1 Front-to-back-end heterogeneous integration processing for III–V-on-silicon lasers. (a) The micro-nano-patterning of a 200 mm SOI wafer is accomplished via CMOS fab tools for the realization of passive and active silicon-based components. (b) In the meanwhile III–V epitaxies are grown prior to bonding, (c) III–V epilayers to SOI bonding and InP substrate removal, (d) processing of the III–V on SOI proceeds for implants and contact definition, and (e, f) the process ends at the 100 mm scale with the metallization of active components, including heaters, modulators, and photodetectors.

Fabrication is then finalized at the 100 mm scale through standard III–V optoelectronics technological steps, including III–V mesa etching, and n- and p-type ohmic contacts definition.

Concerning more specifically the silicon fab depicted in Fig. 8.2, the fabrication approach proceeds from shallowly etched structures to deeply etch features, and it usually starts from the definition of the laser cavity optical resonator.

Depending on the laser architecture, the depth of this first etch step varies as a function of the targeted silicon patterning for the definition of the optical resonator. As specified in Fig. 8.2, the silicon patterning can be adapted on the need for realizing DBR reflectors for Fabry–Pérot resonators or equivalently the grating of a DFB laser or even the highly reflecting mirror of a vertical-cavity surface-emitting laser (VCSEL), as will be presented in the proof-of-concept section.

After the resonator definition, the following lithography etch steps include the patterning of waveguide-to-fiber couplers (125 nm

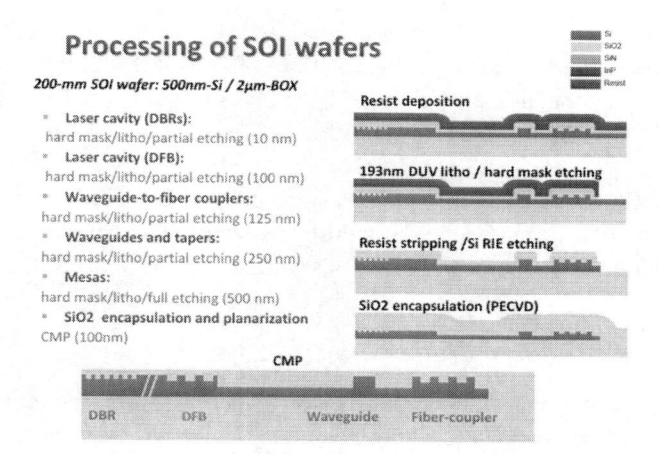

Figure 8.2 Front-end CMOS fab of 200 mm wide, 500 nm thick SOI wafers prior to III–V bonding. Optical cavities of laser resonators are defined through the shallow etch of DBR reflectors (10 nm deep) or more strongly corrugated DFB gratings (100 nm deep etch). Waveguide-to-fiber couplers are realized at the 125 nm etch depth level, while silicon waveguides and tapers are etched down to 250 nm. Finally, silicon is etched till the buried oxide (500 nm deep) for mesa patterning, while a chemical mechanical polishing step planarizes the wafer prior to III–V bonding.

deep), followed then by adiabatic taper transitions (250 nm deep), while the silicon mesa definition is obtained via a full etch to buried oxide. Finally, plasma-enhanced chemical vapor deposition (PECVD) of silica followed by CMP concludes the first phase of the fabrication front end by planarizing the patterned SOI wafer containing silicon passives prior to the bonding of III–V epitaxies for the heterointegration of gain areas.

Concerning the back end of the III–V-on-SOI fab current injection is realized by using n- and p-type intracavity contacts and a proton-implanted 3 μm wide current aperture. To prevent any defect induced by the local H^+ implant from reaching the active region—where it causes nonradiative recombinations—an 1800 nm thick p-doped, graded InP confinement layer is included in the III–V epilayer. The n-type ohmic contact metallurgy consists in the Ni 5 nm/AuGe 30 nm/Ni 5 nm/Au 200 nm stack deposited by electron beam sputtering and is then followed by rapid thermal annealing (RTA) at 380°C under N_2 flux. Ohmic p-type contact metallurgy is instead based on a Pd 10 nm/Ti 30 nm/Pt 80 nm/Au 200 nm stack. A 10 nm thick Ti layer covered by 1 μm of gold has been used for the subsequent contacts' thickening and widening. The III–V mesa definition is obtained with a mix of both wet- (H_2SO_4:H_2O_2:H_2O = 1:1:10 for InGaAsP QWs) and RIE-based (CH_4/H_2 for InP) techniques, resulting in controlled InP surface roughness and improved etching directivity for correct contact insulation and reduced access resistance as low as a few ohms.

8.3.2 *Distributed Bragg Reflector Lasers*

This section reports on DBR hybrid silicon/III–V lasers based on adiabatic mode transformers. The hybrid structure is formed by two vertically superimposed waveguides separated by a $\sim$100 nm thick SiO_2 gap layer. The top waveguide, fabricated in an InP-/InGaAsP-based heterostructure, serves to provide optical gain. The bottom Si waveguide system, which supports all optical functions, is constituted by two tapered rib waveguides (mode transformers), two DBRs, and a surface-grating coupler. The supermodes of this hybrid structure are controlled by an appropriate design of the tapers located at the edges of the gain region. In the middle part of the device almost all the field resides in the III–V waveguide so that the optical mode experiences maximal gain, while in regions near the III–V facets, mode transformers ensure an efficient transfer of the power to silicon waveguides. In brief, such III–V on SOI hybrid laser achitecture is capable to operate under continuous wave regime, showing a room-temperature threshold current of 17 mA, uncooled output power levels of 14 mW, a side-mode suppression ratio (SMSR) well above 50 dB, and a small signal of −3 dB modulation bandwidth above 7.5 GHz.

In detail, an example of the hybrid structure is shown in Fig. 8.3: the Fabry–Pérot laser cavity is formed by two vertically superimposed waveguides, a top III–V waveguide for light amplification, while a bottom Si waveguide system is constituted by two tapered rib waveguides (mode transformers), a wafer-to-fiber coupler, and two DBRs. The optical cavity is defined by the two DBRs spaced 1040 μm apart, while the mode transformers, positioned below the edges of the active waveguide, provide an

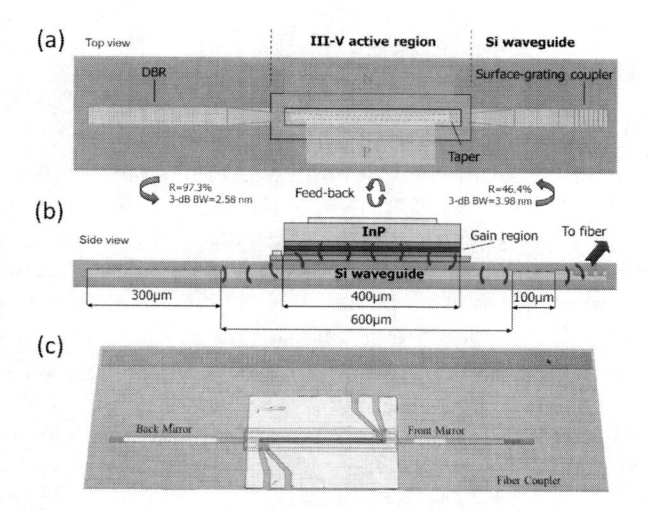

Figure 8.3 Sketch of a III–V-on-SOI DBR laser emitter, including both top (a) and side (b) views. The III–V waveguide meant for light amplification is superimposed over a silicon rib tapered waveguide enabling adiabatic coupling from/to the upper waveguide, light feedback through side DBRs, and light collection to fiber via a surface grating coupler. (c) Contact pads are designed in the typical ground-signal-ground (GSG) configuration for radio frequency (RF) on-off keying (OOK) high-speed modulation.

adiabatic transition by varying the width of the Si rib waveguide. In the middle part of the device, almost all the field resides in the III–V waveguide so that the optical mode experiences maximal gain, while in regions near the III–V facets, mode transformers ensure an efficient coupling between the two levels. The thickness of the bonding interface allows for relaxing fabrication constraints typical of evanescently coupled structures where high control over the thin bonding layer (~5 nm) is required. The 6 μm wide top waveguide is essentially made up of an InGaAsP-/InP-based separate confinement heterostructure (SCH) for carrier confinement and light amplification, while the resonator architecture is defined by a 600 μm long silicon rib waveguide endowed with adiabatic tapers for mode coupling to the III–V waveguide. The Si rib waveguide is terminated at both ends by DBRs with tailored modal reflectivities and a surface grating for fiber-coupling measurements of transverse electric (TE)-polarized light. The shallowly etched 10 μm wide DBRs are characterized by a 10 nm etch depth and a 50% fill factor. Optimized for achieving enhanced modal control and large a SMSR, the two DBRs are 300 μm and 100 μm long with a grating strength [36, 42] $\kappa = 83$ cm^{-1}, showing modal reflectivities of 97.3% and 46.4%, along with 3 dB bandwidths of 2.6 nm and 4 nm, respectively.

The physics of the adiabatic coupling scheme adopted to transfer power between the III–V active waveguide and the silicon waveguide underneath is illustrated in Fig. 8.4. The steady-state finite-element method (FEM) shows the adiabatic coupling kinetics of the supermode inside the hybrid DBR-based emitter when using a 100 nm thick silicon dioxide gap between the active III–V and the silicon waveguide. Figure 8.4 illustrates the electric field patterns (quasi-TE case) for various cross sections of the mode transformer. The tapered waveguide is defined by a rib height and a slab height of 500 nm and 250 nm, respectively. The width of the rib W is widened adiabatically from 0.8 μm to 1.1 μm. This variation of the rib width is carried out on a length of 120 μm. The shape of the mode transformers is optimized to be robust enough with respect to the variations induced by the fabrication processes and particularly the thickness of the oxide separation layer. For

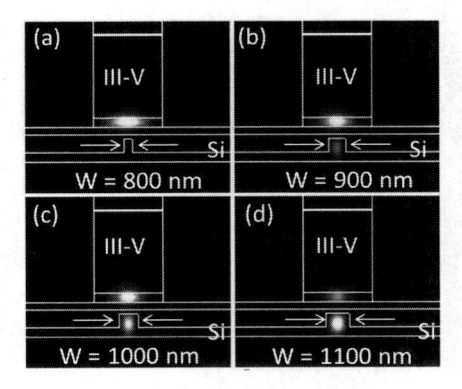

Figure 8.4 Evolution of the lasing supermode inside the hybrid III–V-on-Si Fabry–Pérot cavity. (a) Supermode profile in the III–V section of the structure. (b) Mode transformer: transfer of the supermode power by adiabatically widening the width W of the silicon waveguide. (c) Supermode profile at the phase-matching condition between III–V and Si waveguides. (d) Power transfer to the silicon waveguide toward the end of the taper.

specific configurations, coupling efficiencies higher than 97% were calculated. In addition, if we take into account fabrication process tolerances, the lower limit is found to be ~90%. As illustrated in the Fig. 8.4a,d insets, more than 90% of modal power is coupled back and forth from/to the silicon waveguide to/from the III–V SCH. This ensures efficient amplification of optical modes in the active region aiming at minimizing lasing thresholds, while keeping constant outcoupled power levels in the silicon level.

Enlarging now the view to a modal analysis of the DBR-based resonator, the intrinsic length (above 1 mm) of such a DBR-based Fabry–Pérot cavity brings in the additional issue of controlling the spectral purity of the lasing emission. In detail, the Fabry–Pérot phase condition for resonance depends on the resonator physical length L, the optical mode group index n_g, and the emission wavelength λ, and it is expressed as follows:

$$T_{FP} = \frac{1 - R_1 R_2}{1 + R_1 R_2 - 2\sqrt{R_1 R_2} - 2\sqrt{R_a R_2}\cos\left(\frac{2\pi}{\lambda}2Ln_g\right)}, \quad (8.1)$$

where R_1 and R_2 are, respectively, the power reflectivity of the first and second DBR and T_{FP} is the overall transmission of the Fabry–Pérot.

This results in a dense distribution of longitudinal modes admitted in the cavity as the free spectral range (FSR) reduces for increasing effective cavity lengths, as described in the following equation:

$$FSR = \frac{\lambda^2}{2Ln_g}, \quad (8.2)$$

where c is the light velocity, L is the resonator length, and n_g represents the modal effective index.

The reduced FSR in such architectures (FSR = 0.32 nm computed for an average group index of 3.62) translates into a dense modal frequency comb. Moreover, due to the wideband spectrum over which light amplification is provided—that is, the quantum well photoluminescence emission spectrum—as well as the similar confinement factors shared by all these longitudinal modes, the DBR emitters are affected by an inherent multimode behavior, therefore impinging on the modal selectivity of the laser. However, spectral purity achieved through modal selection in laser sources

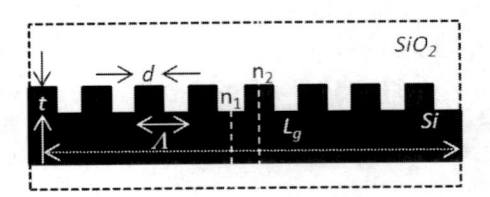

Figure 8.5 Architecture of a high-index-contrast grating realized in silicon and surrounded by a silica matrix realized for waveguided operation. The corrugation strength κ constitutes a key physical constant governing the optical behavior of the grating, including its reflectivity and bandwidth.

is an essential feature for telecom and datacom systems requiring SMSRs well above 30 dB as well as stable mod-hop-free operation. Therefore, the filtering of the dense spectral comb arising from longitudinal modes appearing on the lasing spectrum of such emitters is a mandatory objective to be achieved to get laser sources compatible with telecom standards, and then featuring commercial product compatibility.

As mentioned earlier in the chapter, the high-index contrast of an SOI material system combined with the precise patterning enabled by CMOS planar technology offers us a powerful tool to achieve fine tailoring of the light wave at the wavelength scale. As illustrated in Fig. 8.5, though the geometry of high-index-contrast gratings is essentially defined by the grating pitch (d), period (Λ), and etch depth (t), their optical behavior can be fully described by the so-called grating strength or corrugation factor κ. Such a general constant can be expressed as

$$\kappa = \frac{\pi n_{eff}}{\lambda_0} \frac{\int \int_\Omega (n_1^2 n_2^2 E^2 dx dy)}{\int \int_\infty^\infty E^2 dx dy}, \tag{8.3}$$

where $n_{eff} = (n_1 - n_2)d\Lambda + n_2$ corresponds to the grating effective index and E is the electric field component in the unperturbed region. The κ factor governs the definition of the reflectivity (Eq. 9.4) and bandwidth (Eq. 9.5) exhibited by such high-contrast corrugation for a guided mode travelling across the grating and they can be written as

$$R = \frac{|\kappa|^2 \sinh^2 s L_g}{s^2 \cosh^2 s L_g + \left(\frac{\Delta\beta}{2}\right)^2 \sinh^2 s L_g} \tag{8.4}$$

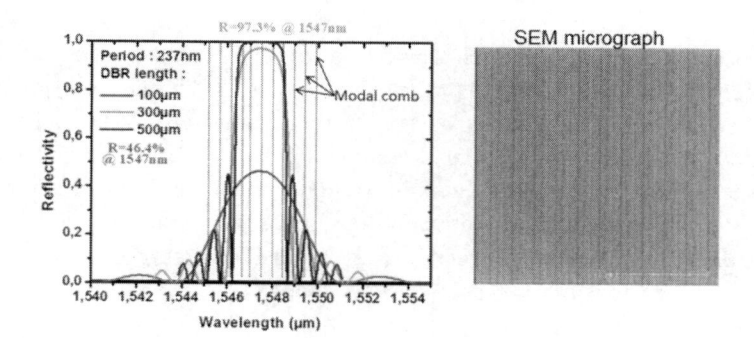

Figure 8.6 (Left) Reflectivity modal response of first-order 1550 nm centered distributed Bragg reflectors (DBRs) using a 10 nm deep shallow etch resulting in a κ factor of 83 cm^{-1}. It can be noted how longitudinal modes in the resonator build up into a frequency comb with a free spectral range of 0.32 nm: the DBR's spectral response is thus tailored through its length to filter out the comb selecting one mode only, preventing multimode lasing. (Right) A scanning electron microscope (SEM) view of the shallowly etched DBRs.

$$\delta\omega_{DBR} = 2v_g\sqrt{|\kappa|^2 + \left(\frac{\pi}{L_g}\right)^2} \qquad (8.5)$$

where $\Delta\beta$ indicates the propagation constant mismatch between the etched and unetched regions of the grating, $s = \sqrt{\kappa^2 - \Delta\beta^2}$ expresses the coupling constant between unperturbed and perturbed modes via the corrugation strength, L_g is the grating length, and v_g is the group velocity.

In this architecture, the κ corrugation factor as well as the frequency response of the DBR can be adjusted at will in order to implement a meticulous filtering of the laser emission spectral comb. This results in single-mode emitters featuring SMSR values well above 50 dB, thus complying with datacom standards and industrial roadmap needs, followed later by product development and commercialization. As shown in Fig. 8.6, the DBRs are designed to be asymmetric in length so that their superposed modal response results in the filtering of a single longitudinal mode out of the spectral comb supported by the Fabry–Pérot resonator.

It is worth to point out once again that such capability is fully relying on the capability provided by CMOS fabrication tools to precisely pattern silicon both in vertical as well as in azimuthal orientations, as DBR etch depths as low as 10 nm ensure a full control over their modal reflectivity and bandwidth for a given grating length L_g. In conclusion, starting from a given corrugation strength of the DBR grating, by choosing the appropriate length of the reflectors, this approach allows high modal selectivity through asymmetric DBR bandwidths. In the same way, by finely varying the DBR grating period a precise wavelength addressing of the DBR-based hybrid sources is achieved.

Static and dynamic characterization of the hybrid DBR laser via electrical probing was performed, collecting the optical signal through a multimode fiber aligned over the laser grating coupler and connected to an optical spectrum analyzer and a power meter. As reported in Figs. 8.7 and 8.8, the III–V-on-SOI source shows continuous-wave lasing up to 65°C with a minimum current threshold (I_{th}) of 17 mA and a maximum uncooled output power above 15 mW at 160 mA of driving current, resulting in a remarkable differential quantum efficiency of 13.3%. In addition, the laser diodes are characterized by a turn-on voltage of 1 V and a series resistance of 7.5 Ω. An SMSR greater than 50 dB at high driving current is achieved, as shown in Fig. 8.9 (left).

Concerning dynamic performances, the electro-optic (EO) small-signal modulation response obtained for driving currents between $6I_{th}$ and $8I_{th}$ indicates a 3 dB bandwidth above 7.2 GHz (Fig. 8.10a), while open-eye diagrams reported in Fig. 8.10b,c confirm 5 Gb/s and 12.5 Gb/s OOK direct modulation at room temperature and 160 mA of bias current in a back-to-back configuration for a $2^{15} - 1$ pseudorandom bit sequence (PRBS) with 17 mW RF power [42].

The need for higher aggregate bandwidths through coarse and dense wavelength division multiplexing (WDM) demands the realization of wavelength-selective lasers as well as on-chip tunable sources [19, 42]. In this sense, hybrid III–V-on-SOI laser arrays have been conceived for a four-channel 12 nm spaced course wavelength division multiplexing (CWDM) in the 1.55 μm band. A fine tailoring of DBRs period is used for selecting longitudinal modes and achieving precise wavelength addressing. Experimental results reported in Fig. 8.9 (right) show a 4-λ array with an average SMSR over 40 dB. A richer wavelength availability on the chip can

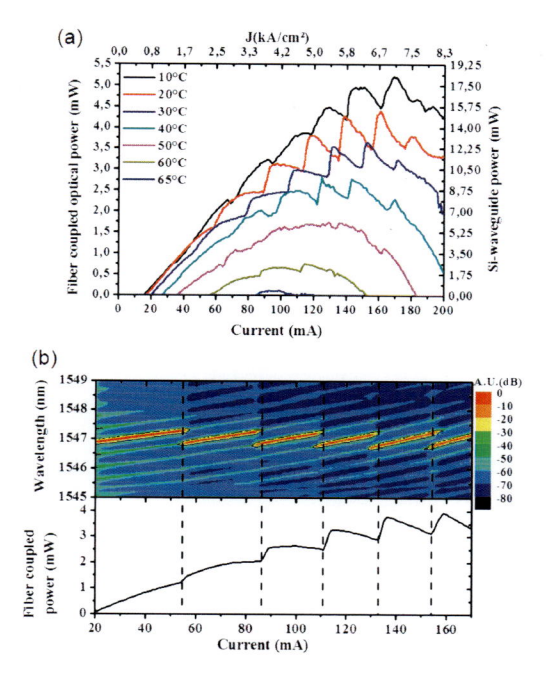

Figure 8.7 (a) *L–I–V* curves at different stage temperatures reporting optical power in both the silicon waveguides (right) and the optical fiber (left) coupled via a surface grating. Values for the waveguide-coupled output power have been normalized with respect to a 500 μm long and 10 μm wide reference waveguide endowed with test grating couplers. (b) Fiber-coupled *L–I* characteristics and lasing spectra behavior contour map. Mode hopping between the different longitudinal modes at higher driving current is due to thermo-optic effect taking place in the laser diode.

be obtained also via emission frequency tuning. By adding on the highly reflective DBR a resistive NiCr film as a heater element, peak reflectivity and phase-matching condition of longitudinal modes are varied, resulting in a continuous-wavelength tuning range of over 20 nm, as shown in Fig. 8.11 (right). Besides the satisfactory performance of these early stage demonstrators further design improvements are needed to enhance tuning span and reduce thermal budget to target maximum operation temperatures at 80 °C and beyond.

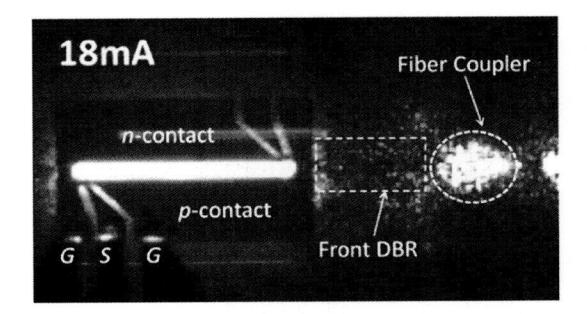

Figure 8.8   Near-infrared camera image of the hybrid laser just above the threshold (I_{th} = 18 mA), showing the fiber coupler (FC), front DBR, and III–V regions. Ground-signal-ground probes were used for RF measurements.

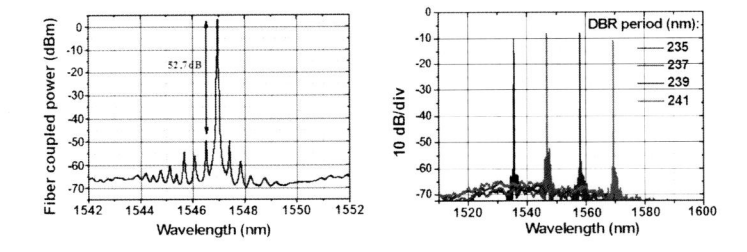

Figure 8.9   (Left) Fiber-coupled lasing spectrum of a DBR-based hybrid III–V-on-Si emitter showing more than 50 dB of SMSRs. Such modal selectivity has been achieved by using two asymmetric DBRs exhibiting the same grating corrugation strength but different lengths in order to pick up one of the modes in the laser frequency comb. (Right) Wavelength addressing through a fine DBR period tuning: a 4 × λ 12 nm spaced array of DBR-based emitters is targeted for course wavelength division multiplexing (CWDM) applications.

8.3.3 *Distributed Feedback Lasers*

The high-index-contrast patterning of silicon can be declined over various architectures of the silicon photonics toolbox, such as waveguides, splitters, couplers, and gratings. In this sense, we have seen that high-index-contrast, shallowly corrugated gratings can be used as DBR reflectors in hybrid Fabry–Pérot lasers, ensuring high modal selectivity, low threshold components, and high-throughput performance by the use of standard CMOS pilot lines.

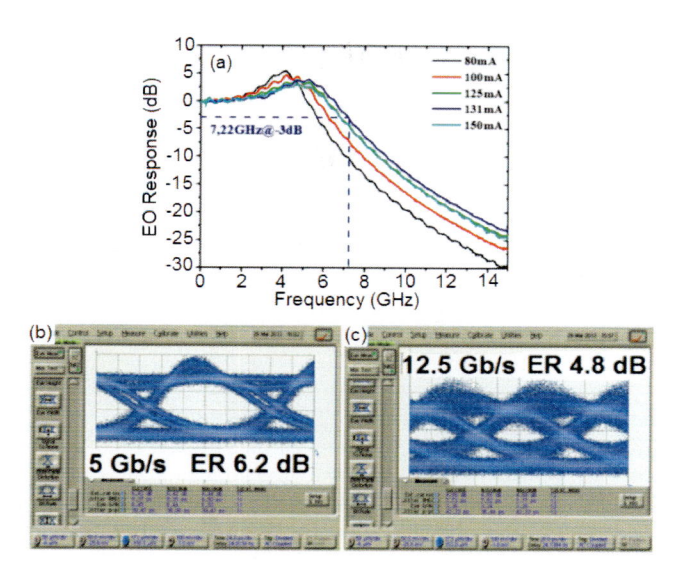

Figure 8.10 (a) Small-signal modulation response of the hybrid DBR laser at RT under different bias conditions. A 3 dB bandwidth above 7.2 GHz is obtained for a driving current of 131 mA. Open-eye diagrams in back-to-back configurations at 5 Gb/s (b) and 12.5 Gb/s (c) are achieved with extinction ratios (ERs) of 6.2 dB and 4.8 dB, respectively.

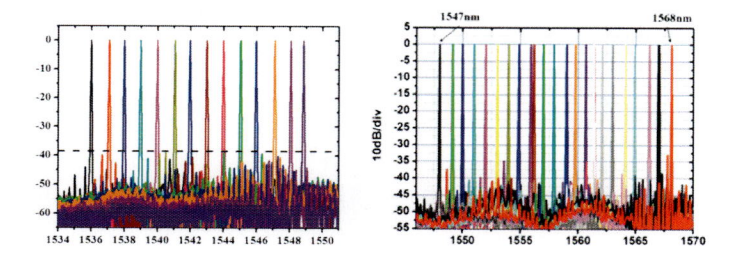

Figure 8.11 (Left) Wavelength tuning has been also implemented in early-stage demonstrators by adding a resistive NiCr film heater on the back DBR. A 21 nm wide continuous tuning range is achieved (right).

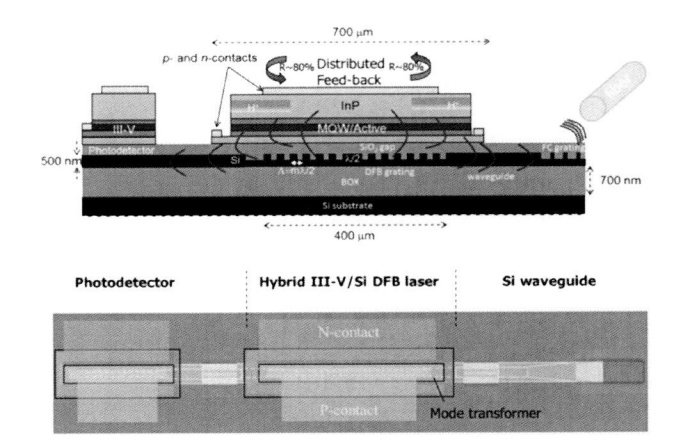

Figure 8.12 Cross-sectional and top views of a III–V-on-SOI distributed feedback (DFB) hybrid laser. Relying on the adiabatic coupling scheme, the optical mode is amplified in the III–V active region, while the Si/SiO$_2$ DFB grating provides modal reflectivity to reach the lasing threshold. A fiber coupler (FC) and a III–V photodiode are also integrated for wafer-to-fiber and wafer-to-waveguide test measurements, respectively.

Analogously, more strongly corrugated silicon gratings can be used for implementing more compact resonators designs such as, for example, DFB lasers as well as VCSELs, endowing the hybrid III–V/SOI laser integration platform with access to power-efficient sources endowed with a fully scalable footprint.

As illustrated in Fig. 8.12, a DFB grating is etched into the bottom-level silicon waveguide, providing enough modal reflectivity to achieve lasing at relatively low threshold, while a $\lambda/4$-long defect at the center of the DFB grating ensures modal selection and precise wavelength addressing. Moreover, by relying on the aforementioned adiabatic coupling scheme, the optical mode is amplified in the III–V active region, being only evanescently coupled to the Si/SiO$_2$ DFB grating. A fiber coupler (FC) and a III–V photodiode are also integrated, respectively, for wafer-to-fiber and wafer-to-waveguide test measurements. The optical behavior of the laser as well as its perfomance are strongly determined by the DFB grating design and its coupling to the supermode in the III–V region. It follows that grating dimensions and corrugation strength factors, as well as the silicon dioxide gap between the silicon and the III–V waveguides, are crucial for the laser design, as these parameters strongly impact the coupling constant between the two superimposed waveguides, therefore shaping the taper transition region and the outcoupled power efficiency down to the silicon level.

In detail, the DFB grating corrugation brings in two different effective indices for the coupled III–V/Si waveguides system. Such effective index patterning of the supermode at a given target wavelength λ results in a DFB grating coupling constant that can be expressed as follows:

$$\kappa_{DFB} \approx \frac{2\Delta n_{eff}}{\lambda}, \tag{8.6}$$

where Δn_{eff} indicates the perturbation in the effective index of the coupled waveguide system supermode implemented by the grating. On the other hand, the reflectivity R provided by the grating at the Bragg wavelength is described as

$$R = \tan h^2(\kappa_{DFB} L_{DFB}), \tag{8.7}$$

with L_{DFB} representing the DFB grating length. By considering typical grating lengths between 500 μm and 1000 μm leading to κ_{DFB} values ranging from 10 cm^{-1} to 30 cm^{-1}, and modal reflectivities between 65% and 80%, we may conclude that a feasible value for the κL factor is approximately comprised between 1 and 1.4.

In Fig. 8.13 (left), we report the DFB grating reflectivity as a function of both grating length and coupling strength, the coupling strength being dependent on the III–V/Si supermode overlap over the grating when taking into account a 100 nm thick silica gap. In Fig. 8.13 (right), the coupling strength factor of a DFB grating as a function of the silicon waveguide width and thickness highlights how a trade-off condition between enough reflectivity, *modal selection*, and architecture footprint is found for κL between 1 and 1.4.

The fabrication process flow follows the one outlined in Section 8.3.1, while DFB gratings have been defined by both 193 nm deep DUV and electron beam lithography (EBL), reaching similar results. Scanning electron microscope (SEM) images of the DFB grating patterning are reported in Fig. 8.14: It can be noted how a very good side-wall verticality is achieved, although the grating etch

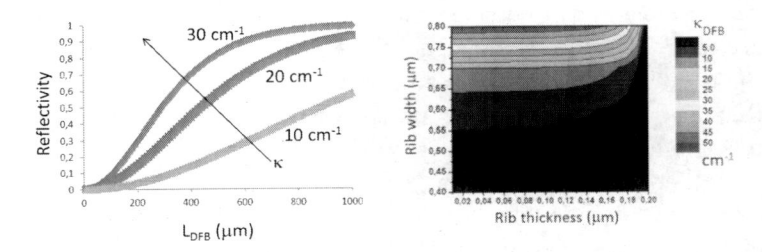

Figure 8.13 DFB grating reflectivity (R) as a function of its length and corrugation strength expressed in cm^{-1}. It can be noted how a good trade-off between enough reflectivity and the architecture footprint is usually stated in the interval $\kappa L = 1 - 1.4$. (Right) Corrugation strength of the DFB grating as a function of both silicon waveguide rib thickness and width, when using a silica gap of 100 nm. A good compromise in this case is found for a Si rib above 0.7 µm in width.

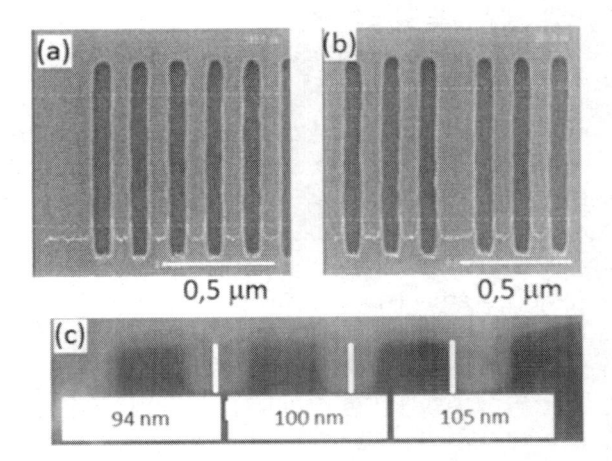

Figure 8.14 Scanning electron microscope image of the first-order DFB grating patterning into silicon after electron beam lithography and 100 nm deep dry etch. (a) Proximity effects at the end as well as at the center (b) of the grating have been optimized. (c) Focused ion beam (FIB) cross section shows a slightly varying etch depth along the grating length, while a very good side-wall verticality is achieved.

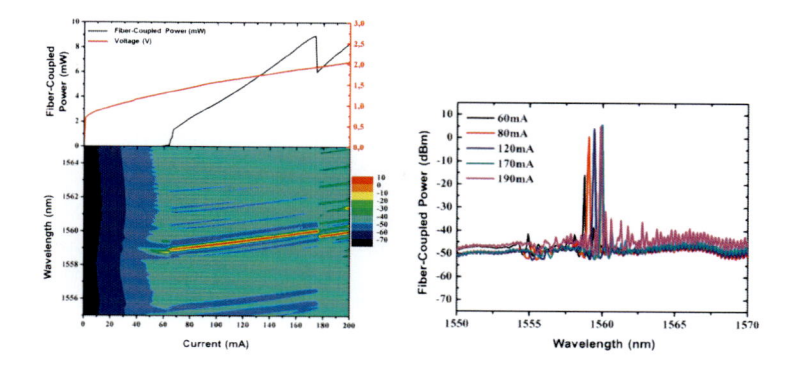

Figure 8.15 $L-I-V$ characteristics of the III–V-on-SOI DFB laser. (Left) The device shows room-temperature continuous-wave single-mode operation with current thresholds below 60 mA and output power levels coupled to test fiber above 6.5 mW for 120 mA current drive. (Right) SMSRs well above 30 dB are achieved, as reported in the right uppermost inset, while the III–V photodetectors (bottom inset) measure photocurrents above 25 mA when the DFB laser is driven at 150 mA.

depth slowly varies across the grating length, modifying the III–V/Si supermode coupling strength.

Static measurements of first-run hybrid III–V-on-SOI DFB lasers are reported in Fig. 8.15. By exploiting the collective approach to active devices fabrication made possible by the molecular wafer bonding, both a III–V-on-silicon EAM and a photodetector were introduced in the device architecture, as illustrated in Fig. 8.16. Tests show that the lasers operate in continuous waves at room temperature, showing current thresholds below 60 mA. Output power levels are instead estimated via the photocurrent generated in the photodiode: in detail, the one-sided power emitted by the DFB driven at 150 mA generates a photocurrent of 27 mA. By conservatively assuming that the internal quantum efficiency of the photodiode is 100% and that all light is coupled into the photodetector, we can infer that the total two-sided power emitted by the DFB laser at 150 mA stays well above 40 mW.

The presence of the III–V EAM makes possible it to enhance the small-signal f-3 dB bandwidth of the stand-alone DFB laser, bringing it to 12 GHz, providing a wider bandwidth for high-speed modulation. Static and dynamic characteristics of the EAM

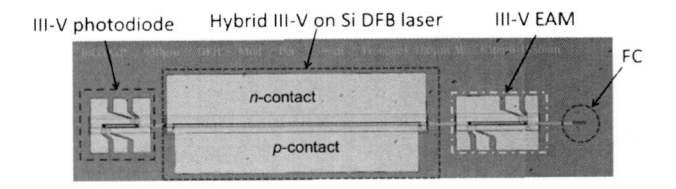

III-V photodiode   Hybrid III-V on Si DFB laser   III-V EAM

FC

Figure 8.16 Microscope image of a hybrid III–V-on-silicon laser modulator photodetector block. A III–V-based electroabsorption modulator (EAM) has been added to the laser building block to improve its aggregate bandwidth by extending it to f-3 dB = 12 GHz.

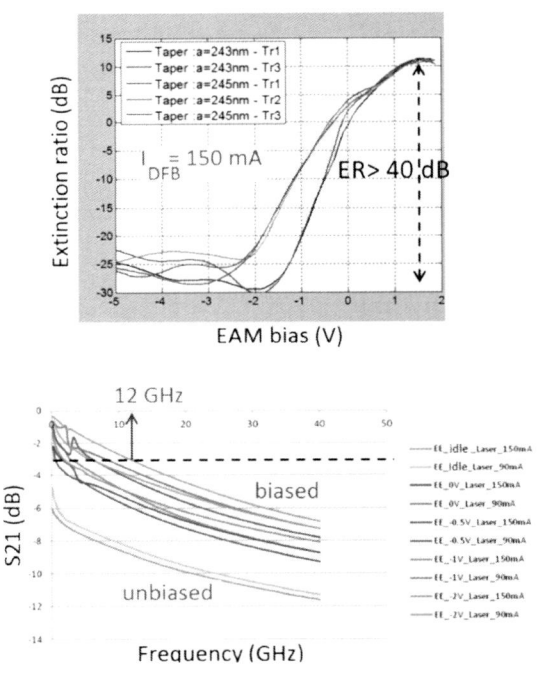

Figure 8.17 Static (top) and dynamic (bottom) analysis of the III–V-on-Si electroabsorption modulator (EAM). On-off keying extinction ratios above 40 dB are achieved for a DFB laser current drive of 150 mA, while a small-signal modulation f-3 dB bandwidth of 12 GHz is obtained when the EAM is polarized at $V_{bias} = -2$ V while the laser drive is set at 90 mA.

modulators polarized with $V_{bias} = -2$ V are reported in Fig. 8.17, highlighting OOK extinction ratios above 35 dB at a DFB laser current of 150 mA and a small-signal cutoff frequency of 12 GHz with the DFB laser current drive set at 90 mA.

8.4 Innovative Proofs of Concepts

As we saw in the previous sections, the heterogeneous integration of group III–V direct-bandgap semiconductors on silicon offers a win-win approach where efficient light amplification is achieved within the III–V direct-bandgap layers, while optical functions are implemented in the passive silicon architectures fabricated by high-throughput CMOS processing tools. Several milestones in such sense have been achieved using direct coupling from III–V waveguides to silicon passive circuits as well as adiabatic mode transforming. However, although such silicon-compatible laser building blocks represent a first viable path toward siliconizing photonics, such sources suffer from several drawbacks such as a still-too-large architecture footprint, moderate bandwidth limitation, and power consumption, as well as the mandatory increase in the passive silicon complexity to achieve the mode-hop-free spectral purity.

An answer to such intrinsic limitations comes from microdisk laser sources. Microdisk resonators are compact ($\sim$10 μm in diameter), thereby enabling small power consumption, while the light wave evanescent coupling to silicon microwaveguides for optical routing proves to be effective as well [43]. Nevertheless, such devices are characterized by an inherent lack of modal selection, demanding to pattern 1D photonic crystals (PhCs) in order to filter out a single color of light [44].

In the next two sections, we will comment on and investigate recent promising solutions consisting of the III–V heterointegration of HCG-VCSELs on SOI.

8.4.1 Integrating VCSEL and Silicon Photonics

VCSELs constitute an almost ideal solution to the aforementioned drawbacks. Since the device was firstly conceived by K. Iga et al. in

1979 [45] VCSEL photonics has conquered the world [46]. VCSELs are nowadays used in a wide-ranging set of developing research fields [47–49] as well as steadily present in several mature industrial applications within consumer optoelectronics, datacoms, sensing, and medicals. Such success is intrinsically related to the several benefits that can be drawn from such photonic architecture: long device lifetime, on-chip wafer testing, cheap modules and package costs, wavelength control, power efficiency, low consumption, high-speed modulation capability, and directive radiation pattern allowing easy coupling to single-mode fibers (SMFs) through thick multimode fibers (MMFs) are just some of the key advantageous features characterizing this device class.

Nevertheless, the small modal gain typical of VCSELs requires the use of highly reflecting mirrors for the vertical confinement in the optical cavity, with typical reflectivity levels well above 99%. However, the monolithic growth of active regions lattice-matched to InP for the near-infrared emission (NIR) requires the growth of very thick, bulky, small-index-contrast ($\Delta n \approx 0.2$–0.3) DBRs characterized by a long photon penetration depth, resulting in reduced modal confinement and higher pumping thresholds. Higher-index-contrast wafer-fused GaAs-based DBRs [50] as well as the use of dielectric materials such as CaF_3/ZnS ($\Delta n \approx 1.1$) [51] represented a viable alternative to replace monolithic InP-based mirrors. Yet, irrespective of the index contrast and materials chosen, the classical DBR architecture was still unable to provide effective solutions for a stable transverse modal and polarization control in this device class. Concerning the first issue, effective-index antiguiding obtained through DBR lateral etching [52] and low-index regrowth [53] results in additional diffraction losses on the optical mode, thereby inducing higher thresholds. As well as that, shallow surface gratings fabricated on the outcoupling region of DBRs [54, 55] provide polarization-mode suppression ratio up to 40 dB. However, such pinning of the emitted polarization is too sensible to the grating architecture parameters and laser operating conditions. Finally, the use of EBL limits the fabrication to a mere prototyping medium-scale paradigm.

Within the past years, subwavelength high-index-contrast metastructures such as 1D and 2D PhCs—or, equivalently, high-contrast gratings (HCGs)—have come to the fore in VCSEL photonics, owing to their unique optical features and enhanced light-harnessing capabilities. Following this trend, the first-ever demonstration of a VCSEL using a double set of HCGs has been achieved by our group in late 2011 [56]. In the past year, tunable VCSEL devices with remarkable spectral purity and output powers operating in the 1550 nm range at 10 Gb/s over 100 km long fibers have been obtained by replacing the top DBR with an InP-/air-based HCG [57]. In mid-2012, surface-emitting optical cavities using a double set of 2D Si-/air-based PhC reflectors [58] have been realized by cost-effective nanoimprint lithography, while experimental results claiming the signature of lasing emission—although when cooled down to cryogenic temperatures—were reported as well [59].

More specifically, our research efforts are oriented to the realization of ultracompact VCSELs as next-generation emitters for silicon photonics applications. This silicon-based photonic building block has been conceived within a large-scale CMOS-compatible processing technology mind-set. Thus, the silicon patterning of HCGs on 200 mm wide SOI wafers aims at both an efficient light harnessing and optimal optical confinement, while the III–V epitaxial layers providing light amplification are wafer-bonded to SOI by state-of-the-art molecular bonding to ensure laser performances, CMOS compatibility, and large-scale low-cost fabrication.

Such novel VCSEL silicon-based architectures prove to be very promising, owing to excellent optical features shown by first-ever optically pumped demonstrators [56]. The device operates in the continuous-wave regime well above room temperature with power thresholds in the submilliwatt range. Moreover, the VCSEL is endowed with remarkable spectral purity enabling single-mode single-polarization lasing, even at very high pumping powers ($\times 10$ P_{th}) over large pumped surfaces (10×10 μm^2). Remarkably, the modal control is exclusively molded by the silicon patterning of HCGs realized by DUV lithography over 200 mm wide SOI wafers [60, 61].

8.4.2 *III–V on SOI HCG-VCSELs*

CMOS-compatible VCSELs using a double set of silicon-made HCGs are sketched up in Fig. 8.18a, while a SEM cross-sectional view of

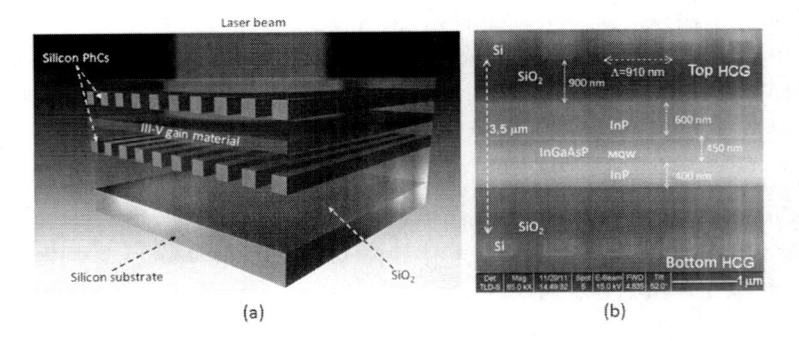

(a)   (b)

Figure 8.18 (a) Schematic view of a III-/V-on-Si VCSEL cavity employing a double set of Si/SiO$_2$ 1D high-contrast gratings (period Λ of 910 nm). The proposed approach makes use of III–V layers for light amplification wafer-bonded to 200 mm wide SOI wafers, the optical cavity being defined by the micro-nano-patterning of silicon layers via 248 nm DUV lithography. (b) Cross-sectional SEM view of the double HCG-VCSEL cavity measuring 3.3 µm × 25 µm in thickness and width, respectively. The bottom HCG is made of crystalline silicon, while the top mirror is obtained from a deposited amorphous silicon layer tailored for minor absorption losses in the wavelength range of interest.

fabricated devices is shown in Fig. 8.18b. Briefly, an InGaAsP-based three-quantum-well active region is embedded between two silica spacers measuring 900 nm each, while the VCSEL cavity is vertically terminated by two 290 nm thick 1D Si/silica HCGs characterized by a 50% fill factor and a 910 nm lattice period, respectively. The 25 µm wide HCG has been designed to enhance its spectral response both in terms of modal reflectivity and optical confinement, aiming at low-threshold VCSEL devices.

The high-index-contrast patterning of the optical medium in such deep-etched gratings allows an almost complete molding of the light wave, which is fully adjustable in both space and spectral domains. Starting from the basics, the light wave is vertically confined within HCGs due to the refractive index contrast between silicon and the surrounding low-index material (silica). More importantly, such strongly corrugated waveguides when operating on band edges of the dispersion curves situated above the light line allow coupling of incoming light to so-called *heavy* photon states, thus enabling an enhanced slowdown of waveguided modes propagating across

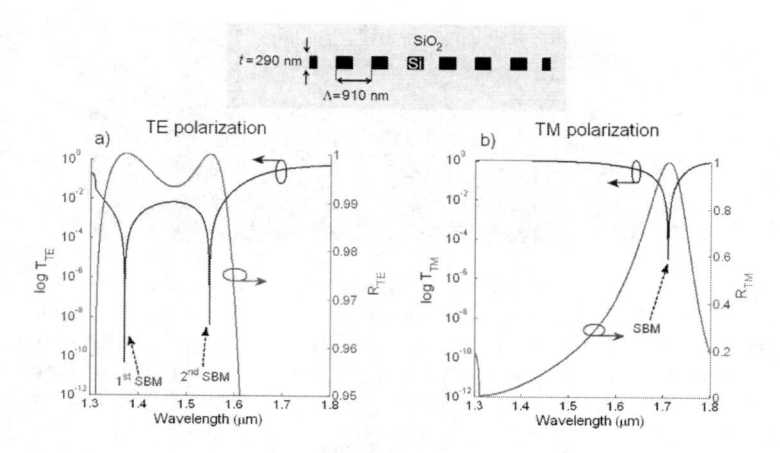

Figure 8.19 (a) The Si/SiO$_2$ HCG is 290 nm thick and shows a period of 910 nm and a silicon fill factor of 50%. (b) RCWA-computed reflectivity and transmission spectra of HCGs show two guided-mode resonances (GMRs) supporting a high-reflection band of 160 nm (R$_{TE}$ >0.99) for TE-polarized waves, while a rather weak transverse magnetic TM-polarized GMR (b) appears far from the 1.55 μm wavelength region.

the HCG. Such resonant coupling can be tailored to achieve power-efficient reflectors endowed with the uttermost compactness—both vertical and lateral—and wide stopbands, also in the order of several hundreds of nanometers, which can be adopted in VCSELs to provide vertical and lateral confinement, thus improving their performance. These concepts are illustrated by means of both rigorous coupled-wave analysis (RCWA) and finite-difference time domain (FDTD) simulations in Figs. 8.19 and 8.20.

Static light-in-light-out (LL) analysis of optically pumped III–V-on-Si HCG-VCSEL is reported in Fig. 8.21. Devices show continuous-wave operation at 1.55 μm up to 45°C, with room-temperature threshold power in the submilliwatt range. Concerning spectral purity, the heterostructure-confined HCG architecture, as shown in Fig. 8.20, enhances the laser modal selection to almost 30 dB of the SMSR. Moreover, the fabricated HCG mirrors are inherently polarization sensitive, providing the VCSEL with a full TE polarization selection, as illustrated in Fig. 8.21c.

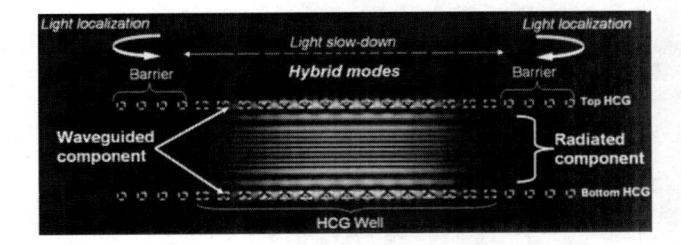

Figure 8.20 Optical hybrid mode intensity of a double HCG-VCSEL computed by 3D FDTD. Hybrid optical modes sustained by HCG-VCSEL cavities are partly radiated in between the mirrors, as well as waveguided within HCGs. Joint action of light localization and slowdown in heterostructure-confined HCGs achieved by locally thinning the silicon bars at the edge (barriers).

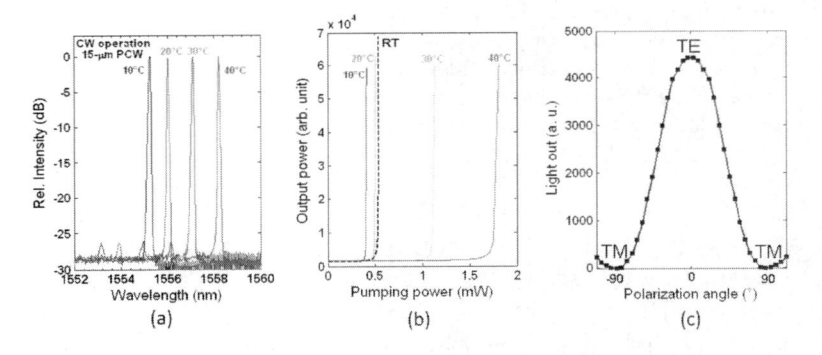

Figure 8.21 Modal, thermal, and polarization features of double HCG-VCSELs under optical pumping excitation. (a) Continuous-wave lasing behavior of double HCG-VCSELs at different pumping powers and stage temperatures. Single-mode emission with a 27 dB of transverse SMSR is obtained up to a stage temperature of 43°C. The estimated thermal tuning coefficient is 0.06 nm/K. (b) Light-in-light-out curves corresponding to different stage temperatures. (c) Polarization-resolved spectra denoting a pinned TE-polarized emission.

8.5 Conclusions

Meeting the ever-growing interconnectivity challenge posed by a conscious data-driven society is setting a closer horizon over the viable exploitation of present copper-based links in our telecommunication backbone and computational machines.

Silicon lies at the heart of a forthcoming paradigm shift that is set to revolutionize datacom infrastructures nowadays by triggering a photonics-based computational era. In this context, the maturity and cost-effectiveness of CMOS technology—so far the mainstay of the electronic industry—will be leveraged into photonics-based devices and systems, endowed with the inherent capability to transfer and compute data at very high bitrates ($\sim$Tb/s), over a small footprint with minor power dissipation, showing considerable integration potential with micro-nano-electronic nodes. In other words, silicon photonics operates within an energy-efficient communication framework, where the scaling of IT infrastructures from micronscale microprocessors to macroscale data centers and high-performance computing infrastructures will be achieved by the clever integration of a photonics-enhanced transceiving platform featuring high scalability and dense functionality.

The chapter shed light over the recent achievements recorded in silicon photonics, notably concerning the heterointegration of III–V materials on silicon aiming at realizing small-footprint, power-efficient, cheap laser emitters onto PICs.

Section 8.2 introduced the reader to the reasons driving the current efforts in both research and industrial environments for integrating laser sources on silicon photonic chips. Considering the inherent costs of flip-chipping lasers onto PICs in terms of time-consuming small-throughput active alignment, the heterogeneous integration of III–V epilayers on micro-nano-patterned silicon and SOI represents a viable and cost-effective path to bridge laser optoelectronics performance to silicon fab maturity, featuring a key step toward a cheap, reliable photonics–electronics convergence.

In Section 8.3 the authors described the main technological steps of the III–V-on-Si heterogeneous integration for both front- and back-end processing phases. The section also provided a wide overview of the latest results concerning electrically driven DBR- and DFB-based III–V hybrid lasers on 200 mm SOI wafers. Such components show cutting-edge performance in terms of wall-plug efficiency, power consumption, spectral purity, and high-speed direct modulation. As well as that, the section highlighted how the III–

V-on-silicon integration potential can be easily and smartly declined over other key building blocks such as EAMs and photodetectors, therefore enlarging the scope of the III–V heterointegration toward the circuit level. Moreover, the deterministic control of light that can be obtained at a small footprint by mature CMOS technology endows active devices with greater flexibility over spectral and modal features, enabling full wavelength addressing and coarse as well as dense WDM over main optical telecom and datacom bands for higher system aggregate bandwidth capacity.

In the bridging movement of III–V optoelectronics to silicon maturity, innovative proofs of concepts aiming at further miniaturizing present integrated laser technology play a key role for reducing power and thermal budgets of future silicon photonics-powered links. Section 8.4 described the recent developments obtained in heterointegrating III–V-on-SOI HCG-VCSELs. Such demonstrators show a room-temperature power threshold level in the submilliwatt range, while continuous-wave operation up to 45°C has been achieved in optically pumped devices exhibiting high modal selection and full polarization purity. Although these preliminary results bode well for the future of this new laser architecture, the efficient electrical drive of this device class on silicon, however, remains a topical objective to be met in order to prove the compatibility of such a novel approach with product prototyping, as well as for reducing its research-to-market transfer time.

In this sense, round-the-clock perspective work is now focused on active device miniaturization, performance consolidation through the improvement of both thermal and power budgets, and novel proof-of-concept introduction in the research and development pipelines for next-generation silicon photonics transceiver chips. This challenging task is shaping present efforts toward CMOS-compatible silicon photonics-enhanced cards and modules, thus paving the way to meet the future needs of a relentlessly growing data-driven society.

References

1. Cisco. (2014). *"Entering the Zettabyte Era" Is a Document Part of the Cisco Visual Networking Index, an Ongoing Initiative to Track and*

Forecast the Impact of Visual Networking Applications, http://www.cisco.com/en/US/solutions/collateral/ns341/ns525/ns537/ns705/ns827/VNI_Hyperconnectivity_WP.html.

2. Reed, G.T. (2004). The optical age of silicon. *Nature*, **427**, pp. 595–596.

3. Asghari, M., and Krishnamoorthy, A.V. (2011). Silicon photonics: energy efficient communication, *Nat. Photonics*, **5** pp. 268–270.

4. Krishnamoorthy, A.V., Ho, R., Zheng, X., Schwetman, H., Lexau, J., Koka, P., Li, G., Shubin, I., and Cunningham, J.E. (2009). Computer systems based on silicon photonic interconnects, *Proc. IEEE*, **97**, 7, pp. 1337–1361.

5. Dell, J.O. (2013). *Intel Takes Silicon Photonics Technology out of Research, into Production, & Open Source*, SantaClara OpenSummit, VentureBeat, http://venturebeat.com/2013/01/16/silicon-photonics/.

6. Intel. (2013a). *World's First Intel-Based Optical PCI Express Server*, Fujitsu Annual Forum, ICM Munich, Germany, 2013, https://www-ssl.intel.com/content/www/us/en/research/intel-labs-silicon-photonics-optical-pci-express-server.html.

7. Boyraz, O., and Jalali, B. (2004). Demonstration of a silicon Raman laser, *Opt. Express*, **12**, p. 5269.

8. Rong, H., Jones, R., Liu, A., Cohen, O., Hak, D., Fang, A., and Paniccia, M. (2005). A continuous-wave Raman silicon laser, *Nature*, **433**, pp. 725–728.

9. Pavesi, L., Dal Negro, L., Mazzoleni, C., Franzó, G., and Priolo, F. (2000). Optical gain in silicon nanocrystals, *Nature*, **408**, pp. 440–444.

10. Navarro-Urrios, D., Lebour, Y., Jambois, O., Garrido, B., Pitanti, A., Daldosso, N., Pavesi, L., Cardin, J., Hijazi, K., Khomenkova, L., Gourbilleau, F., and Rizk, R. (2009). Optically active Er^{3+} ions in SiO_2 codoped with Si nanoclusters, *J. Appl. Phys.*, **106**, 9, pp. 0931071–0931075.

11. Gelloz, B., Kojima, A., and Koshida, N. (2005). Highly efficient and stable luminescence of nanocrystalline porous silicon treated by high-pressure water vapor annealing, *Appl. Phys. Lett.*, **87**, p. 031107.

12. Solehmainen, K., Kapulainen, M., Heimala, P., and Polamo, K. (2004). Erbium doped waveguides fabricated with atomic layer deposition method, *IEEE Photonics Tech. Lett.*, **16**, pp. 194–196.

13. Camacho-Aguilera, R.E., Cai, Y., Patel, N., Bessette, J.T., Romagnoli, M., Kimerling, L.C., and Michel, J. (2012). An electrically pumped germanium laser, *Opt. Express*, **20**, pp. 11316–11320.

14. P. De Dobbelaere. (2013). *Silicon Photonics Technology Platform for High Speed Interconnect*, Semiconwest, Oct. 2013.

15. Zilkie, A.J., Seddighian, P., Bijlani, B.J., Qian, W., Lee, D.C., Fathololoumi, S., Fong, J., Shafiiha, R., Feng, D., Luff, B.J., Zheng, X., Cunningham, J.E., Krishnamoorthy, A.V., and Asghari, M. (2012). Power-efficient III-V/silicon external cavity DBR lasers, *Opt. Express*, **20**, pp. 2345–2346.

16. Fang, A.W., Jones, R., Park, H., Cohen, O., Raday, O., Paniccia, M.J., and Bowers, J.E. (2007). Integrated AlGaInAs-silicon evanescent race track laser and photodetector, *Opt. Express*, **15**, pp. 2315–2322.

17. Ben Bakir, B., Descos, A., Olivier, N., Bordel, D., Grosse, P., Augendre, E., Fulbert, L., and Fedeli, J.-M. (2011). Electrically driven hybrid Si/III-V Fabry-Pérot lasers based on adiabatic mode transformers, *Opt. Express*, **19**, 11, pp. 10317–10325.

18. Dahlem, M.S., Holzwarth, C.W., Khilo, A., Kärtner, F.X., Smith, H.I., and Ippen, E.P. (2011). Reconfigurable multi-channel second-order silicon microring-resonator filterbanks for on-chip WDM systems, *Opt. Express*, **19**, pp. 306–331.

19. Ben Bakir, B., de Gyves, A.V., Orobtchouk, R., Lyan, P., Porzier, C., Roman, A., and Fedeli, J.-M. (2010). Low-loss (<1 dB) and polarization-insensitive edge fiber couplers fabricated on 200-mm silicon-on-insulator wafers, *IEEE Photonics Technol. Lett.*, **22**, 11, pp. 739–741.

20. Antelius, M., Gylfason, K.B., and Sohlström, H. (2011). An apodized SOI waveguide-to-fiber surface grating coupler for single lithography silicon photonics, *Opt. Express*, **19**, pp. 3592–3598.

21. Liao, L., Samara-Rubio, D., Morse, M., Liu, A., Hodge, D., Rubin, D., Keil, U., and Franck, T. (2005). High speed silicon Mach-Zehnder modulator, *Opt. Express*, **13**, pp. 3129–3135.

22. Van Campenhout, J., Pantouvaki, M., Verheyen, P., Selvaraja, S., Lepage, G., Yu, H., Lee, W., Wouters, J., Goossens, D., Moelants, M., Bogaerts, W., and Absil, P. Low-voltage, low-loss, multi-Gb/s silicon micro-ring modulator based on a MOS capacitor. In *Optical Fiber Communication Conference (OFC2012)*, Los Angeles, USA, paper OM2E.4.

23. Vivien, L., Rouvié re, M., Fé dé li, J.-M., Marris-Morini, D., Damlencourt, J.-F., Mangeney, J., Crozat, P., El Melhaoui, L., Cassan, E., Le Roux, X., Pascal, D., and Laval, S. (2007). High speed and high responsivity germanium photodetector integrated in a silicon-on-insulator microwaveguide, *Opt. Express*, **15**, pp. 9843–9848.

24. Bogaerts, W., Selvaraja, S.K., Dumon, P., Brouckaert, J., De Vos, K., Van Thourhout, D., and Baets, R. (2010). Silicon-on-insulator spectral filters fabricated with CMOS technology, *IEEE J. Sel. Top. Quant. Electron.*, **16**, 1, pp. 33–44.

25. Intel. (2013b). *Intel Showcases 50 Gb/s Silicon Link*, http://optics.org/news/1/2/19/.

26. Graydon, O. (2011). View from group IV photonics: hope from hybrids, *Nat. Photon.*, **5**, pp. 718–719.

27. Kromer, H., Liu, T.-Y., and Petroff, P.M. (1989). GaAs on Si and related systems: problems and prospects, *J. Cryst. Growth*, **95**, pp. 96–102.

28. Kato, K., and Tohmori, Y. (2000). PLC hybrid integration technology and its application to photonic components, *IEEE J. Sel. Top. Quant. Electron.*, **6**, pp. 4–13.

29. Sasaki, J., Itoh, M., Tamanuki, T., Hatakeyama, H., Kitamura, S., Shimoda, T., Kato, T. (2001). Multiple-chip precise self-aligned assembly for hybrid integrated optical modules using Au–Sn solder bumps, *IEEE Trans. Adv. Packag.*, **24**, pp. 569–575.

30. Fedeli, J.-M., Di Cioccio, L., Marris-Morini, D., Vivien, L., Orobtchouk, R., Rojo-Romeo, P., Seassal, C., and Mandorlo, F. (2008). Development of silicon photonics devices using microelectronic tools for the integration on top of a CMOS wafer, *Adv. Opt. Technol.*, p. 412518.

31. Roelkens, G., Liu, L., Liang, D., Jones, R., Fang, A.W., Koch, B.R., and Bowers, J.E. (2010). III-V/Silicon photonics for on-chip and intra-chip optical interconnects, *Laser Photonics Rev.*, **4**, pp. 751–779.

32. Fang, A.W., Park, H., Cohen, O., Jones, R., Paniccia, M.J., and Bowers, J.E. (2006). Electrically pumped hybrid AlGaInAs-silicon evanescent laser, *Opt. Express*, **14**, pp. 9203–9210.

33. Sun, X., Zadok, A., Shearn, M.J., Diest, K.A., Ghaffari, A., Atwater, H.A., Scherer, A., and Yariv, A. (2009). Electrically pumped hybrid evanescent Si/InGaAsP lasers, *Opt. Lett.*, **34**, pp. 1345–1347.

34. Fang, A.W., Jones, R., Park, H., Cohen, O., Raday, O., Paniccia, M.J., and Bowers, J.E. (2007). Integrated AlGaInAs-silicon evanescent race track laser and photodetector, *Opt. Express*, **15**, pp. 2315–2322.

35. Fang, A.W., Lively, E., Kuo, Y.-H., Liang, D., and Bowers, J.E. (2008). A distributed feedback silicon evanescent laser, *Opt. Express*, **16**, pp. 4413–4419.

36. Heck, M.J.R., Hui-Wen, C., Fang, A.W., Koch, B.R., Liang, D., Park, H., Sysak, M.N., and Bowers, J.E. (2011). Hybrid silicon photonics for optical interconnects, *IEEE J. Sel. Top. Quant. Electron.*, **17**, 2, pp. 333–346.

37. Koch, B.R., Norberg, E.J., Kim, B., Hutchinson, J., Shin, J., Fish, G., and Fang, A. (2013). Integrated silicon photonic laser sources for telecom and Datacom. In *Optical Fiber Communication Conference (OFC2013)*, Anaheim, USA, paper PDP5C.8.

38. Keyvaninia, S., Roelkens, G., Van Thourhout, D., Jany, C., Lamponi, M., Le Liepvre, A., Lelarge, F., Make, D., Duan, G.-H., Bordel, D., and Fedeli, J.-M. (2013). Demonstration of a heterogeneously integrated III-V/SOI single wavelength tunable laser, *Opt. Express*, **21**, pp. 3784–3792.

39. Yariv, A., and Sun, X.K. (2007). Supermode Si/ III–V hybrid lasers, optical amplifiers and modulators: a proposal and analysis, *Opt. Express*, **15**, pp. 9147–9151.

40. Sun, X., and Yariv, A. (2008). Engineering supermode silicon/III–V hybrid waveguides for laser oscillation, *J. Opt. Soc. Am. B*, **25**, pp. 923–926.

41. Sun, X., Liu, H.-C., and Yariv, A. (2009). Adiabaticity criterion and the shortest adiabatic mode transformer in a coupled-waveguide system, *Opt. Lett.*, **34**, pp. 280–282.

42. Streifer, W., Scifres, D.R., and Burnham, R. (1975). Coupling coefficients for distributed feedback single- and double-heterostructure diode lasers, *IEEE J. Quant. Electron.*, **11**, 11, pp. 867–873.

43. Mandorlo, F., Rojo-Romeo, P., Ferrier, L., Olivier, N., Orobtchouk, R., Fedeli, J.-M., and Letartre, X. (2011). CMOS CW tunable III-V microdisk LASERs for optical interconnects in integrated circuits. In *8th IEEE International Conference on Group IV Photonics, GFP '11*, pp. 178–180, Sep. 14–16, 2011.

44. Mandorlo, F., Rojo Romeo, P., Olivier, N., Ferrier, L., Orobtchouk, R., Letartre, X., Fedeli, J.-M., and Viktorovitch, P. (2012). Controlled multi-wavelength emission in full CMOS-compatible micro-lasers for on-chip interconnections, *IEEE J. Lightwave Technol.*, **30**, pp. 3073–3080.

45. Soda, H., Iga, K., Kitahara, C., and Suematsu, Y. (1979). GaInAsP/InP surface emitting injection lasers. *Jpn. J. Appl. Phys.*, **18**, pp. 2329–2330.

46. Iga, K. (2000). Surface-emitting laser – Its birth and generation of new optoelectronics field, *IEEE J. Sel. Top. Quant. Electron.*, **6**, 6, pp. 1201–1215.

47. Kögel, B., Halbritter, H., Jatta, S., Maute, M., Böhm, G., Amann, M.-C., Lackener, M., Schwarzott, M., Winter, F., and Meissner, P. (2007). Simultaneous spectroscopy of NH_3 and CO using a > 50 nm continuously tunable MEMS-VCSEL, *IEEE Sensor J.*, **7**, 11, pp. 1483–1489.

48. Svensson, T., Andersson, M., Rippe, L., Svanberg, S., Andersson-Engels, S., Johansson, J., and Folestad, S. (2008). VCSEL-based oxygen spectroscopy for structural analysis of pharmaceutical solids, *Appl. Phys. B: Laser Opt.*, **90**, 2, pp. 345–354.

49. Bosman, E., Missinne, J., Van Hoe, B., Van Steenberge, G., Kalathimekkad, S., Van Erps, J., Milenkov, I., Panajatov, K., Van Gijseghem, T., Dubruel, P., Thienpont, H., and Daele, P. (2011). Ultrathin optoelectronic device packaging in flexible carriers, *IEEE J. Sel. Top. Quant. Electron.*, **17**, 3, pp. 617–628.

50. Mereuta, A., Suruceanu, G., Caliman, A., Iakovlev, V., Sirbu, A., and Kapon, E. (2009). 10-Gb/s and 10-km error-free transmission up to 100°C with 1.3-μm wavelength wafer-fused VCSELs, *Opt. Express*, **17**, 15, pp. 12981–12986.

51. Hofmann, W., Müller, M., Nadtochiy, A., Meltzer, C., Mutig, A., Böhm, G., Rosskopf, J., Bimberg, D., Amann, M.-C., and Chang-Hasnain, C. (2009). 22 Gb/s long wavelength VCSELs, *Opt. Express*, **17**, 20, pp. 17547–17553.

52. Martinsson, H., Vukusic, J., Grabherr, M., Michalzik, R., Jager, R., Ebeling, K., and Larsson, A. (1999). Transverse mode selection in large-area oxide-confined vertical-cavity surface-emitting lasers using a shallow surface relief, *IEEE Photonics Technol. Lett.*, **11**, 12, pp. 1536–1538.

53. Oh, T.H., McDaniel, M.R., Huffaker, D.L., and Deppe, D.G. (1998). Cavity-induced antiguiding in a selectively oxidized vertical-cavity surface-emitting laser, *IEEE Photonics Technol. Lett.*, **10**, 1, pp. 12–14.

54. Gustavsson, J., Haglund, Å., Vukušić, J., Bengtsson, J., Jedrasik, P., and Larsson, A. (2005). Efficient and individually controllable mechanisms for mode and polarization selection in VCSELs, based on a common, localized, sub-wavelength surface grating, *Opt. Express*, **13**, pp. 6626–6634.

55. Debernardi, P., Ostermann, J.M., Feneberg, M., Jalics, C., and Michalzik, R. (2005). Reliable polarization control of VCSELs through monolithically integrated surface gratings: a comparative theoretical and experimental study, *IEEE J. Sel. Top. Quant. Electron.*, **11**, 1, pp. 107–116.

56. Sciancalepore, C., Ben Bakir, B., Letartre, X., Harduin, J., Olivier, N., Seassal, C., Fedeli, J.-M., and Viktorovitch, P. (2012). CMOS-compatible ultra-compact 1.55-μm emitting VCSEL using double photonic crystal mirrors, *IEEE Photonics Technol. Lett.*, **24**, 6, pp. 455–457.

57. Rao, Y., Chase, C., Huang, M.C.Y., Khaleghi, S., Chitgarha, M.R., Ziyadi, M., Worland, D.P., Willner, A., and Chang-Hasnain, C. (2012). Continuous tunable 1550-nm high contrast grating VCSEL. In *CLEO: Applications and Technology, OSA Technical Digest* (online), Optical Society of America, 2012, paper CTh5C.3.

58. Zhao, D., Yang, H., Chuwongin, S., Seo, J.H., Ma, Z., Zhou, W. (2012). Design of photonic crystal membrane-reflector-based VCSELs, *IEEE Photonics J.*, **4**, 6, pp. 2169–2175.

59. Yang, H., Zhao, D., Chuwongin, S., Seo, J.-H., Yang, W., Shuai, Y., Berggren, J., Hammar, M., Ma, Z., and Zhou, W. (2012). Transfer-printed stacked nanomembrane lasers on silicon, *Nat. Photonics*, **6**, pp. 615–620.

60. Sciancalepore, C., Ben Bakir, B., Letartre, X., Fedeli, J.-M., Olivier, N., Bordel, D., Seassal, C., Rojo-Romeo, P., Regreny, P., and Viktorovitch, P. (2011). Quasi-3D light confinement in double photonic crystal reflectors VCSELs for CMOS-compatible integration, *IEEE J. Lightwave Technol.*, **29**, 13, pp. 2015–2024.

61. Sciancalepore, C., Ben Bakir, B., Seassal, C., Letartre, X., Harduin, J., Olivier, N., Fedeli, J.-M., Viktorovitch, P. (2012). Thermal, modal and polarization features of double photonic crystal vertical-cavity surface-emitting lasers, *IEEE Photonics J.*, **4**, 02, pp. 399–410.

Chapter 9

Application of Organic Semiconductors toward Transistors

Yoshihiro Kubozono,[a,b] Xuexia He,[a] Shino Hamao,[a] Eri Uesugi,[a] Yuma Shimo,[c] Takahiro Mikami,[c] Hidenori Goto,[a] and Takashi Kambe[d]

[a] *Research Laboratory for Surface Science, Okayama University, Okayama 700-8530, Japan*
[b] *Research Center of New Functional Materials for Energy Production, Storage and Transport, Okayama University, Okayama 700-8530, Japan*
[c] *Department of Electrical and Electronic Engineering, Okayama University, Okayama 700-8530, Japan*
[d] *Department of Physics, Okayama University, Okayama 700-8530, Japan*
kubozono@cc.okayama-u.ac.jp

We introduce here the use of thin films and single crystals of organic molecules in field-effect transistors (FETs) and survey recent progress in organic-based FETs, with special emphasis on the excellent p-channel FET characteristics obtainable with thin films and single crystals of phenacene. The highest-reported field-effect mobility, μ, in an organic thin-film FET is 21 cm^2 V^{-1} s^{-1} at room temperature [1] in an FET based on a thin film of alkyl-substituted picene. Such a high μ value offers promise of practical applications

Nanodevices for Photonics and Electronics: Advances and Applications
Edited by Paolo Bettotti
Copyright © 2016 Pan Stanford Publishing Pte. Ltd.
ISBN 978-981-4613-74-3 (Hardcover), 978-981-4613-75-0 (eBook)
www.panstanford.com

for organic FETs. This chapter is based on the PhD thesis of Dr. Xuexia He of our research group [2] and our recent papers on organic FETs [3–20]. Organic FETs based on phenacene are used to illustrate typical FET characteristics in this chapter, and the active research concerning such FETs is fully reported in our recent review article [21].

9.1 Introduction to Field-Effect Transistors

A transistor, derived from "transfer resistor," is a three-terminal electronic device for controlling an electric current. In this device, the electric current flowing between two terminals is controlled by the third terminal, that is, the electric current flowing between terminals can be amplified by the other terminal. A field-effect transistor (FET) is one of the transistors in which an electric field controls the conductivity of a semiconductor material. As seen from the schematic picture of an FET device in Fig. 9.1, an electric current flows through the active layer between two electrodes when a bias voltage is applied. The active layer may be a semiconductor such as Si, inorganic material (III–V and II–VI compounds), or organic material. The active layer and the two electrodes are called the channel, the source electrode, and the drain electrode. The source and drain electrodes are structurally indistinguishable in the device, but the source electrode is always grounded when applying a bias voltage V_D, which is applied to the drain electrode. The electric current is controlled by an adjustable bias voltage, which is applied to the third electrode, called the gate electrode. This voltage is called the gate voltage, or V_G.

As seen in Fig. 9.1, voltage V_G is applied to the gate electrode, which is in contact with the active layer across the gate dielectric. Generally, SiO_2 is used as the gate dielectric in an FET because of its strong protection against the leakage of gate current, which might occur when applying V_G, and the ease of fabrication of SiO_2, which can be formed by thermal oxidation of Si. When V_G is applied to the gate electrode, positive and negative electric charges are separated. Consequently, a charge-enriched region is produced in the active layer underneath the gate electrode, known as the channel region.

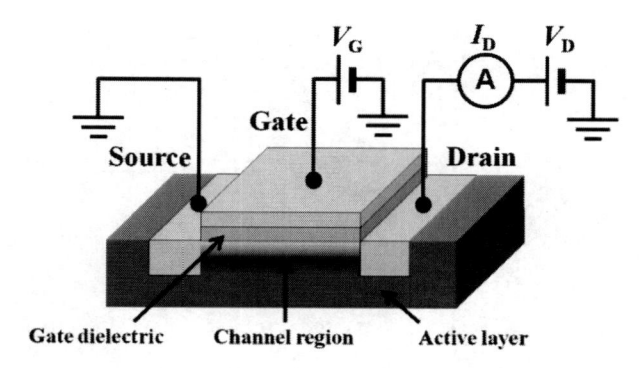

Figure 9.1 Typical FET structure.

This charge-enriched region can be regarded as a low-resistance region in which the electric current can flow. The negative charges refer to electrons, while the positive charges are holes. Both are light and move easily through the channel region between source and drain electrodes with the application of V_D. On the basis of the polarity of the applied V_G, the enriched charges will be either negative or positive, that is, either electrons or holes. When a negative V_G is applied, the charge that is enriched consists of holes, while applying a positive V_G enriches the electron population. In the former case, moving holes comprise the electric current flow, a situation called p-channel operation, while in the latter case it is electrons that flow and the operation is called n-channel operation. These charges are called carriers because the electric current is carried by these charges (holes and electrons).

In a Si metal-oxide semiconductor field-effect transistor (MOSFET), which is a fundamental device in current electronics, Si is the active layer. In this FET, the Si crystals contain as a dopant an element located in either column III (B) or column V (P or As). The former dopant can accept electrons from Si atoms, while the latter can donate electrons to Si atoms. Therefore, the hole concentration is higher in the former than in nondoped Si crystals, making this a p-type semiconductor. In contrast, the electron concentration is higher in the latter than in nondoped Si crystals, making this an n-type semiconductor. The actual dopant concentration is 1 atom of dopant for each 10^6 Si atoms. Thus, the semiconductor (active

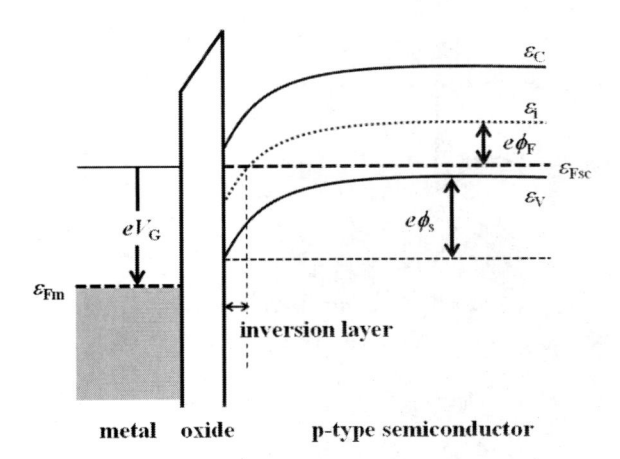

Figure 9.2 The formation of an inversion layer in a typical MOSFET; $V_G > 0$.

layer) used in Si MOSFETs is called an extrinsic semiconductor. In Si MOSFETs, a hole (electron)-enriched region should be formed in the Si semiconductor when applying a negative (positive) V_G. However, p-channel operation is observed only in Si MOSFETs with an n-type Si semiconductor, while n-channel operation is observed in Si MOSFETs with a p-type Si semiconductor. In other words, the carrier that is more concentrated (majority carrier) cannot carry the electric current, even if a suitable V_G is applied, but a carrier with a lower concentration (minority carrier) can carry the electric current. This is because the hole (electron) concentration increases in the channel region when a negative (positive) V_G is applied to an n-type (p-type) semiconductor through the gate dielectric. Therefore, the channel is formed by the minority carrier, in what is called an inversion layer. Thus, a Si MOSFET is a unipolar device. The formation of an inversion layer in a p-type semiconductor is shown in the energy diagram depicted in Fig. 9.2.

The first FET device was proposed by Lilienfeld [22], and he received the first patent in 1930. The second was presented in 1933 [23]. The patents attracted very little attention from most scientists because of the unreliability of the devices. However, advances in technology allowing the fabrication of pure Si crystals opened an avenue for the application of FETs in practical electronics. In 1960,

Atalla and Kahng succeeded in creating the first practical FET device [24–26]. This FET is the direct ancestor of the present FET device, that is, this invention led to the birth of modern electronics.

An FET device is a very effective electric current amplifier because it can increase electric current by 10^7–10^{12} times. Furthermore, an FET can realize a lower noise level than other types of transistors, such as the bipolar junction transistor (BJT), which means a highly stable operation. From a manufacturing perspective, an FET device has advantages over a BJT, because an FET requires fewer steps to build than do BJTs and other transistors. In addition, FETs can be integrated into an electronic circuit, that is, many FET chips can be integrated on one substrate. The integration can lead to low power consumption and high-speed operation. Currently, the development of advanced (functional) FETs has become a major focus in physics, chemistry, and engineering research because of the high expectations regarding future applications The most recent and advanced FET devices include carbon nanotube FETs [27–30], DNA FETs [31–33], graphene FETs [34–37], and the vertical organic triode [38, 39]. In particular, graphene FETs are attracting significant attention not only for potential applications based on their high operation speed but also in fundamental physics because their transport properties directly reflect relativistic quantum physics [34–37]. A new type of graphene FET [40] and some fundamental physics of graphene FETs [41, 42] have recently been reported by our group.

Organic FETs use organic materials as the active layer. The study of organic FETs is currently attracting a great deal of interest from scientists and engineers because of their flexibility, light weight, large area coverage, and ease of design [43–56]. Also, organic FETs can be fabricated with low-cost and low-energy processes because the active layer can easily be formed by a solution process. In contrast to organic FETs, a high-temperature process is indispensable for the fabrication of Si MOSFETs since temperatures greater than 1000°C are required to obtain pure single crystals of Si. Organic FETs can be used in various types of equipment, such as identification tags, drivers of active matrix displays, and smart cards, because of their flexibility and light weight. Furthermore, organic FETs are quite compatible with plastic substrates, enabling

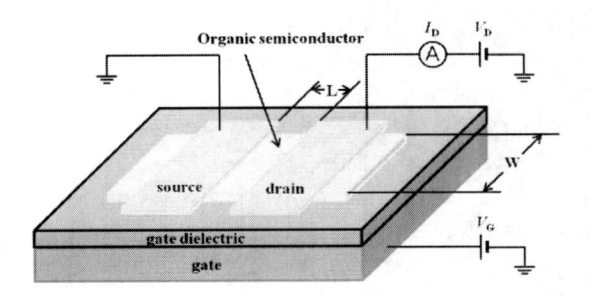

Figure 9.3 Typical structure of an organic FET device.

the formation of integrated circuits on flexible plastic substrates. This leads to the concept of ubiquitous electronics that people can use freely at any time, in any place.

The typical structure of an organic FET is shown in Fig. 9.3. This structure is the same as that of other FET devices such as Si MOSFETs. The FET consists of three electrodes (source, drain, and gate), a gate dielectric, and an active layer. The active layer is an organic material instead of Si or other inorganic materials. As described later, thin films or single crystals of organic materials are used as the active layer. The most important point is that organic materials form van der Waals crystals, which is different from Si and other inorganic materials. Si atoms are organized into crystals by the formation of covalent bonds between them, which can produce a wide band owing to the strong transfer integral between Si atoms. In contrast, in organic crystals, the band is very narrow because of the weak transfer integral between organic molecules. The characteristics of these materials affect the operation speed of FETs. The operation speed can be evaluated by a single important FET parameter, the field-effect mobility (μ). The larger the μ, the faster the operation. In a Si MOSFET, the typical μ value is greater than 1000 cm^2 V^{-1} s^{-1} [57–66], while the μ values in organic thin-film FETs are generally lower than 10 cm^2 V^{-1} s^{-1} [43–56]. The low μ of organic FETs is easily explained by the narrow band mentioned above. Increasing μ values is one of the most important missions in organic FET research. Low-voltage operation is also significant in organic FETs. For this purpose, SiO$_2$, with its dielectric constant ε

(=3.9) can be replaced by other gate dielectrics with high ε (high-k gate dielectrics), such as HfO_2 and Ta_2O_5.

Here, it is important to say that the mechanism of operation in organic FETs is different from that in Si MOSFETs. In Si MOSFETs, the inversion layer in which minority carriers are enriched is formed in the extrinsically doped Si semiconductor. Therefore, a Si MOSFET operates with the minority carrier and it is in principle a unipolar transistor. In an organic FET, no impurity is doped into the active layer. Therefore, in an organic FET, carriers are accumulated in an intrinsic (pure) organic semiconductor without any dopant impurities. Consequently, the transistor operation is not explained by the formation of an inversion layer. Such a difference can lead to unique and interesting physics quite different from that of a Si MOSFET.

9.2 Historical View of the Development of Organic Transistors

The first description of an organic FET appeared in the early 1970s. Barbe and Westgate reported field-effect measurements on single crystals of metal-free phthalocyanine under atmospheric conditions [67]. Modification of channel conductance induced by the variation of gate potential was observed in this system, and the paper clarified the surface parameters of single-crystal phthalocyanine using this technique. This may have been the first organic FET device. A realistic organic FET device was fabricated by Tsumura et al. in 1986 [68] in a study that used polythiophene in the active layer. This device showed normally-off-type FET characteristics. The drain current, I_D, was increased by 2 to 3 orders of magnitude by the application of V_G. The highest μ observed was at the most $\sim 10^{-5}$ cm^2 V^{-1} s^{-1} [68]. In 1993, the same group improved the μ value to 0.22 cm^2 V^{-1} s^{-1} using a polythienylenevinylene FET [69].

In 1992, Horowitz et al. reported a series of organic thin-film FETs using various π-conjugated oligomers, such as the oligothiophene series and two linear polyacenes (tetracene and pentacene; Fig. 9.4) [70]. This was the first use of pentacene in an FET device. Presently, pentacene is being used extensively in the study of organic

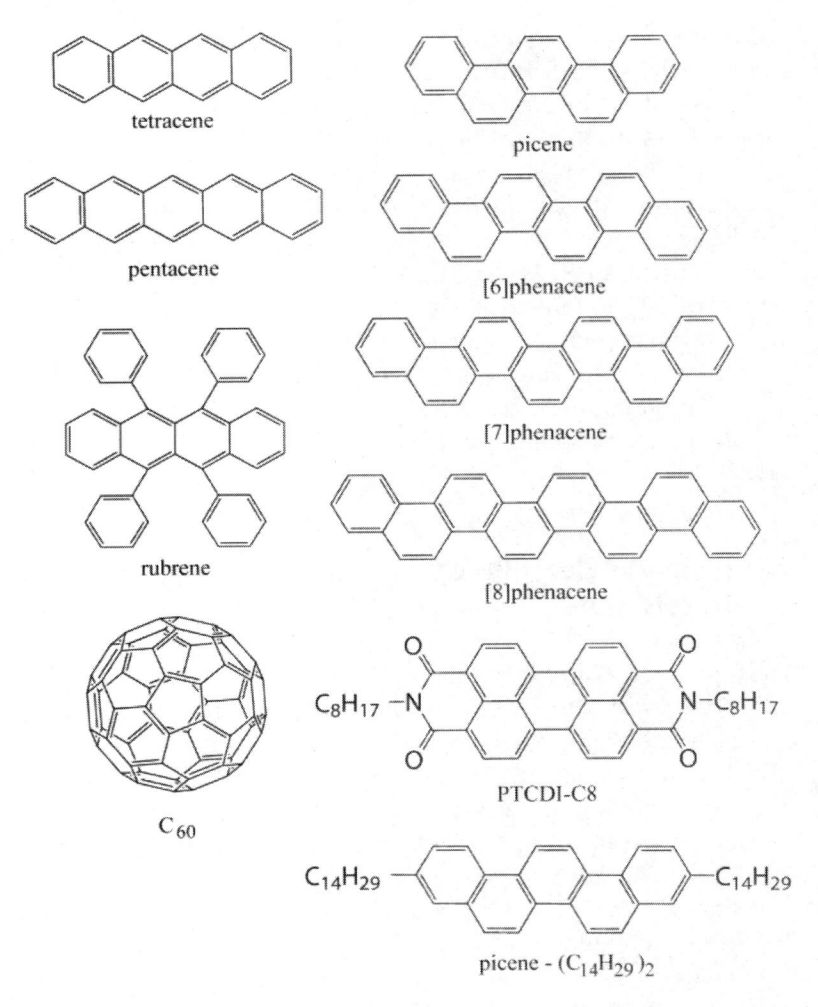

Figure 9.4 Molecular structures of organic materials appearing in this chapter.

FETs because of its potential for practical applications. Dodabalapur et al. made an FET device with a thiophene-type oligomer (α-hexathienylene). The on–off ratio of drain current I_D reached $>10^6$ [55]. The current density and switching speed realized in this device were sufficient for practical electronic circuits. Subsequently, Dimitrakopoulos et al. fabricated a pentacene thin-film FET in which the thin film was made by molecular beam deposition [71]. The maximum μ of that FET was recorded as 0.038 cm^2 V^{-1} s^{-1}. Furthermore, Lin et al. reported pentacene thin-film FETs with μ values as high as 1.5 cm^2 V^{-1} s^{-1} [72]. These pentacene FETs operated in p-channel mode, that is, holes carry the electric current in these FETs. In 2004, Yasuda et al. reported the impressive result that a pentacene FET can operate in both p- and n-channel modes [73]. In this device, the source and drain electrodes in the FET device were formed with Ca. The Fermi level of Ca is close to the lowest unoccupied molecular orbital (LUMO) in pentacene; the LUMO is related to the conduction of electrons (n-channel). The μ value in p-channel mode (hole mobility) was 4.5 $\times$ 10^{-4} cm^2 V^{-1} s^{-1}, while the μ in n-channel mode (electron mobility) was 2.7 $\times$ 10^{-5} cm^2 V^{-1} s^{-1}. This result shows clearly that the operation of an organic FET is based on the active layer of the intrinsic (nondoped) semiconductor, that is, an organic FET operates at the Mott–Schottky limit.

These organic FETs are made of thin films of organic materials. An FET device based on an organic single crystal was fabricated for the first time with rubrene by Podzorov et al. in 2003 [74]. The molecular structure of rubrene is shown in Fig. 9.4. The single crystal of rubrene was grown by physical vapor transport under a flow of hydrogen gas. The FET device demonstrated p-channel characteristics and a μ up to 1 cm^2 V^{-1} s^{-1} with an on–off ratio of $\sim$10^4. An FET device fabricated with a pentacene single crystal exhibited p-channel FET characteristics, with a μ as high as 0.30 cm^2 V^{-1} s^{-1} and an on–off ratio as high as 5 $\times$ 10^6 [75]. Furthermore, a tetracene single-crystal FET was also fabricated, with a μ value as high as 0.4 cm^2 V^{-1} s^{-1} [76]. Takeya et al. reported a very high μ value in a rubrene single-crystal FET [77]. The maximum μ value reached was 18 cm^2 V^{-1} s^{-1} in two-terminal measurement mode and 40 cm^2 V^{-1} s^{-1} in four-terminal measurement mode. Presently, the highest-reported μ value is 94 cm^2 V^{-1} s^{-1} for κ-(BEDT-TTF)$_2$Cu[N(CN)$_2$]Br (BEDT-TTF = bis(ethylenedithio)tetrathiafulvalene) in an organic single-crystal FET in four-terminal measurement mode [78].

9.3 Operation Mechanism of Organic Transistors

An FET device makes it possible to amplify the drain current, which flows between source and drain electrodes, by using the gate voltage. The operation mechanism in organic FET devices can be expressed with mathematical formulae, but these actually refer to the operation mechanism of Si MOSFETs. Strictly speaking, the operation of an organic FET is different from that of a Si MOSFET. The operation in the former type of FET is based on the accumulation of carriers, and an intrinsic semiconductor acts as the active layer, while the operation of the latter FET is explained by the formation of an inversion layer. In the latter FET, an n-type or p-type Si semiconductor is used as the active layer of the FET, that is, an extrinsic semiconductor is the key material in a Si MOSFET. The operation of an FET has already been fully explained in many textbooks [57–66]. Technical terms necessary for understanding FET operation are explained in this section.

The field-effect mobility, μ, shows how strongly a hole or an electron is influenced in the channel region of an FET when V_D is applied; the channel is opened by applying V_G. If the carrier responsible for channel conduction in an FET is a hole (electron), the field-effect mobility is called hole (electron) field-effect mobility. Thus, the mobility in an FET is called field-effect mobility, while the word "mobility" is generally used for the parameter that shows how strongly carriers in the semiconductor are influenced by an electric field. Therefore, the field-effect mobility can be determined from the FET characteristics, and the value expressed in units of cm^2 V^{-1} s^{-1}. In an organic FET, the field-effect mobility is affected not only by the intrinsic nature of the channel material, such as the existence of an effective π-conduction network, but also by extrinsic factors such as defects, grain boundaries, impurities, and contact resistance. The defects, grain boundaries, and impurities are formed extrinsically in the channel region, and the contact resistance relates to carrier injection between the active layer and source/drain electrodes. Therefore, the field-effect mobility is an important parameter that is directly related to the device performance.

In a Si MOSFET, a strong inversion layer can be formed by applying V_G so that the band curvature at the surface (interface between the active layer and the gate dielectric) becomes equal to twice the value, $2e\varphi_F$ ($=2\,|\varepsilon_{Fsc} - \varepsilon_i|$), of the Fermi potential energy, $e\varphi_F$, where ε_{Fsc} and ε_i refer to Fermi energy and intrinsic energy, respectively (see Fig. 9.2). The V_G that forms the strong inversion layer is called the threshold voltage. In a Si MOSFET, the channel starts to open at V_{TH}. In an organic FET, the threshold voltage, V_{TH}, is the voltage required to accumulate enough carriers to form an accumulation layer. Actually, $|V_{TH}|$ is strongly influenced by interface states such as trap states formed in the channel region, that is, $|V_{TH}|$ increases with an increase in trap states.

The on–off ratio is the ratio of the switch-on drain current to the off current. The minimum absolute current below $|V_{TH}|$ is regarded as the off current, while the maximum absolute current is regarded as the on current. The on–off ratio is important in the integration of FETs, that is, the formation of logic gate circuits. Suppressing the off current is a significant way to decrease power consumption. The enhancement of on drain current is indispensable for high-power FETs that are required for mobile communication systems, communication satellite relays, and compact radars.

The subthreshold swing, S, is the parameter indicating how much V_G is required to increase $|I_D|$ by a factor of 10. This value is normally called the S factor because another name for subthreshold swing is subthreshold slope. This value is given in V decade^{-1}. A smaller S provides more efficient/rapid switching. The lowest S is theoretically 60 mV decade^{-1} at 300 K. For an organic FET, in which the accumulation layer forms the channel region, the minimum S is theoretically evaluated to be more than 60 mV decade^{-1} [60].

In the organic FET device shown in Fig. 9.3, the I_D can be controlled by applying V_G. A positive V_G attracts electrons to the interface between the gate dielectric and the organic active layer, while a negative V_G attracts holes to the interface between the gate dielectric and the organic active layer. These electrons or holes form a conducting channel between the source and the drain. Typical $I_D - V_D$ curves at different V_Gs in FET devices are shown in Fig. 9.5. The curves are called output curves, and the $I_D - V_D$ in the low V_D range ($V_D << V_G - V_{TH}$) is called the linear regime, while the $I_D - V_D$ in the high V_D range ($V_D > V_G - V_{TH}$) is called the saturation regime. The I_D is constant in this range. Figure 9.6 shows an $I_D - V_G$ curve

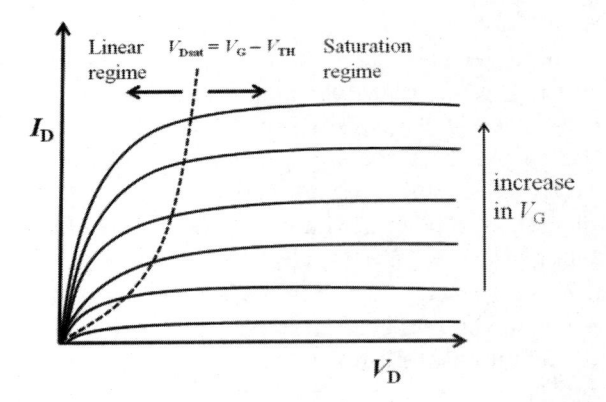

Figure 9.5 $I_D - V_D$ characteristics (output curves) for different V_G values. The dashed parabolic line ($V_D = V_G - V_{TH}$) separates the linear regime and the saturation regime.

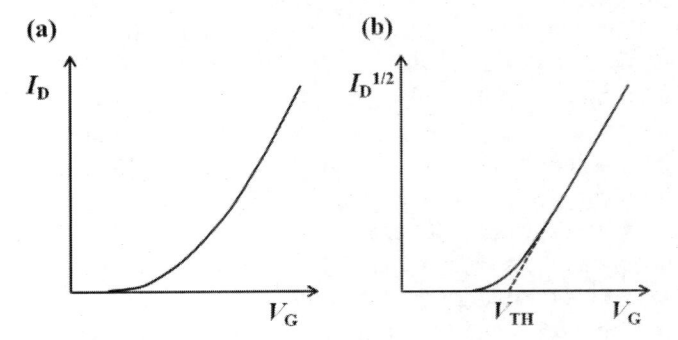

Figure 9.6 (a) $I_D - V_G$ characteristic (transfer curve) measured in the saturation regime. (b) The saturation drain current is proportional to ($V_G - V_{TH})^2$, indicating a linear relation between $I_D^{1/2}$ and V_G.

at fixed V_D. This curve is called the transfer curve. The theoretical formula for the $I_D - V_G$ curve differs between linear and saturation regimes. The μ, V_{TH}, on–off ratio, and S are evaluated from the transfer curve. Normally, in FET experiments, the output curves are measured to determine the linear and saturation regimes before measuring transfer curves.

In a Si MOSFET, the channel can be formed by minority carriers, a phenomenon that is called an inversion layer. However, in organic FETs the channel is formed by accumulated carriers in the accumulation layer. The formulae based on the formation of an inversion layer in Si MOSFETs must be modified for organic FETs because the I_D in Si MOSFETs is directly related to the inversion charge, $Q_{inv}(y)$, modified by the depletion charge, $Q_{dep}(y)$. The basic concept is applicable to organic FETs by simply replacing the $Q_{inv}(y)$ by the accumulation charge, $Q_{acc}(y)$. In this section, the mathematical formulae are derived by assuming the inversion layer.

The I_D in the inversion layer is given by the following equation [57]:

$$\int_0^L I_D dy = -\mu W \int_{V_S}^{V_D} Q_{inv}(y)\, dV(y), \tag{9.1}$$

where W and L are channel width and length, respectively, as shown in Fig. 9.6. $V(y)$ is the local electrostatic potential satisfying the boundary conditions, $V(0) = V_S$ and $V(L) = V_D$. Here, the general theory that covers nonvanishing V_S is presented. The $Q_{inv}(y)$ induced by V_G is estimated from the parallel-plate capacitor model; then Eq. 9.1 is expressed as

$$I_D = \mu C_{ox} \frac{W}{L} \int_{V_S}^{V_D} (V_G - [V_{THo} + V(y)]) dV(y), \tag{9.2}$$

where C_{ox}, V_{THo}, and $V(y)$ refer to the capacitance per area of the gate dielectric, the nonideal threshold voltage, and the voltage at y, respectively. This equation can also be applied to an organic FET in which the channel is formed by carrier accumulation. Considering the dependence of the threshold voltage on the source voltage V_s,

$$V_{TH} = V_{THo} + V_s. \tag{9.3}$$

Therefore, Eq. 9.2 reduces to

$$I_D = \frac{\mu C_{ox} W}{L} \left\{ (V_G - V_{TH}) V_{DS} - \frac{V_{DS}^2}{2} \right\}. \tag{9.4}$$

Here, $V_{DS} \equiv V_D - V_S$. In the linear regime, as $V_{DS} << V_G - V_{TH}$, the second term can be ignored Consequently, we can write the following equation:

$$I_D = \frac{\mu C_{ox} W}{L} \left\{ (V_G - V_{TH}) V_{DS} \right\}. \tag{9.5}$$

In the saturation regime, $dI_D/dV_{DS} = 0$ in Eq. 9.4. Therefore, V_D in the saturation regime can be expressed as follows:

$$V_D = V_G - V_{TH} + V_s. \tag{9.6}$$

Substituting Eq. 9.6 into Eq. 9.4 then allows the I_D in the saturation regime to be derived:

$$I_D = \mu C_{ox} W \frac{1}{L} \frac{(V_G - V_{TH})^2}{2} \tag{9.7}$$

This formula is generally used to evaluate the μ value from the $I_D - V_G$ curve in the saturation regime.

The S factor can be expressed as

$$S = \frac{dV_G}{d\log I_D}. \tag{9.8}$$

Therefore, the S factor can be evaluated from the $\log I_D - V_G$ curve. The S factor is related to the density of trap states. If the trap states are few in the active layer, the S factor decreases [57]. The S factor in the accumulation regime is fully discussed in Ref. [79].

9.4 Transistor Device Fabrication

Thin films or single crystals are used for the active layer in organic FETs. The organic thin films are generally formed using two methods, thermal deposition and deposition from solution (solution process). In the former method, the sample is thermally evaporated at high temperature under vacuum to form a thin film. The thickness of the film is monitored and a film suitable for an FET device can be fabricated easily. In the latter method, the sample is dissolved in a suitable solvent and the solution is coated on the substrate by spin-coating or casting. The general process for fabricating FET devices with thin films is as follows

Generally, a substrate of SiO_2/Si is used in organic FETs because the SiO_2/Si substrate is easily obtained commercially and SiO_2 is a good gate dielectric material with low gate leakage current. The surface of the SiO_2/Si substrate should be washed with acetone, methanol, distilled water, and H_2O_2/H_2SO_4 (1:4 by volume) to remove any organic and inorganic impurities on the SiO_2 surface

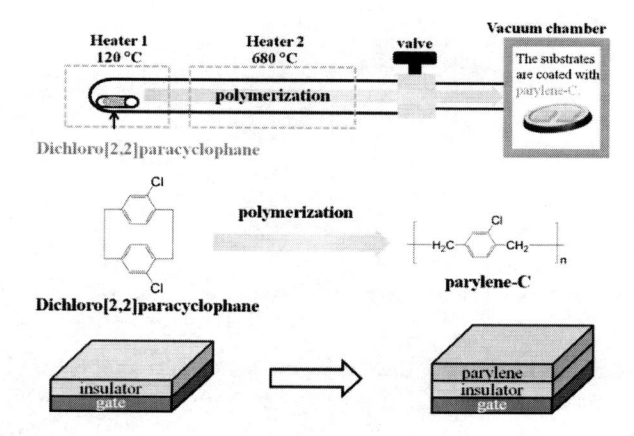

Figure 9.7 A schematic representation of the parylene-coating process.

that may disturb smooth carrier transport and finally the substrate must be cleaned thoroughly with distilled water. This washing process is called sulfuric acidhydrogen peroxide mixture (SPM) washing The washing of SiO_2/Si substrates is performed in an ultrasonic bath. Other gate dielectrics such as Ta_2O_5, HfO_2, ZrO_2, and $PbZr_{0.52}Ti_{0.48}O_3$ (PZT) can be also used as gate dielectrics in organic FETs. These gate dielectrics show a higher ε_x than SiO_2, providing the advantage that higher carrier densities can be accumulated at lower V_G values; such a gate dielectric is called a high-k gate dielectric. These substrates are generally washed with methanol and isopropanol.

The surfaces of these gate dielectrics are covered with parylene or hexamethyldisilazane (HMDS) to create hydrophobic stages that reduce the trap states for carriers. The hysteresis of transport properties shrinks significantly when a hydrophobic surface is formed [9]. Furthermore, covering a surface with parylene or HMDS suppresses gate current leakage. The experimental capacitance per area, C_o, for the hydrophobic-treated gate dielectric is obtained by extrapolating to 0 Hz the capacitance recorded at any alternating current (AC) amplitude using an LCR meter. In this way, the C_o at 0 Hz is used in the evaluation of FET parameters.

The parylene-coating process is shown in Fig. 9.7. In parylene coating, the precursor material, dichloro[2,2]paracyclophane, was

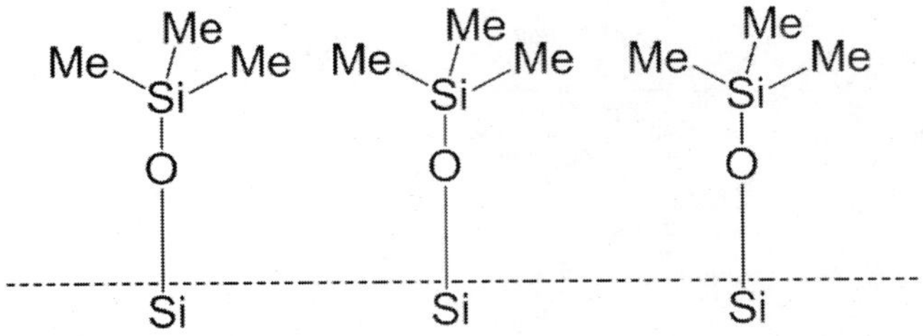

Figure 9.8 Hydrophobic surface of HMDS-coated SiO_2.

sublimed at 120°C and thermally polymerized at 680°C. During parylene coating, the vacuum in the parylene-coating system was maintained at 10^{-2} Torr. As shown in Fig. 9.7, the SiO_2/Si substrate was placed in a vacuum chamber. Poly-(2-chloroxylene) (parylene-C) is reported to be mechanically stable from −200°C to +150°C. In forming an HMDS-coated SiO_2 surface, HMDS is dissolved in hexane (HMDS:hexane = 1:9 by volume), and the SiO_2/Si substrate is immersed in the solution for 12 hours. The SiO_2 surface treated with HMDS provides a hydrophobic surface layer on the SiO_2 surface, as shown in Fig. 9.8.

The thin film is formed on the substrates using thermal deposition or the solution process described above, and the channel is formed with a shadow mask, that is, the source and drain electrodes are formed with Au or other noble metals by thermal deposition. The source and drain electrodes can also be fabricated using a solution process, that is, a conducting polymer such as poly(3,4-ethylenedioxythiophene)-poly(styrenesulfonate) (PEDOT:PSS) is used for the formation of electrodes in the solution process. The thickness of source/drain electrodes is typically 50 nm. A typical device structure (a top-contact structure) is shown in Fig. 9.9.

In the fabrication of a single-crystal FET, a single crystal is carefully placed on the hydrophobic-treated gate dielectric. The single crystals are normally produced by physical vapor transport, which provides a solvent-free single crystal. A typical synthesis method is described in Ref. [80]. The channel is prepared on the single crystal through the formation of source/drain electrodes by thermal deposition of Au or noble metals on the top-contact

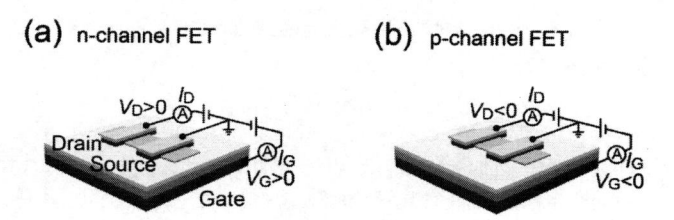

(a) n-channel FET   **(b)** p-channel FET

Figure 9.9 Top-contact structures and measurement modes for n-channel and p-channel FETs.

structure. The source and drain electrodes are made in the same way as on a thin-film FET. The contact resistance between electrodes and a single crystal tends to be higher than that between electrodes and a thin film. Therefore, we tried to insert various electron acceptors in the space between the electrodes and the single crystal [20]. We found that an electron acceptor with a high redox potential, such as tetrafluoro-7,7,8,8-tetracyano-p-quinodimethane (F4TCNQ), usefully reduced the Schottky barrier height (contact resistance). The process of device fabrication is shown in Fig. 9.10.

The measurement modes for n-channel and p-channel FETs are depicted in Fig. 9.9. The source electrode is always grounded in the measurement of FET characteristics, which is termed "source grounded." In an n-channel FET, the electric current flows from drain to source electrodes, meaning that electrons are provided from the source. In p-channel mode, the current flows from the source to the drain electrodes, meaning that holes are provided from the source. In n-channel and p-channel FETs, electrons and holes (respectively) accumulate to form the channel when V_G is applied.

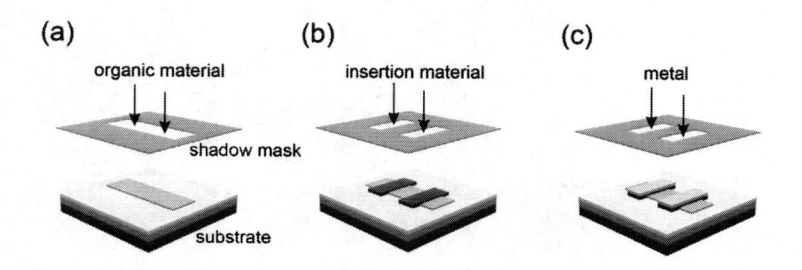

(a)   **(b)**   **(c)**

Figure 9.10 Fabrication of an organic FET device.

9.5 Typical Examples of N-Channel and P-Channel Organic Transistors

The typical FET characteristics of n-channel organic thin-film FETs are shown in Figs. 9.11 and 9.12. Figure 9.11 shows the characteristics of a C_{60} thin-film FET with a Ta_2O_5 gate dielectric (high-k gate dielectric). The I_D increases with an increase in V_G at a fixed V_D, suggesting n-channel operation. The C_{60} thin-film FET operates in an Ar-filled glove box or vacuum, but the FET characteristics are not observed under atmospheric conditions. The μ value of this device is 3.5×10^{-2} cm^2 V^{-1} s^{-1}, and $|V_{th}|$ is 17 V, providing low-voltage operation. The highest μ value achieved so far in a C_{60} thin-film FET is 0.3–0.5 cm^2 V^{-1} s^{-1} [3, 4, 81, 82]. Thus, the C_{60} FET is one of the better n-channel organic FETs. However, n-channel operation is more difficult than p-channel operation because the electrons are easily captured by –OH groups on the gate dielectric [83]. Therefore, finding organic materials that can contribute to good n-channel operation is of paramount importance in this field. In particular, accomplishing n-channel operation under atmospheric conditions would be a very significant advance.

N,N'-dioctyl-3,4,9,10-perylenedicarboximide (PTCDI-C8) is also a promising material for n-channel operation. Figure 9.12 shows the typical FET characteristics of a PTCDI-C8 thin-film FET with a SiO_2 gate dielectric. In the same manner as the C_{60} thin-film FET, the I_D increases with an increase in V_G. The positive V_G and V_D can operate the PTCDI-C8 thin-film FET, suggestive of n-channel operation. The

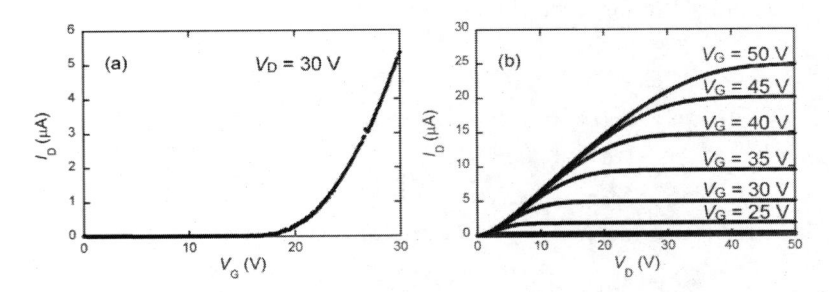

Figure 9.11 Transfer and output curves of a C_{60} thin-film FET with Ta_2O_5 gate dielectric.

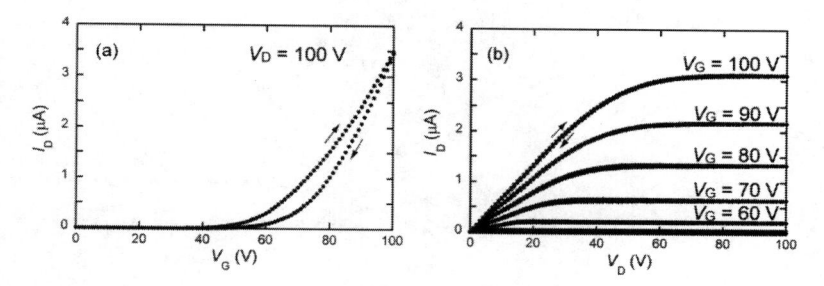

Figure 9.12 Transfer and output curves of a PTCDI-C8 thin-film FET with SiO$_2$ gate dielectric.

μ value is evaluated to be 2.6×10^{-1} cm^2 V^{-1} s^{-1} from the forward transfer curve. The FET characteristics were recorded in an Ar-filled glove box. The molecular structures of C$_{60}$ and PTCDI-C8 are shown in Fig. 9.4.

One of the best-investigated p-channel thin-film FETs is the pentacene FET. The μ value reaches 1.5 cm^2 V^{-1} s^{-1} [72]. The properties of pentacene thin-film FETs have been extensively studied, but they are very sensitive to the presence of air because the pentacene molecule oxidizes to pentaquinone under atmospheric conditions [21]. Recently, the application of phenacene in thin-film FETs has been reported and is noteworthy because of the stability of phenacene [6, 12, 19]. The phenacene molecule shows a W-shaped structure consisting of fused benzene rings (see Fig. 9.4). The electronic structure is shown in our review papers [21, 84]. The wide bandgap (more than 3 eV) and deep highest occupied molecular orbital (HOMO) level are characteristic of this type of molecule and account for its chemical stability.

Figure 9.13 shows the typical FET characteristics of a [6]phenacene thin-film FET with a SiO$_2$ gate dielectric. The molecular structure of [6]phenacene (W-shaped and six benzene rings) is shown in Fig. 9.4. The $|I_D|$ increases with an increase in $|V_G|$ at the fixed negative V_D, suggestive of p-channel FET characteristics. The electric current flows from the source to the drain. From the forward transfer curve, the μ value was evaluated to be 3.7 cm^2 V^{-1} s^{-1}. Since we recently observed the μ to be as high as 7.4 cm^2 V^{-1} s^{-1} [19], it can be concluded that the [6]phenacene thin-

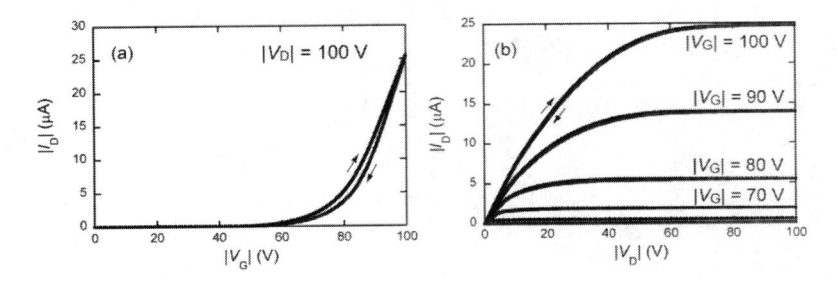

Figure 9.13 Transfer and output curves of a [6]phenacene thin-film FET with a SiO_2 gate dielectric.

film FET is promising as a p-channel organic FET. We previously fabricated a picene thin-film FET that showed a μ value as high as 1 cm^2 V^{-1} s^{-1} [5, 6]. The molecular structure of picene is shown in Fig. 9.4. The FET showed O_2-sensing properties, which implies that the μ and absolute I_D increase under an O_2 atmosphere. Thus, excellent FET characteristics are observed with the use of picene and [6]phenacene. Furthermore, a [7]phenacene thin-film FET has also been fabricated that displayed μ values as high as 0.71 cm^2 V^{-1} s^{-1} [12]. Details of FET properties (O_2 sensing, hysteresis, device flexibility, bottom-contact-type device, devices with high-k gate dielectric electric-double-layer [EDL] capacitor, etc.) based on picene, [6]phenacene, and [7]phenacene thin-film FETs are reported in Refs. [6], [19], and [12], respectively. All the FET characteristics suggest that phenacene thin-film FETs are promising candidates for future applications.

Very recently, we investigated the FET characteristics of alkyl-substituted phenacenes. Our group synthesized 3,10-ditetradecylpicene, picene-$(C_{14}H_{29})_2$ [1]. The molecular structure is shown in Fig. 9.4. The FET devices were fabricated with picene-$(C_{14}H_{29})_2$ thin films [1]; the thin films were fabricated by either thermal deposition or a solution process. The FET characteristics of picene-$(C_{14}H_{29})_2$ with a SiO_2 gate dielectric are shown in Fig. 9.14. The typical p-channel properties are observed in the FET with a thin film prepared by thermal deposition, showing that the μ value determined from the forward transfer curve (Fig. 9.14a) is as high as 8.3 cm^2 V^{-1} s^{-1}. We made six picene-$(C_{14}H_{29})_2$ FETs with

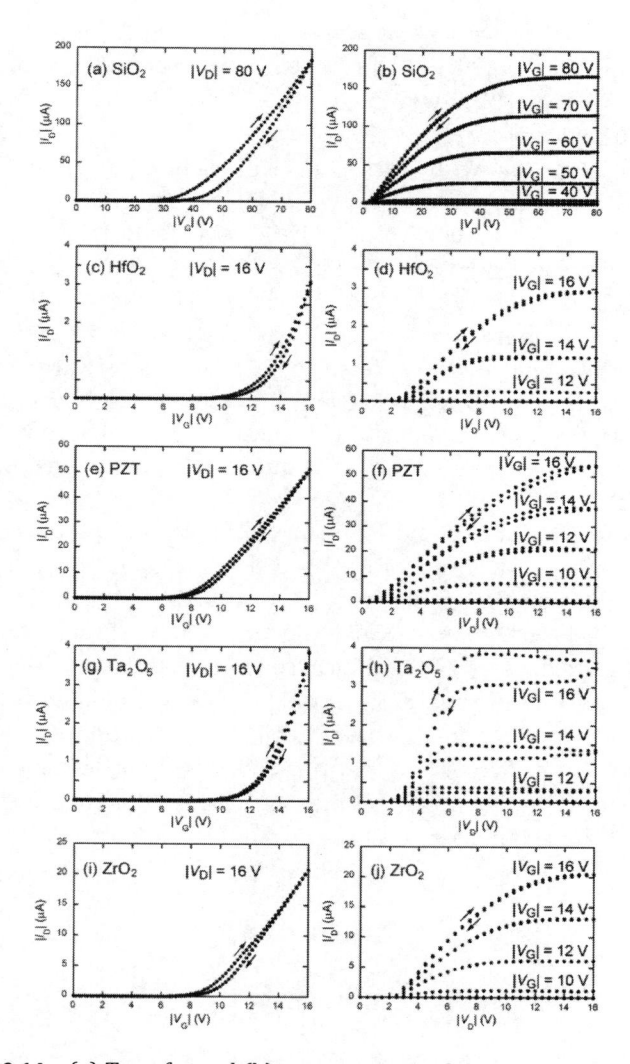

Figure 9.14 (a) Transfer and (b) output curves of a picene-$(C_{14}H_{29})_2$ thin-film FET with a SiO_2 gate dielectric. (c) Transfer and (d) output curves of a picene-$(C_{14}H_{29})_2$ thin-film FET with a HfO_2 gate dielectric. (e) Transfer and (f) output curves of a picene-$(C_{14}H_{29})_2$ thin-film FET with a PZT gate dielectric. (g) Transfer and (h) output curves of a picene-$(C_{14}H_{29})_2$ thin-film FET with a Ta_2O_5 gate dielectric. (i) Transfer and (j) output curves of a picene-$(C_{14}H_{29})_2$ thin-film FET with a ZrO_2 gate dielectric.

SiO_2 gate dielectrics. The average μ value was 7(2) cm^2 V^{-1} s^{-1}, exhibiting excellent FET characteristics. Furthermore, we fabricated picene-$(C_{14}H_{29})_2$ FETs with high-k gate dielectrics. The typical FET characteristics of picene-$(C_{14}H_{29})_2$ thin-film FETs with high-k gate dielectrics are shown in Fig. 9.14. The μ values determined from the forward transfer curves shown in Fig. 9.14 are 4.2, 13, 4.7, and 8.6 cm^2 V^{-1} s^{-1}, respectively, for HfO$_2$, PZT, Ta$_2$O$_5$, and ZrO$_2$ gate dielectrics. The $|V_{TH}|$ values are 11, 6.7, 10, and 7.4 V, respectively, for HfO$_2$, PZT, Ta$_2$O$_5$, and ZrO$_2$ gate dielectrics, indicative of low-voltage operation. We fabricated some FET devices with each gate dielectric, and the average μ values determined from the forward transfer curves were 5(1), 14(4), 11(1), and 9(2) cm^2 V^{-1} s^{-1}, respectively, with HfO$_2$, PZT, Ta$_2$O$_5$, and ZrO$_2$ gate dielectrics [1]. Thus, it has been verified that picene-$(C_{14}H_{29})_2$ is a quite promising material for FET devices.

Figure 9.15 shows the FET characteristics of a [7]phenacene single-crystal FET with a SiO$_2$ gate dielectric, displaying typical p-channel FET characteristics. As seen in Fig. 9.15, a 3 nm thick layer of F4TCNQ is inserted into the space between source/drain electrodes and the single crystal to reduce contact resistance. The output curves show a little concavity in the low $|V_D|$ regime, but basically linear behavior. From the forward transfer curve, the μ value was evaluated to be 3.9 cm^2 V^{-1} s^{-1}. We fabricated single-crystal FETs with picene, [6]phenacene, [7]phenacene, and [8]phenacene. The μ value of the [7]phenacene single-crystal FET is the highest among these phenacene single-crystal FETs. This is due to the high quality

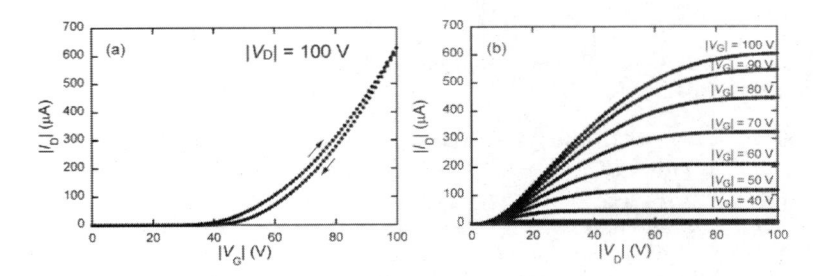

Figure 9.15 (a) Transfer and (b) output curves of a [7]phenacene single-crystal FET with a SiO$_2$ gate dielectric.

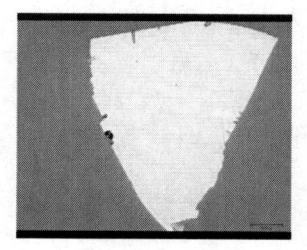

Figure 9.16   Optical micrograph image of a [7]phenacene single crystal.

of the crystals, visible in the optical micrograph images in Fig. 9.16. Since the [6]phenacene thin-film FET provides the highest μ value, the μ value is not directly affected by the intrinsic nature of the molecule (electronic structure and transfer integral) but extrinsic factors (defects, impurities and grain boundaries).

9.6   Future Prospects for Organic Transistors

Organic FETs offer many advantages, including flexibility, light weight, low-cost/low-energy fabrication, and ease of design for future electronics, but their low field-effect mobility and low reliability are serious problems that must be solved prior to practical application. Organic thin-film FETs made of new organic materials exhibiting higher μ values than 15 cm^2 V^{-1} s^{-1} have been developed during the past five years [85, 86]. We have also developed organic thin-film FETs with phenacene-type molecules, which show very high μ values. In particular, the picene-$(C_{14}H_{29})_2$ thin-film FET displayed μ values as high as 21 cm^2 V^{-1} s^{-1} ($< \mu >= 14(4)$ cm^2 V^{-1} s^{-1}) [1], which may be the highest μ value yet seen in organic thin-film FETs. The high μ values achieved in organic FETs should provide the next step toward practical transistors because high μ values are directly related to the operation speed.

Furthermore, the reliability and durability of organic FET devices are important problems that need to be resolved prior to their practical application. We clarified that phenacene molecules are stable under atmospheric conditions [21, 84], where their FETs also operate stably. However, the phenacene FETs cannot be used at high

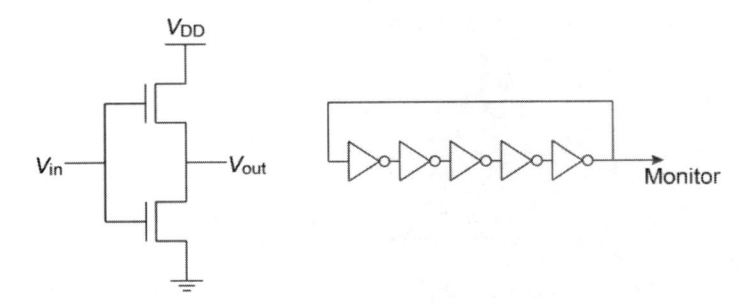

Figure 9.17 Schematic representation of CMOS inverter and ring oscillator.

temperatures because phenacene molecules sublime from the thin films above $100°C$. The stability of phenacene thin films at high temperatures remains a significant difficulty to be overcome.

Complementary metal-oxide semiconductor (CMOS) logic gate circuits (NOT, NAND, OR, NOR) can be fabricated using both n-channel and p-channel operations [87]. For example, the CMOS inverter (NOT circuit) shown in Fig. 9.17 consists of n-channel and p-channel FETs. In the ideal inverter, the logic value of 1 (high voltage V_{DD}) is observed when $V_{in} < V_{DD}/2$, while the logic value of 0 (low voltage 0) is observed when $V_{in} > V_{DD}/2$. C_{60} and PTCDI-C8 are suitable for n-channel FETs in CMOS logic gate circuits, while phenacene is promising for p-channel FETs in CMOS circuits. Currently, our group is beginning to develop a ring oscillator circuit by combining some CMOS inverters composed of PTCDI-C8 and picene-$(C_{14}H_{29})_2$ thin-film FETs. The ring oscillator will be constructed on a flexible plastic substrate, leading to flexible logic gate circuits.

The development of solution-processed organic FETs is a most important subject in this research field because no heating process is required in the fabrication, which means low energy consumption. This technique can open the way to the manufacture of printable organic FETs without any high-temperature processes. We are currently also working to synthesize substituted phenacene molecules that can be dissolved in organic solvents at useful concentrations, that is, we hope to find a way of preparing phenacene ink as a step on the path to a printable FET. Because of the high stability of phenacene FETs, this may lead to solution-processed organic FETs that can operate under ambient conditions. Logic gate circuits could also be produced using solution-processed FETs [88]. Furthermore, the solution process could allow the fabrication of large-scale transistor devices for sensor and medical applications.

Phenacene thin-film and single-crystal FETs are fabricated with EDL capacitors [13, 89], which provide low-voltage operation because of their very high capacitance. These devices may also be suitable for medical sensors. From the viewpoint of basic science, the EDL FET is very attractive because of the high carrier density that is accumulated in the channel region of the active layer, suggesting the possibility of controlling the electronic structure in the active layer. Since some phenacene solids have shown superconductivity when doped with alkali metal atoms [90], the high electron density attainable in solid phenacene using FET techniques may trigger the emergence of novel physical properties such as metallic behavior and superconductivity. Actually, such field-induced superconductivity has already been achieved in some inorganic 2D layered materials [91–93]. Thus, the current study of organic FETs addresses not only transistor performance but also exploring material design directed at the emergence of novel physical properties.

References

1. Okamoto, H., Hamao, S., Goto, H., Sakai, Y., Izumi, M., Gohda, S., Kubozono, Y., and Eguchi, R. (2014). Transistor application of alkyl-substituted picene, *Sci. Rep.*, **4**, p. 5048.

2. Doctor thesis prepared by Dr. Xuexia He of our research group.

3. Kubozono, Y., Haas, S., Kalb, W.L., Joris, P., Meng, F., Fujiwara, A., and Batlogg, B. (2008). High-performance C60 thin-film field-effect transistors with parylene gate insulator, *Appl. Phys. Lett.*, **93**, p. 033316.

4. Kawasaki, N., Ohta, Y., Kubozono, Y., Konishi, A., and Fujiwara, A. (2008). Transport properties in C(60) field-effect transistor with a single Schottky barrier, *Appl. Phys. Lett.*, **92**, p. 163307.

5. Okamoto, H., Kawasaki, N., Kaji, Y., Kubozono, Y., Fujiwara, A., and Yamaji, M. (2008). Air-assisted high-performance field-effect transistor with thin films of picene, *J. Am. Chem. Soc.*, **130**, p. 10470.

6. Kawasaki, N., Kubozono, Y., Okamoto, H., Fujiwara, A., and Yamaji, M. (2009). Trap states and transport characteristics in picene thin film field-effect transistor, *Appl. Phys. Lett.*, **94**, p. 043310.

7. Kaji, Y., Mitsuhashi, R., Lee, X., Okamoto, H., Kambe, T., Ikeda, N., Fujiwara, A., Yamaji, M., Omote, K., and Kubozono, Y. (2009). High-performance C-60 and picene thin film field-effect transistors with conducting polymer electrodes in bottom contact structure, *Org. Electron.*, **10** p. 432.

8. Kaji, Y., Kawasaki, N., Lee, X., Okamoto, H., Sugawara, Y., Oikawa, S., Ito, A., Okazaki, H., Yokoya, T., Fujiwara, A., and Kubozono, Y. (2009). Low voltage operation in picene thin film field-effect transistor and its physical characteristics, *Appl. Phys. Lett.*, **95**, p. 183302.

9. Kawasaki, N., Kalb, W. L., Mathis, T., Kaji, Y., Mitsuhashi, R., Okamoto, H., Sugawara, Y., Fujiwara, A., Kubozono, Y., and Batlogg, B. (2010). Flexible picene thin film field-effect transistors with parylene gate dielectric and their physical properties, *Appl. Phys. Lett.*, **96**, p. 113305.

10. Nouchi, R., and Kubozono, Y. (2010). Anomalous hysteresis in organic field-effect transistors with SAM-modified electrodes: structural switching of SAMs by electric field, *Org. Electron.*, **11**, p. 1025.

11. Lee, X., Sugawara, Y., Ito, A., Oikawa, S., Kawasaki, N., Kaji, Y., Mitsuhashi, R., Okamoto, H., Fujiwara, A., Omote, K., Kambe, T., Ikeda, N., and Kubozono, Y. (2010). Quantitative analysis of O-2 gas sensing characteristics of picene thin film field-effect transistors, *Org. Electron.*, **11**, p. 1394.

12. Sugawara, Y., Kaji, Y., Ogawa, K., Eguchi, R., Oikawa, S., Gohda, H., Fujiwara, A., and Kubozono, Y. (2011). Characteristics of field-effect transistors using the one-dimensional extended hydrocarbon [7]phenacene, *Appl. Phys. Lett.*, **98**, p. 013303.

13. Kaji, Y., Ogawa, K., Eguchi, R., Goto, H., Sugawara, Y., Kambe, T., Akaike, K., Gohda, S., Fujiwara, A., and Kubozono, Y. (2011). Characteristics of conjugated hydrocarbon based thin film transistor with ionic liquid gate dielectric, *Org. Electron.*, **12** p. 2076.

14. Kawai, N., Eguchi, R., Goto, H., Akaike, K., Kaji, Y., Kambe, T., Fujiwara, A., and Kubozono, Y. (2012). Characteristics of single crystal field-effect transistors with a new type of aromatic hydrocarbon, picene, *J. Phys. Chem. C*, **116**, p. 7983.

15. Komura, N., Goto, H., He, X., Mitamura, H., Eguchi, R., Kaji, Y., Okamoto, H., Sugawara, Y., Gohda, S., Sato, K., and Kubozono, Y. (2012). Characteristics of [6]phenacene thin film field-effect transistor, *Appl. Phys. Lett.*, **101**, p. 083301.

16. Sugawara, Y., Ogawa, K., Goto, H., Oikawa, S., Akaike, K., Komura, N., Eguchi, R., Kaji, Y., Gohda, S., and Kubozono, Y. (2012). O-2-exposure and light-irradiation properties of picene thin film field-effect transistor: A new way toward O-2 gas sensor, *Sens. Actuators, B*, **171–172**, p. 544.

17. He, X., Eguchi, R., Goto, H., Uesugi, E., Hamao, S., Takabayashi, Y., and Kubozono, Y. (2013). Fabrication of single crystal field-effect transistors with phenacene-type molecules and their excellent transistor characteristics, *Org. Electron.*, **14**, p. 1673.

18. Nishihara, Y., Kinoshita, M., Hyodo, K., Okuda, Y., Eguchi, R., Goto, H., Hamao, S., Takabayashi, Y., and Kubozono, Y. (2013). Phenanthro[1,2-b : 8,7-b']dithiophene: a new picene-type molecule for transistor applications, *RSC Adv.*, **3**, p. 19341.

19. Eguchi, R., He, X., Hamao, S., Goto, H., Okamoto, H., Gohda, S., Sato, K., and Kubozono, Y. (2013). Fabrication of high performance/highly functional field-effect transistor devices based on [6]phenacene thin films, *Phys. Chem. Chem. Phys*, **15**, p. 20611.

20. He, X., Hamao, S., Eguchi, R., Goto, H., Yoshida, Y., Saito, G., and Kubozono, Y. (2014). Systematic control of hole-injection barrier height with electron acceptors in [7]phenacene single-crystal field-effect transistors, *J. Phys. Chem. C*, **118**, p. 5284.

21. Kubozono, Y., He, X., Hamao, S., Teranishi, K., Goto, H., Eguchi, R., Kambe, T., Gohda, S., and Nishihara, Y., (2014). Transistor Application of Phenacene Molecules and Their Characteristics, *Eur. J. Inorg. Chem.*, **2014**, p. 3806.

22. Lilienfeld, J. E., Method and Apparatus for Controlling Electric Currents, US Patent 1, 745, 175, filed October 8, 1926, patented January 28, p. 1930.

23. Lilienfeld, J. E., Device for Controlling Electric Current, US Patent 1, 900, 018, filed March 28, 1928, patented March 7, p. 1933.

24. Kahng, D., and Atalla, M. M. (1960). Silicon-silicondioxidesurfacedevice. In IRE Device Research Conference, Pittsburgh, p. 1960.

25. Kahng, D., Electric Field Controlled Semiconductor Device, US Patent 3, 102, 230, filed May 31, 1960, patented August 27, p. 1963.

26. Sze, S. M. (2002). *Semiconductor Devices: Physics and Technology*, John Wiley & Sons, New York.

27. Cleuziou, J.-P., Wernsdorfer, W., Bouchiat, V., Ondarçuhu, T., and Monthioux, M. (2006). Carbon nanotube superconducting quantum interference device, *Nat. Nanotech.*, **1**, p. 53.

28. Deshpande, V. V., Chandra, B., Caldwell, R., Novikov, D. S., Hone, J., and Bockrath, M. (2009). Mott insulating state in ultraclean carbon nanotubes, *Science*, **323**, p. 106.

29. Liang, W., Bockrath, M., Bozovic, D., Hafner, J. H., Tinkham, M., and Park, H. (2001). Fabry-Perot interference in a nanotube electron waveguide, *Nature*, **411**, p. 665.

30. Sahoo, S., Kontos, T., Furer, J., Hoffmann, C., Gräber, M., Cottet, A., and Schönenberger, C. (2005). Electric field control of spin transport, *Nat. Phys.*, **1**, p. 99.

31. Porath, D., Bezryadin, A., Vries, S., and Dekker, C. (2000). Direct measurement of electrical transport through DNA molecules, *Nature*, **403**, p. 635.

32. Yoo, K.-H., Ha, D. H., Lee, J.-O., Park, J. W., Kim, J., Kim, J. J., Lee, H.-Y., Kawai, T., and Choi, H. Y. (2001). Electrical conduction through poly(dA)-poly(dT) and poly(dG)-poly(dC) DNA molecules, *Phys. Rev. Lett.*, **87**, p. 198102.

33. Hwang, J. S., Kong, K. J., Ahn, D., Lee, G. S., Ahn, D. J., and Hwang, S. W. (2002). Electrical transport through 60 base pairs of poly(dG)-poly(dC) DNA molecules, *Appl. Phys. Lett.*, **81**, p. 1134.

34. Novoselov, K. S., Geim, A. K., Morozov, S. V., Jiang, D., Katsnelson, M. I., Grigorieva, I. V., Dubonos, S. V., and Firsov, A. A. (2005). Two-dimensional gas of massless Dirac fermions in graphene, *Nature*, **438**, p. 197.

35. Zhang, Y. Tan, Y.-W. Stormer, H. L., and Kim, P. (2005). Experimental observation of the quantum Hall effect and Berry's phase in graphene, *Nature*, **438**, p. 201.

36. Young, A. F., and Kim, P. (2009). Quantum interference and Klein tunnelling in graphene heterojunctions, *Nat. Phys.*, **5**, p. 222.

37. Heersche, H. B., Herrero, P. J., Oostinga, J. B., Vandersypen, L. M. K., and Morpurgo, A. F. (2007). Bipolar supercurrent in graphene, *Nature*, **446**, p. 56.

38. Fischer, A., Scholz, R., Leo, K., and Lüssem, B. (2012). An all C-60 vertical transistor for high frequency and high current density applications, *Appl. Phys. Lett.*, **101**, p. 213303.

39. Fischer, A., Siebeneicher, P., Kleemann, H., Leo, K., and Lüssem, B. (2012). Bidirectional operation of vertical organic triodes, *J. Appl. Phys.*, **111**, p. 044507.

40. Goto, H., Uesugi, E., Eguchi, R., Fujiwara, A., and Kubozono, Y. (2013). Edge-dependent transport properties in graphene, *Nano Lett.*, **13**, p. 1126.

41. Uesugi, E., Goto, H., Eguchi, R., Fujiwara, A., and Kubozono, Y. (2013). Electric double-layer capacitance between an ionic liquid and few-layer graphene, *Sci. Rep.*, **3**, p. 1595.

42. Goto, H., Uesugi, E., Eguchi, R., and Kubozono, Y. (2013). Parity effects in few-Layer graphene, *Nano Lett.*, **13**, p. 5153.

43. Sun, Y., Liu, Y., and Zhu, D. (2005). Advances in organic field-effect transistors, *J. Mater. Chem.*, **15**, p. 53.

44. Murphy, A. R., and Frechet, J. M. J. (2007). Organic semiconducting oligomers for use in thin film transistors, *Chem. Rev.*, **107**, p. 1066.

45. Anthony, J. E. (2008). The larger acenes: versatile organic semiconductors, *Angew. Chem., Int. Ed.*, **47**, p. 452.

46. Allard, S., Forster, M., Souharce, B., Thiem, H., and Scherf, U. (2008). Organic semiconductors for solution-processable field-effect transistors (OFETs), *Angew. Chem., Int. Ed.*, **47**, p. 4070.

47. Liu, S., Wang, W. M., Briseno, A. L., Mannsfeld, S. C. B., and Bao, Z. (2009). Controlled deposition of crystalline organic semiconductors for field-effect-transistor applications, *Adv. Mater.*, **21**, p. 1217.

48. Wang, S., Kappl, M., Liebewirth, I., Müller, M., Kirchhoff K., Pisula, W., and Müllen, K. (2012). Organic Field-Effect Transistors based on highly ordered single polymer fibers *Adv. Mater.*, **24**, p. 417.

49. Ie, Y., Ueta, M., Nitani, M., Tohnai, N., Miyata, M., Tada, H., and Aso, Y. (2012). Air-Stable n-type organic field-effect transistors based on 4,9-dihydro-s-indaceno[1,2-b:5,6-b ']dithiazole-4,9-dione unit, *Chem. Mater.*, **24**, p. 3285.

50. Cavallini, M., D'Angelo, P., Criado, V. V., Gentili, D., Shehu, A., Leonardi, F., Milita, S., Liscio, F., and Biscarini, F. (2011). Ambipolar multi-stripe organic field-effect transistors, *Adv. Mater.*, **23**, p. 5091.

51. Keil, C., and Schlettwein, D. (2011). Development of the field-effect mobility in thin films of F16PcCu characterized by electrical in situ measurements during device preparation, *Org. Electron.*, **12**, p. 1376.

52. Lv, A., Puniredd, S. R., Zhang, J., Li, Z., Zhu, H., Jiang, W., Dong, H., He, Y., Jiang, L., Li, Y., Pisula, W., Meng, Q., Hu, W., and Wang, Z. (2012). High mobility, air stable, organic single crystal transistors of an n-type diperylene bisimide, *Adv. Mater.*, **24**, p. 2626.

53. Bao, Z., Lovinger, A.J., and Dodabalapur, A. (1996). Organic field-effect transistors with high mobility based on copper phthalocyanine, *Appl. Phys. Lett.*, **69**, p. 3066.

54. Yi, H. T., Payne, M. M., Anthony, J. E., and Podzorov, V. (2012). Ultra-flexible solution-processed organic field-effect transistors, *Nat. Commun.*, **3**, p. 1259.

55. Dodabalapur, A., Torsi, L., and Katz, H. E. (1995). Orgnic transistors - 2-dimensional transport and improved electrical characteristics, *Science*, **268**, p. 270.

56. Zschieschang, U., Kang, M. J., Takimiya, K., Sekitani, T., Someya, T., Canzler, T. W., Werner, A., Nimoth, J. B., and Klauk, H. (2012). Flexible low-voltage organic thin-film transistors and circuits based on C-10-DNTT, *J. Mater. Chem.*, **22**, p. 4273.

57. Grahn, H. T. (1999) *Introduction to Semiconductor Physics*, World Scientific, Singapore.

58. Colinge, J. P., and Colinge, C. A. (2006). *Physics of Semiconductor Devices*, Springer, Massachusetts.

59. Schroder, D. K. (2006) *Semiconductor Material and Device Characterization*, John Wiley & Sons, New Jersey.

60. Sze, S. M. (2007). *Physics of Semiconductor Devices*, John Wiley & Sons New Jersey.

61. Mishra, U., and Singh, J. (2008). *Semiconductor Device Physics and Design*, Springer, Dordrecht.

62. Liou, J. J., Conde, A. O., and Sanchez, F. G. (1998). *Analysis and Design of Mosfets: Modeling, Simulation, and Parameter Extraction*, Kluwer Academic, Massachusetts.

63. Kymissis, I. (2009). *Organic Field Effect Transistors: Theory, Fabrication and Characterization*, Springer, New York.

64. Bao, Z., and Locklin, J. (2007). *Organic Field-Effect Transistors* CRC Press, Taylor & Francis Group, Boca Raton.

65. Hummel, R. E. (2012). *Electronic Properties of Materials*, Springer, New York.

66. Singh, Y., and Agnihotri, S. (2009). *Semiconductor Devices*, I.K. International, New Delhi.

67. Barbe, D. F., and Westgate, C. R. (1970). Surface state parameters of metal-free phthalocyanine single crystals *J. Phys. Chem. Solids*, **31**, p. 2679.

68. Tsumura, A., Koezuka, H. and Ando, T. (1986). Macromolecular electronic device - field-effect transistor with a polythiophene thin-film, *Appl. Phys. Lett.*, **49**, p. 1210.

69. Fuchigami, H., Tsumura, A., and Koezuka, H. (1993). Polythienylenevinylene thin-film-transistor with high carrier mobility, *Appl. Phys. Lett.*, **63**, p. 1372.

70. Horowitz, G., Peng, X., Fichou, D., and Garnier, F. (1992). Role of the semiconductor/insulator interface in the characteristics of π-conjugated-oligomer-based thin-film transistors, *Synth. Met.*, **51**, p. 419.

71. Dimitrakopoulos, C. D., Brown, A. R., and Pomp, A. (1996). Molecular beam deposited thin films of pentacene for organic field effect transistor applications, *J. Appl. Phys.*, **80**, p. 2501.

72. Lin, Y.-Y., Gundlach, D. J., Nelson, S. F., and Jackson, T. N. (1997). Stacked pentacene layer organic thin-film transistors with improved characteristics, *Electron Device Lett.*, **18**, p. 606.

73. Yasuda, T., Goto, T., Fujita, K., and Tsutsui, T. (2004). Ambipolar pentacene field-effect transistors with calcium source-drain electrodes, *Appl. Phys. Lett.*, **85**, p. 2098.

74. Podzorov, V., Pudalov, V. M., and Gershenson, M. E. (2003). Field-effect transistors on rubrene single crystals with parylene gate insulator *Appl. Phys. Lett.*, **82**, p. 1739.

75. Butko, V. Y., Chi, X., Lang, D. V., and Ramirez, A. P. (2003). Field-effect transistor on pentacene single crystal, *Appl. Phys. Lett.*, **83**, p. 4773.

76. de Boer, R. W. I., Klapwijk, T. M., and Morpurgo, A. F. (2003). Field-effect transistors on tetracene single crystals, *Appl. Phys. Lett.*, **83**, p. 4345.

77. Takeya, J., Yamagishi, M., Tominari, Y., Hirahara, R., Nakazawa, Y., Nishikawa, T., Kawase, T., Shimoda, T., and Ogawa, S. (2007). Very high-mobility organic single-crystal transistors with in-crystal conduction channels, *Appl. Phys. Lett.*, **90**, p. 102120.

78. Kawasugi, Y., Yamamoto, H. M., Hosoda, M., Tajima, N., Fukunaga, T., Tsukagoshi, K., and Kato, R.(2008). Strain-induced superconductor/insulator transition and field effect in a thin single crystal of molecular conductor*Appl. Phys. Lett.*, **92**, p. 243508.

79. Kawasaki, N., Ohta, Y., Kubozono, Y., Konishi, A., and Fujiwara, A. (2008). An investigation of correlation between transport characteristics and trap states in n-channel organic field-effect transistors, *Appl. Phys. Lett.*, **92**, p. 163307.

80. Jiang, H., and Kloc, C. (2013). Single-crystal growth of organic semiconductors, *MRS Bull.*, **38**, p. 28.

81. Kanbara, T., Shibata, K., Fujiki, S., Kubozono, Y., Kashino, S., Urisu, T., Sakai, M., Fujiwara, A., Kumashiro, R., and Tanigaki, K. (2003). N-channel field effect transistors with fullerene thin films and their application to a logic gate circuit, *Chem. Phys. Lett.*, **379**, p. 223.

82. Kobayashi, S., Takenobu, T., Mori, S., Fujiwara, A., and Iwasa, Y., (2003). Fabrication and characterization of C-60 thin-film transistors with high field-effect mobility, *Appl. Phys. Lett.*, **82**, p. 4581.

83. Chua, L.-L., Zaumseil, J., Chung, J.-F., Ou, E.C.-W., Ho, P.K.-H., Sirringhaus, H., and Friend, R. H. (2005). General observation of n-type field-effect behaviour in organic semiconductors, *Nature*, **438**, p. 194.

84. Kubozono, Y., Mitamura, H., Lee, X., He, X., Yamanari, Y., Takahashi, Y., Suzuki, Y., Kaji, Y., Eguchi, R., Akaike, K., Kambe, T., Okamoto, H., Fujiwara, A., Kato, T., Kosugigh, T., and Aoki, H. (2011). Metal-intercalated aromatic hydrocarbons: a new class of carbon-based superconductors, *Phys. Chem. Chem. Phys.*, **13**, p. 16476.

85. Kurihara, N., Yao, A., Sunagawa, M., Ikeda, Y., Terai, K., Kondo, H., Saito, M., Ikeda, H., and Nakamura, H. (2013). High-mobility organic thin-film transistors over 10 cm^2 v^{-1} s^{-1} fabricated using bis(benzothieno)naphthalene polycrystalline films, *Jpn. J. Appl. Phys.*, **52**, p. 05DC11.

86. Amin, A. Y., Khassanov, A., Reuter, K., Meyer-Friedrichsen, T., and Halik, M. (2012). Low-voltage organic field effect transistors with a 2-tridecyl[1]benzothieno[3,2-b][1]benzothiophene semiconductor layer, *J. Am. Chem. Soc.*, **134**, p. 16548.

87. Kang, S.-M., and Leblebici, Y. (2003). *CMOS Digital Integrated Circuits Analysis and Design*, McGraw-Hill, New York.

88. Ball, J. M., Wöbkenberg, P. H., Colléaux, F., Heeney, M., Anthony, J. E., McCulloch, I., Bradley, D. D. C., and Anthopoulos, T. D. (2009). Solution processed low-voltage organic transistors and complementary inverters, *Appl. Phys. Lett.*, **95**, p. 103310.

89. Okamoto, H., Eguchi, R., Hamao, S., Goto, H., Gotoh, K., Sakai, Y., Izumi, M., Takaguchi, Y., Gohda, S., and Kubozono, Y. (2014). An Extended Phenacene-type Molecule, [8] Phenacene: Synthesis and Transistor Application, *Sci. Rep.*, **4**, p. 5330.

90. Mitsuhashi, R., Suzuki, Y., Yamanari, Y., Mitamura, H., Kambe, T., Ikeda, N., Okamoto, H., Fujiwara, A., Yamaji, M., Kawasaki, N., Maniwa, Y., and Kubozono, Y. (2010). Superconductivity in alkali-metal-doped picene, *Nature*, **464**, p. 76.

91. Ueno, K., Nakamura, S., Shimotani, H., Ohtomo, A., Kimura, N., Nojima, T., Aoki, H., Iwasa, Y., and Kawasaki, M. (2008). Electric-field-induced superconductivity in an insulator, *Nat. Mater.*, **7**, p. 855.

92. Ye, J. T., Inoue, S., Kobayashi, K., Kasahara, Y., Yuan, H. T., Shimotani, H., and Iwasa, Y. (2010). Liquid-gated interface superconductivity on an atomically flat film, *Nat. Mater.*, **9**, p. 125.

93. Ueno, K., Nakamura, S., Shimotani, H., Yuan, H.T., Kimura, N., Nojima, T., Aoki, H., Iwasa, Y., and Kawasaki, M. (2011). Discovery of superconductivity in KTaO3 by electrostatic carrier doping, *Nat. Nanotech.*, **6**, p. 408.

Chapter 10

Photonic Ring Resonators for Biosensing

Carlos Errando-Herranz and Kristinn B. Gylfason
KTH Royal Institute of Technology, Stockholm, Sweden
gylfason@kth.se

10.1 Introduction

The rapid growth of new sensors, such as accelerometers, gyro-scopes, and pressure sensors, in modern smartphones owes its success to the miniaturization techniques developed by the micro-electromechanical systems (MEMS) and semiconductor industries.

For biological applications, biosensors are already finding great use in medical diagnostics in hospitals and pharmaceutical research. However, to bring biosensor technology to a wider user base, the magic of miniaturization must be brought to bear.

In this chapter, we present the photonic ring resonator biosensor technology, which, by taking advantage of nanofabrication technologies, is well suited for this kind of miniaturization.

Below, we start by introducing the concept of biosensing and discussing the current challenges in the field, followed by the history

Nanodevices for Photonics and Electronics: Advances and Applications
Edited by Paolo Bettotti
Copyright © 2016 Pan Stanford Publishing Pte. Ltd.
ISBN 978-981-4613-74-3 (Hardcover), 978-981-4613-75-0 (eBook)
www.panstanford.com

of ring resonator biosensor development. We continue by reviewing the theory of biosensing with ring resonators and conclude with an overview of the fabrication methods used.

10.1.1 *Label-Free Biosensing*

In the most general terms, a *biosensor* is a device that combines a biological recognition component with a physical detector component, the transducer, in order to measure a physical property [1]. The recognition component may, for example, be composed of nucleic acids [2], enzymes [3], or antibodies [4]. The transducer could be optical [1, 5–7], electromechanical [8], electrochemical [9], magnetic, or calorimetric. The main advantage of biosensing is that the sensor benefits from the natural selectivity of the biological recognition component. In a sense, it takes advantage of the trial and error process already done by evolution. When a selective recognition element is combined with a sensitive transducer, a powerful analytical tool emerges.

The most established biosensors make use of special reporter molecules or particles, known as labels, to read out the measured quantity. Most often these labels are fluorescent or radioactive, but plasmonic or magnetic nanoparticles are also commonly used. *Label-based* sensors can measure down to the single-molecule level, but they have a number of significant drawbacks that make it worthwhile to explore alternative methods.

One limitation of label-based methods is the time, and thus cost, of the labeling step [10]. For each new class of experiments, a labeling protocol must be developed. Another is the complexity that the labels add to the reaction under study, and thus an increased risk of interference and misinterpretation [11]. Finally, since labels are typically added in a secondary step, most label-based methods can report only on the end point of a reaction and do not provide real-time kinetic information. This information can, however, easily be gained by label-free methods.

Label-free biosensors have the unique advantage of supplying real-time quantitative information in physical units on the progress of (bio)chemical reactions, without the risk of interference from labels. From this information, reaction rates, binding affinities,

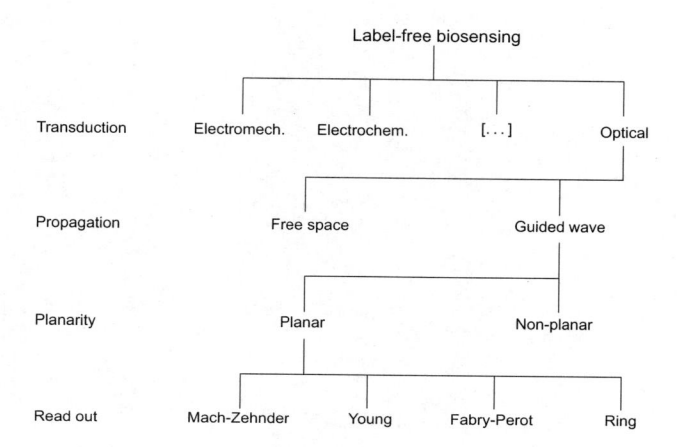

Figure 10.1 A taxonomy of label-free biosensing. The branch under study here is indicated with boldface labels. For a background, consult recent reviews on label-free biosensing: in general [13], electromechanical [8], electrochemical [9], optical [14] guided wave [6], planar waveguide based [5], nonplanar waveguide based [15], ring resonator gased [16].

and even layer densities [12] can be calculated. For other uses, such as simple positive/negative assays where real-time physical information is not needed, well-established label-based methods are usually a better choice.

Figure 10.1 shows one way of classifying label-free biosensing methods. The branch under study here is indicated by boldface labels. The cited references point to recent reviews of the respective topics.

10.1.2 *Biosensing with Planar Optical Waveguides*

Optical biosensors based on planar waveguides have a number of important advantages: Since they are optical, they are immune to electromagnetic interference, and the use of optical fibers allows for simple guidance of light to and from the optical chip. Since they are guided wave systems, the optical path is well defined by etching into the core layer, and thus the adjustment of bulky components, for which free-space optics are notorious, is avoided [5].

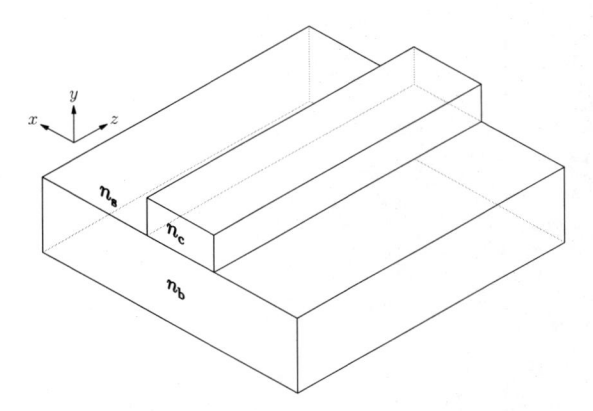

Figure 10.2   A schematic planar strip waveguide. Light propagates along the z direction and $n_c > n_b, n_s$.

Figure 10.2 shows a schematic illustration of a planar strip waveguide, the type most commonly used for biosensing. Light is confined in the high-refractive-index core ($n_c > n_b, n_s$) by total internal reflection.

A fraction of the optical field of the wave propagating in the waveguide, known as the evanescent field, extends a few hundred nanometers[a] into the liquid sample, and thus the sample refractive index close to the surface influences the propagation of light at the interface.

Planar waveguide sensors have two modes of operation, volume sensing and surface sensing. In volume sensing, the refractive index of the bulk sample n_s is of interest, for example, in refractometry of liquids and gases. In surface sensing, however, the interface is covered with a recognition component that binds selectively to the target of interest, and thus only the refractive index of this thin surface layer is of interest. Since the refractive index of aqueous protein solutions is linear with density [17], the density of protein binding on the surface σ_p can be quantitatively determined. This mode is the most relevant for real-time molecular interaction studies. Figure 10.3 shows how the evanescent field extends into

---

[a]The exact penetration depth depends on the wavelength and the refractive index step.

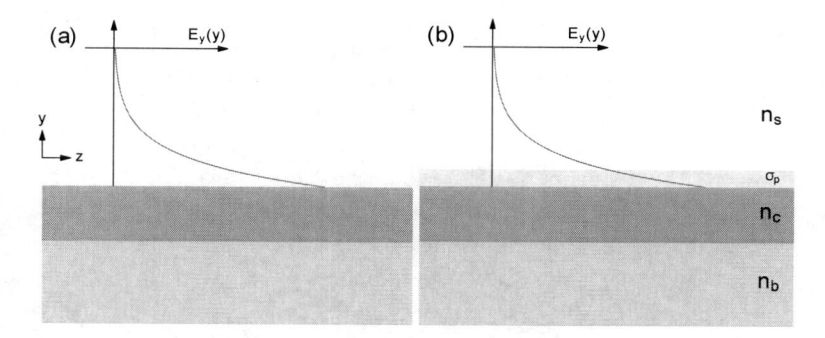

Figure 10.3 The evanescent field of a propagating wave decays exponentially with distance from the refractive index step of a waveguide. (a) In volume sensing, the bulk refractive index n_s of the whole liquid top cladding is of interest. (b) In surface sensing, only the density of a thin surface layer σ_p is of interest. Within this layer, the surface-binding reactions under study take place.

the liquid sample at the surface of waveguides and illustrates the difference between (a) volume sensing and (b) surface sensing.

The transduction principle of surface sensing with label-free planar waveguide biosensors is thus basically that a mass accumulation on the surface changes the refractive index of the surface layer, and thus the effective refractive index of the electromagnetic wave (mode) propagating in the waveguide.

Label-free biosensors based on planar *interferometric waveguide* configurations [12] are already well established in the analytical laboratory. State-of-the-art evanescent-field-based laboratory equipment achieves a volume refractive index limit of detection of the order of 10^{-8} [14] refractive index units (RIUs) and a surface mass density detection limit of 0.2 pg/mm^2 [12].

10.1.3 *Current Challenges*

The application of label-free biosensors based on planar waveguide interferometers in biology and medicine faces challenges such as parallelization of real-time experiments and effective sample transport. As we reason below, ring resonator biosensors address this challenges in a unique way, offering an attractive alternative to interferometric approaches.

10.1.3.1 Parallelization

The complexity of biological systems, and the inherent variability of living things, dictates careful use of statistics and control experiments to yield confidence in biosensing results. The more repeated the experiments, the more the confidence. Furthermore, the large number of possible interaction pathways in biological systems calls for many different experiments in order to map the interaction space. Thus, there is an obvious need to parallelize real-time label-free binding experiments, in the same way as label-based genetic assays have been parallelized by current DNA microarray technology.

Scaling down the sensor addresses this limitation by enabling the fabrication of large sensor arrays in a small space. However, from an optical perspective, it has an important disadvantage: decreased interaction length. The detection limit of optical methods is usually linearly dependent on the length of the optical cell employed. Consequently, a downscaling of 2 to 3 orders of magnitude spells trouble for the detection limit.

Resonators provide a way to effectively extend the interaction length of optical measurements. However, traditional mirror-based resonant cavities are not easily fabricated on a chip.

In planar waveguide *ring resonators*, light resonates inside a compact circular sensing element. These sensors are easily fabricated by standard integrated optics techniques and can be combined with other integrated optical and fluidic support functions in a 2D array to address the parallelization limitation of current systems. The small footprint and fabrication by standard techniques enable mass production at low cost [15].

In ring resonators, light propagates in the form of circulating waveguide modes. As shown in Fig. 10.4a, the circulating waves add constructively at those wavelengths that are divisors of the ring circumference, that is, at the resonance wavelengths. When biological material accumulates on the waveguide surface, the wavelength of light in the ring waveguide shortens and the resonance condition moves to a longer free-space wavelength, as shown in Fig. 10.4b. This wavelength shift is the signal used for biosensing.

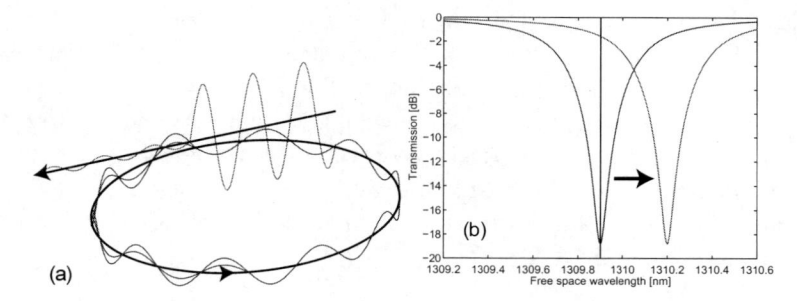

Figure 10.4 (a) Light in ring resonators propagates in the form of circulating waveguide modes. The waves add constructively at those wavelengths that are divisors of the ring circumference (i.e., at resonance). In turn, these waves add destructively with the light traveling along the bus waveguide, yielding an inverted resonance profile in the transmitted power. (b) When biological material accumulates on the waveguide surface, the wavelength of light in the ring shortens and the resonance condition moves to a longer free-space wavelength. This wavelength shift is the signal used for biosensing.

In contrast to, for example, Mach–Zehnder waveguide interferometers, the degree of interaction of light and the sample in ring resonators is not limited by the physical length of the sensing waveguide but rather by the number of rounds the circulating light can complete before decay, characterized by the resonator quality factor Q. To achieve a low detection limit, interferometers need a long sensing arm (usually a few millimeters). This has several consequences: First, changing the sample in a long flow cell takes time. This is a severe limitation for real-time experiments, where the dynamics of the initial response are of particular interest. Second, such a long device cannot be densely integrated in 2D arrays, which are highly desirable both from a cost and an application perspective. Third, conventional laboratory automation systems, such as robotic microarray spotters, deliver circular spots of diameters down to 60 μm and are thus poorly suited for selectively coating long, narrow optical waveguides.

Planar ring resonators made in GeSbS glass have demonstrated a detection limit of 8×10^{-7} RIU [18]. Since the performance-limiting components of planar ring resonators are fabricated by high-precision lithography, replicating such sensors on a chip does not reduce the sensing performance. Furthermore, the microfabrication methods employed excel at replicating identical devices over large surfaces, as can clearly be seen in the fabrication of modern central processing units (CPUs) and computer memory. For practical miniaturized analysis systems, however, one also needs to handle liquid samples on the microscale, and such techniques have only recently reached maturity, with the rapid development of the field of microfluidics [19].

10.1.3.2 Sample transport

A sensor cannot resolve reaction kinetics faster than the time it takes to refresh the sample covering its active sensing area. The active sensing area of many waveguide interferometers is in the mm^2 range [20–22], and thus a high sample flow speed is required in order to resolve fast reaction kinetics [23]. High flow speed, in turn, demands robust fluidics to handle high liquid pressure and can result in a large shear stress at the sensing surface.

The small footprint of ring resonators not only permits the creation of compact sensor arrays but also enables the resolution of fast kinetics. Single-mode ring resonators can have active areas in the 10 μm^2 range [68], and by metering samples with on-chip microfluidic circuits and separating them with air pockets to avoid diffusion during transport a ring-resonator-based sensor should greatly reduce the time response of current long-waveguide interferometers.

10.2 History

From the 1980s untill today, the research on photonic evanescent-field-based sensors has quickly developed from the first planar waveguide sensors, through the first experiments with ring resonator sensing, to highly integrated lab-on-a-chip devices based on arrays of ring resonator biosensors.

In this section, we review the historical development, by discussing notable reports on photonic ring resonator sensing.

10.2.1 *Planar Waveguide Evanescent-Field-Based Sensors*

A great deal of pioneering work on employing the evanescent field of planar optical waveguides for label-free sensing was performed at ETH (Zürich, Switzerland) by Lukosz and Tiefenthaler in the 1980s and the early 1990s [24, 25]. Most of the work was done on sol-gel-formed slab waveguides, with optical confinement only in the out-of-plane dimension. Surface gratings, embossed into the waveguiding layers [26], were used to read out changes in the effective index of the waveguides. Much of the early theoretical analysis of such 1D waveguide sensors [27, 28] remains a useful guideline for the design of lithographically patterned waveguides with optical confinement in two dimensions.

During the 1990s, more groups entered the field. Of particular notice is the work at the MESA Research Institute (University of Twente, Enschede, the Netherlands). Utilizing the microfabrication tools of the semiconductor industry to pattern single-mode waveguides, the group created more complex devices, such as integrated Mach–Zehnder [22, 29] and transverse electric–transverse magnetic (TE-TM) difference interferometers with integrated photodetectors [30].

10.2.2 *Planar Ring Resonator Sensors*

In 1997, Sohlström and Öberg at IMC AB (currently Acreo Swedish ICT AB, Kista, Sweden) reported the use of planar ring resonators for measuring the refractive index of liquids [31]. Their device operated at a wavelength of 1550 nm and used Si_3N_4 strip waveguides on a silicon oxide bottom cladding. It included an integrated splitter, for splitting light into a sensing arm and a reference arm, each of which was coupled to ring resonators of 1 mm radii. The waveguide cores were only 140 nm thick in order to achieve high refractive index sensitivity by pushing optical energy out into the sample region. Repeated injections of 1% sucrose solutions indicated a volume refractive index detection limit of 5×10^{-6} RIU.

In 2002, Krikouv et al. at Twente University demonstrated a similar sensor based on a Si_3N_4 disk resonator with only a 15 μm radius [32]. Since the device used a vertical coupling scheme for better control of the critical coupling distance between the disk and the bus waveguide, it could be fabricated by standard UV lithography. The small disk displayed three radial whispering gallery (WG) modes, and thus sensing required proper interpretation of the multimode transmission spectrum. The devices showed a volume refractive index sensitivity of 23 nm/RIU and a detection limit of 10^{-4} RIU.

In 2003, Chao and Guo at the University of Michigan (Ann Arbor, United States) showed volume refractive index sensing with an imprinted polystyrene (PS) ring sensor on a silicon dioxide cladding [33]. In a follow-up paper in 2006 [34], these devices were shown to have a detection limit of 5×10^{-5} RIU.

In early 2006, two papers were published on biomolecule surface sensing with ring resonators. Chao and Guo showed binding of streptavidin to a biotin-coated surface of a PS ring resonator [34]. On the basis of very limited data, the authors estimated a surface mass density detection limit of 250 pg/mm^2. Yalcin et al. at Boston University and Nomadics Inc. (Massachusetts, United States) also worked with avidin–biotin binding but on a Hydex glass ring [35]. The authors did not estimate the surface mass detection limit, but the rings showed a volume refractive index detection limit of 1.8×10^{-5} RIU.

At about the same time, Ksendzov and Lin at the California Institute of Technology (Pasadena, California, United States) also published on surface sensing of biomolecules [36]. They used a Si_xN_y ring of 2 mm radius on an SiO_2 cladding excited with a He/Ne laser at a 633 nm wavelength. The authors provided rather limited performance data and used a temperature-ramping method to read out the resonance shift, so comparison with the other published results shown in Table 10.1 is difficult. However, the authors estimated an avidin volume mass density detection limit of 6.8 ng/ml for their setup, but since the mass transport properties of the flow system have a large influence on this figure, comparison with other ring resonator sensors remains difficult.

Because of the high refractive index of silicon, silicon waveguides provide two benefits for sensing: Rings can be made very small without reducing Q by bending loss, and the evanescent electric field at the silicon surface is very high, yielding high sensitivity for surface sensing. In 2007, De Vos et al. at Ghent University (Ghent, Belgium) presented biosensing with silicon rings of only 5 μm radius [68] and showed antibody and protein sensing with a detection limit of 17 pg/mm^2 in follow-up papers in 2009 [37, 38]. A number of groups have shown similar devices [10, 39–41]. In 2010 and 2011, the Ghent group presented cascaded ring resonator biosensors [42, 43]. In cascaded ring resonators, the first ring acts as a filter, allowing wavelength interrogation with a low-cost broadband light source.

The first steps toward ring resonator sensor arrays were taken by Ramachandran et al. at Nomadics Inc. in 2007 [44]. The group integrated five rings of the type used by Yalcin et al., mentioned above, on a single chip. In 2009, the same group reported dynamic monitoring of cell growth under stimulus such as by toxic compounds [45] using the same chip layout. However, instead of an integrated optical approach, individual input and output fibers were coupled to each ring and light splitting was handled off-chip.

In 2007 and 2008, photonic ring resonator biosensors based on a novel waveguide type known as slot waveguides were presented [46], and their biosensing capabilities demonstrated [47] with a limit of detection of 20 pg/mm^2 (limited by the laser wavelength step). In a following work, a multiplexed chip with seven Si$_3$N$_4$ rings with 70 μm radius, on-chip light splitting and referencing, and integrated sample-handling system in polydimethylsiloxane (PDMS) was presented [48]. The integrated chip yielded a volume refractive index detection limit of 5×10^{-6} RIU and a surface mass density detection limit of 0.9 pg/mm^2.

In 2010, Xu et al. [49] at the National Research Council of Canada (Ottawa, Ontario, Canada) presented a silicon-on-insulator (SOI) ring resonator sensor array using wavelength division multiplexing (WDM) to connect 4 rings to the same bus waveguide, and measured simultaneously two protein analytes with an integrated sample-handling system in SU-8, yielding a detection limit of 0.3 pg/mm^2.

In late 2009 and early 2010, Washburn et al. at the University of Illinois (Urbana-Champeign, Illinois, United States) moved closer to applications by demonstrating a multiplexed chip for quantification of cancer biomarkers in complex media [10] and four other protein biomarkers [50]. In follow-up papers in 2011 and 2012, the same group reported biosensing of four cytokines [51, 52], viruses [53], four microRNAs [54], protein biomarkers in human serum [55], and four DNA sequences [56]. Recently, this group has reported enhancement of the sensitivity and specificity of ring resonator biosensors by secondary labeling with antibodies [57] and biofunctionalized submicron beads [58], an enzymatic signal enhancement strategy [59], and protein–protein and protein–lipid interaction screening [60]. The chip layout used in all the papers referred to above consists of 32 SOI rings of 15 μm radius, and light is coupled in and out of the chip with surface-grating couplers. The resonator arrays and the instrumentation for analyzing the ring resonance frequencies were acquired from Genalyte, Inc. (San Diego, California, United States) [101], a start-up company currently commercializing silicon-ring-resonator-based biosensors under the trade name Maverick.

More recently, some groups presented photonic ring resonator biosensors functionalized with aptamers [61, 62]. Aptamers are nucleic acids that bind specifically to certain proteins, thus potentially increasing the robustness of the surface functionalization, compared to using antibodies.

Table 10.1 summarizes the main performance metrics of some notable reports of planar ring resonator sensors.

10.2.3 *Sample Handling on Planar Ring Resonator Sensors*

As we introduced in the first section of this chapter (10.1.3.2), single-mode ring resonators can have active areas in the 10 μm^2 range [68], permitting the creation of compact sensor arrays and the resolution of fast kinetics. To bring sample and biosensor in contact and be able to resolve kinetics, an integrated sample-handling system (lab on a chip) is necessary.

A lab-on-a-chip system combining microfluidic sample handling with photonic ring resonator biosensors has great potential to become a compact and sensitive option for the future of medical diagnostics, and thus several groups have reported advances toward this concept.

In the very first proofs of concept, the sample was brought in contact with the sensor via pipetting [31–34], making a drying

Table 10.1

Year	Main novelty	Wavel. [nm]	Core	Bottom cladding	Sample handling	Radius [µm]	Vol. sens. [nm/RIU]	Vol. lim. [10^{-6}RIU]	Surf. lim. [pg/mm^2]	Ref.
1997	Ring sensor	1550	Si$_3$N$_4$	SiO$_2$	Pipetting	1000		5		[31]
2001	Small disk	780	Si$_3$N$_4$	SiO$_2$	Pipetting	15	23	100		[32]
2003	Polymer ring	1550	PS	SiO$_2$	Pipetting	30		50		[33]
2004	Biosensing	1550	PS	SiO$_2$	Pipetting	45		10	250	[34]
2005	Biosensing	1550	Hydex	Hydex	Immersion	50	140	18		[35]
2007	Small Si ring	1550	Si	SiO$_2$	Flow cell	5	70	10	17	[38, 68]
2007	Multi-sensor chip[a]		Hydex	Hydex	Immersion	50	140	18		[44]
2007	First slot sensor	1310	Si$_3$N$_4$	SiO$_2$	Pipetting	70	210	200[b]	20[b]	[46, 47]
2008	Pulley coupler	1550	GeSbS	SiO$_2$	Integrated PDMS	20	182	0.8		[18]
2008	Digital fluidics	1550	SU-8	SiO$_2$	Flow cell	300	69			[63]
2009	Monitoring of cell growth[a]		Hydex	Hydex	Immersion	50	140	18		[45]
2009	Si slot sensor	1550	Si	SiO$_2$	Flow cell	5	298	40		[64]
2009	Biosensing in serum	1560	Si	SiO$_2$	Mylar	15		3		[10, 65]
2009	Slot sensor array	1310	Si$_3$N$_4$	SiO$_2$	Integrated PDMS	70	246	5	0.9	[48]
2010	Cancer marker array	1560	Si	SiO$_2$	Mylar cell	15				[50]
2010	WDM approach	1570	Si	SiO$_2$	PDMS - SU-8 cell	20	135	2	0.3	[49]
2010	Cascaded resonators	1530	Si	SiO$_2$	Integrated PDMS	~400	2169	8.3		[42]
2011	Four cytokines	1560	Si	SiO$_2$	Mylar cell	15				[52]
2011	Biosensing in plasma	1560	Si	SiO$_2$	Mylar cell	15				[55]
2011	Four microRNA	1560	Si	SiO$_2$	Mylar cel	15				[57]
2012	Digital fluidics	1550	Si	SiO$_2$	Flow cell	5	78			[66]
2013	Protein-protein	1560	Si	SiO$_2$	Mylar cell	15				[60]
2013	Dense integration	1550	Si$_3$N$_4$	SiO$_2$	Integrated OSTE	20	50			[67]

[a]Light splitting by off-chip optical switch.
[b]Limited by laser wavelength step.

step between each measurement necessary, thus precluding the resolution of kinetics.

As the research on ring resonator biosensors developed, this limitation was overcome by the use of flow cells [35, 38, 64, 68]. A similar principle (i.e., Mylar microfluidics) is still being used by some groups [10, 42, 50, 52, 55, 57, 60]. However, to fabricate a functional biosensor device capable of measuring kinetics, an integrated sample-handling system is necessary to achieve better flow control and reduce the minimum sample volume.

In 2008, Hu et al. [18] presented the first integrated ring resonator sensor, in which PDMS was the chosen material for the fabrication of the microfluidic sample-handling system. PDMS microchannels (100×30 μm) were fabricated via replica molding and bonded to the sensor chip by oxygen plasma activation [18].

Since then, PDMS has been widely used for integrating microfluidic systems on photonic chips [37, 42, 48]. Carlborg et al. [48] presented a chip containing seven slot waveguide ring resonator biosensors with an integrated PDMS microfluidic system fabricated by molding and bonded to the photonic chip via oxygen plasma activation. However, access holes were punched manually, substantially increasing the wafer footprint of the chip.

However, in biological and medical applications, PDMS suffers from high adsorption and consequent leaching of small molecules [69, 70] and biocompatibility problems [71]. Moreover, the available techniques for bonding PDMS to silicon substrates, the most common being bonding by oxygen plasma activation, make PDMS incompatible with surface biofunctionalization on an industrial scale [72]. These limitations are apparent in [37], in which stamping with an epoxy glue results in channel clogging. For these reasons, to commercially manufacture biosensors with microfluidics, PDMS has to be replaced and the concept redeveloped [73].

A lab-on-a-chip system based on PDMS often takes orders of magnitude the wafer space occupied by the photonic biosensor system. Thus, the compactness, and the economical attractiveness, of photonic biosensors is lost.

To overcome the drawbacks of PDMS, in 2013, Errando-Herranz et al. presented a densely integrated lab-on-a-chip system with photonic interferometers [74] and ring resonators [67]. The wafer footprint of photonics and microfluidics was matched by combining grating couplers with microfluidics fabricated in a novel off-stoichiometric thiol-ene (OSTE) polymer [70] in a single-step process. The microfluidic layer was then dry-bonded to a ring resonator sensor chip in a bonding process compatible with biofunctionalization. Moreover, photolithography of vias in OSTE avoided squeeze film formation and enabled fabrication of tightly packed integrated ring resonator sensor chips.

Photonic ring resonator biosensing can potentially benefit from the emerging technology of digital microfluidics [75]. Digital microfluidics technology reduces the sample volume to a minimum, avoids cross talk between samples, and allows flexibility in sample preparation on a chip. In 2008 the first system integrating digital microfluidics and polymer photonic ring resonators was presented by Luan et al. [63]. However, the complexity of the system made it unsuitable for biosensing in practical applications. The concept was recently simplified and applied to multiplexed silicon photonic ring resonators by Lerma-Arce et al. [66] at Ghent University.

Table 10.1 summarizes the sample-handling systems used in some notable reports of planar ring resonator sensor devices.

10.3 Theory

We now review the theory necessary for interpreting biosensing experiments using ring resonators. We start by deriving an expression for the transmission spectrum of a ring resonator and then discuss the main performance metrics relevant for sensing.

10.3.1 *Scattering Matrix for a Side-Coupled Waveguide Ring Resonator*

To construct a frequency domain model of a ring resonator, side-coupled to a single-bus waveguide, we follow the treatment of [76] and [77] and consider the abstraction of Fig. 10.5.

The complex mode amplitudes B, b, D, and d are normalized such that their squared magnitude corresponds to the modal power. The symbol case distinguishes between bus modes (uppercase) and

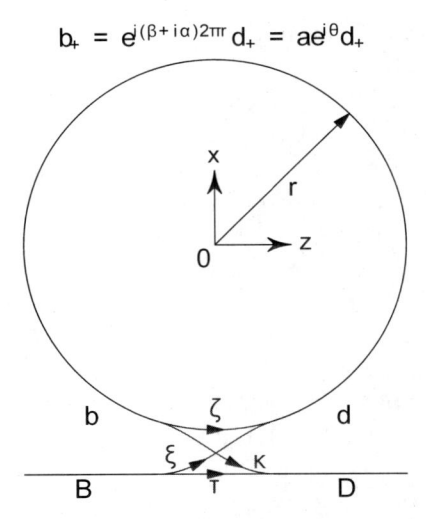

$$b_+ = e^{j(\beta + i\alpha)2\pi r} d_+ = ae^{j\theta}d_+$$

Figure 10.5 An abstraction of a waveguide ring resonator side-coupled to a bus waveguide. B, b, D, and d are the complex mode amplitudes. Ring modes are lowercase and bus modes uppercase. A subscript of $+$ or $-$ denotes a mode propagating toward the positive or negative z axis, respectively. The complex coupling coefficients ξ and κ, and transmission coefficients ζ and τ, describe the coupling.

cavity modes (lowercase). A subscript of plus $(+)$ or minus $(-)$ denotes a mode propagating toward the positive or negative z axis, respectively.

The following assumptions are made:

- All waveguides are unimodal per polarization orientation and do not couple between polarizations.
- The various kinds of losses are incorporated in the attenuation constant α of the individual modes.
- Back reflections are negligible.
- Interaction between modes is negligible outside the coupling region.

The operation of the coupler can now be described in terms of a scattering matrix that establishes a linear relation between the amplitudes B_-, b_-, D_+, and d_+ of the outgoing waves and the amplitudes B_+, b_+, D_-, and d_- of the incoming waves. The values of the complex coupling coefficients ξ and κ, and transmission coefficients ζ and τ, depend on the nature of the coupling.

$$\begin{bmatrix} B_- \\ b_- \\ D_+ \\ d_+ \end{bmatrix} = \begin{bmatrix} 0 & 0 & \tau & \xi \\ 0 & 0 & \kappa & \zeta \\ \tau & \kappa & 0 & 0 \\ \xi & \zeta & 0 & 0 \end{bmatrix} \begin{bmatrix} B_+ \\ b_+ \\ D_- \\ d_- \end{bmatrix} \tag{10.1}$$

The zeros in the matrix implement the assumption of negligible back reflections.

This system can be divided into two subsystems of parallel equations:

$$\begin{bmatrix} B_- \\ b_- \end{bmatrix} = \begin{bmatrix} \tau & \xi \\ \kappa & \zeta \end{bmatrix} \begin{bmatrix} D_- \\ d_- \end{bmatrix} \tag{10.2}$$

and

$$\begin{bmatrix} D_+ \\ d_+ \end{bmatrix} = \begin{bmatrix} \tau & \kappa \\ \xi & \zeta \end{bmatrix} \begin{bmatrix} B_+ \\ b_+ \end{bmatrix} \tag{10.3}$$

If we assume lossless coupling, the coupling matrices must be unitary. By making the matrices unitary ($\mathbf{M}^{*\mathrm{T}}\mathbf{M} = \mathbf{I}$), and solving the system, we have $\zeta = \tau^*$, $\xi = -\kappa^*$ and $|\tau|^2 + |\kappa|^2 = 1$ [78]. The scattering matrix of the waves traveling along the positive z axis (10.3) (and similarly for the waves traveling along the negative z axis) thus reduces to

$$\begin{bmatrix} D_+ \\ d_+ \end{bmatrix} = \begin{bmatrix} \tau & \kappa \\ -\kappa^* & \tau^* \end{bmatrix} \begin{bmatrix} B_+ \\ b_+ \end{bmatrix} \tag{10.4}$$

The single-mode ring waveguide of length $L = 2\pi r$ supports a cavity mode with a complex propagation constant $\gamma = \beta + i\alpha$, that is, with a phase propagation constant β (real) and an attenuation constant α (real, positive). For propagation along the cavity loop, with s measuring the propagation distance, the fields evolve proportionally to $e^{i\gamma s}$, leading to the relation

$$b_+ = e^{i(\beta+i\alpha)2\pi r} d_+ = e^{-\alpha L} e^{i\beta L} d_+ = a e^{i\theta} d_+, \tag{10.5}$$

where a and θ are the real mode amplification factor and the phase shift, respectively, for one round trip. For zero loss $a = 1$. The round-trip phase shift θ is sometimes referred to as detuning [79].

10.3.1.1 Power transfer

Due to the symmetry and linearity of the device, it is sufficient to consider an excitation of one of the ports. We assume $B_+ = \sqrt{P_{in}} = 1$ and $D_- = 0$ and solve (10.4) and (10.5) for the transmitted power $P_T = |D_+|^2$. Our neglection of reflections implies that there is no back-reflected power $B_- = 0$.

Solving (10.4) for the transmitted mode gives

$$D_+ = \frac{-a + \tau e^{-i\theta}}{-a\tau^* + e^{-i\theta}}, \tag{10.6}$$

and for the circulating mode

$$b_+ = \frac{-a\kappa^*}{-a\tau^* + e^{-i\theta}}. \tag{10.7}$$

Setting $\tau = |\tau| e^{i\psi}$, yields the transmitted power

$$P_T = |D_+|^2 = \frac{a^2 + |\tau|^2 - 2a|\tau|\cos(\theta + \psi)}{1 + a^2|\tau|^2 - 2a|\tau|\cos(\theta + \psi)}, \tag{10.8}$$

and the power circulating in the ring cavity

$$P_C = |b_+|^2 = \frac{a^2(1 - |\tau|^2)}{1 + a^2|\tau|^2 - 2a|\tau|\cos(\theta + \psi)}. \tag{10.9}$$

10.3.1.2 Resonances

On resonance, when $\cos(\theta + \psi) = 1$ or, equivalently, when

$$\beta = \frac{2\pi m - \psi}{L} =: \beta_m, \tag{10.10}$$

where m is the integer longitudinal mode number, we obtain

$$P_T = \frac{(a - |\tau|)^2}{(1 - a|\tau|)^2}, \tag{10.11}$$

and

$$P_C = \frac{a^2(1 - |\tau|)^2}{(1 - a|\tau|)^2}. \tag{10.12}$$

Critical coupling, $P_T = 0$, is achieved when $a = |\tau|$, that is, when internal cavity field losses equal the noncoupled field fraction [80].

10.3.1.3 Free spectral range

The distance in free-space wavelength between resonances, or *free spectral range* (FSR), is of interest for most practical applications. The resonant configuration next to β_m is

$$\beta_{m-1} = \frac{2(m-1)\pi - \psi}{L} = \beta_m - \frac{2\pi}{L} \approx \beta_m + \frac{\partial \beta}{\partial \lambda} \Delta \lambda, \qquad (10.13)$$

where $\Delta\lambda$ is the difference between the vacuum wavelengths corresponding to the two resonant configurations. By expressing the first-order dispersion approximation on the right-hand side in terms of the group index n_g using

$$\beta(\lambda) \simeq \beta_0 - n_g \frac{k}{\lambda}(\lambda - \lambda_0), \qquad (10.14)$$

we find

$$\Delta\lambda \approx -\frac{2\pi}{L}\left(\frac{\partial \beta}{\partial \lambda}\right)^{-1} = \frac{\lambda^2}{n_g L} \qquad (10.15)$$

In Fig. 10.6, we plot (10.8) and (10.9) as functions of free-space wavelength. When drawing the graph, we account only for first-order dispersion of the phase propagation constant β by setting $\theta = 2\pi L n_g/\lambda$ and ignore dispersion of the coupling coefficients. Furthermore, we assume $\psi = 0$. In the most general case, however, all the parameters in (10.8) and (10.9) can depend on wavelength.

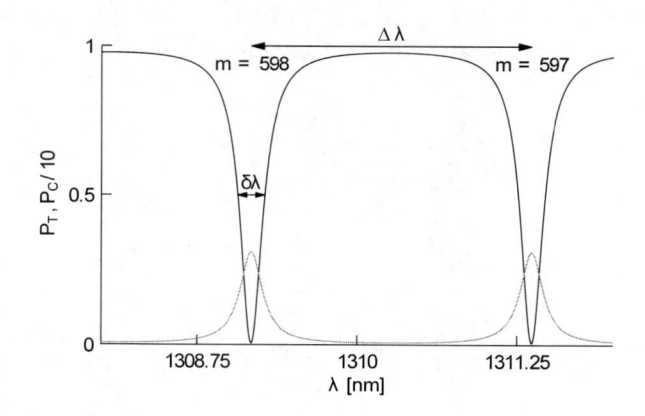

Figure 10.6 The transmitted power P_T (black line) and circulating power P_C (gray line) of a side-coupled ring resonator as functions of the free-space wavelength λ, for $a = 0.87$, $\tau = 0.85$, $L = 2\pi \times 70$ μm, and $n_g = 1.78$.

10.3.1.4 Resonator finesse and quality factor

By employing the approximation of (10.14) and expanding the cosine term of (10.9) one can find the full spectral width at half maximum [76, 81, 82]:

$$\delta\lambda = \frac{\lambda^2}{\pi n_g L} \frac{1 - |\tau|e^{-\alpha L}}{|\tau|^{\frac{1}{2}} e^{\frac{-\alpha L}{2}}} \tag{10.16}$$

The finesse $\mathcal{F}$ of a resonator is defined as the ratio of the FSR and the full width at half maximum. The resonator finesse determines how many channels fit in a wavelength division multiplexing scheme. From (10.15) and (10.16) we find

$$\mathcal{F} = \frac{\Delta\lambda}{\delta\lambda} = \pi \frac{|\tau|^{\frac{1}{2}} e^{\frac{-\alpha L}{2}}}{1 - |\tau|e^{-\alpha L}} \tag{10.17}$$

The quality factor of the resonator is the ratio of the wavelength and the resonance width:

$$Q = \frac{\lambda}{\delta\lambda} = \frac{n_g L}{\lambda}\mathcal{F} = \frac{n_g 2\pi r}{\lambda}\mathcal{F} = n_g k r \mathcal{F} \tag{10.18}$$

10.3.1.5 Sensing

When employed as sensors, a perturbation p, to be measured, changes the phase propagation constant β of the ring waveguides. Thus the resonance at λ_m shifts to a new wavelength $\tilde{\lambda}_m$. A linear approximation in p and the wavelength shift $\Delta_p \lambda_m = \tilde{\lambda}_m - \lambda_m$ yields [76]

$$\beta(p, \tilde{\lambda}_m) \approx \beta(0, \lambda_m) + \frac{\partial\beta}{\partial p}p + \frac{\partial\beta}{\partial\lambda}(\tilde{\lambda}_m - \lambda_m) = \beta_m, \tag{10.19}$$

and thus

$$\Delta_p \lambda_m = -p\frac{\partial\beta}{\partial p}\left(\frac{\partial\beta}{\partial\lambda}\right)^{-1} = p\frac{\partial\beta}{\partial p}\frac{\lambda_m^2}{2\pi n_g} = p\frac{\partial n_e}{\partial p}\frac{\lambda_m}{n_g} \approx \frac{\Delta n_e}{n_g}\lambda_m. \tag{10.20}$$

The form

$$\frac{\Delta_p \lambda_m}{\lambda_m} \approx \frac{\Delta n_e}{n_g} \tag{10.21}$$

is convenient for comparing experimental results of $\Delta\lambda_m/\lambda_m$ to predictions of $\Delta n_e/n_g$ obtained by numerical eigenmode solvers.

10.3.1.6 Volume sensing

For volume sensing $p = \Delta n_s$, and we obtain

$$\frac{\Delta \lambda_m}{\lambda_m} = \Delta n_s \frac{\partial n_e}{\partial n_s} \frac{1}{n_g},$$
(10.22)

and the volume sensitivity can thus be expressed as

$$S_n = \frac{\partial n_e}{\partial n_s} \frac{\lambda_m}{n_g} \quad [\text{nm/RIU}].$$
(10.23)

10.3.1.7 Surface sensing

For surface sensing $p = \Delta \sigma_p$, where σ_p is the surface density of the sensor surface, and we obtain

$$\frac{\Delta \lambda_m}{\lambda_m} = \Delta \sigma_p \frac{\partial n_e}{\partial \sigma_p} \frac{1}{n_g},$$
(10.24)

and a surface sensitivity

$$S_\sigma = \frac{\partial n_e}{\partial \sigma_p} \frac{\lambda_m}{n_g} \quad [\text{nm/(pg/mm}^2)].$$
(10.25)

10.3.2 *Sensor Figure of Merit and Detection Limit*

The performance of a resonant sensor is determined by two properties, its sensitivity, that is, how far the resonant condition moves for a certain change in the sensed parameter, and the sharpness of its resonance, that is, how accurately one is able to determine the resonant condition. These two properties are equally important, and thus a suitable *sensor figure of merit* gives them equal weight. A commonly used figure of merit is the ratio of the sensitivity (bigger is better) and the resonance width (smaller is better) [83]

$$\text{FOM} = \frac{\partial \lambda_m}{\partial p} \frac{1}{\delta \lambda} = \frac{S}{\delta \lambda}.$$
(10.26)

This figure of merit is quite abstract and general and permits quantitative comparison of very different resonant sensor realizations. It is, however, also a somewhat naive, since it doesn't take into account many practical limitations that arise when complete sensor systems are assembled. For example, interrogating a sensor with an extremely high Q requires a laser with a very narrow line width and fine wavelength stepping.

The sensor *detection limit* is a closely related concept [84], although more practically inspired. It compares the sensitivity and noise level of an actual sensor system implementation. Thus it includes noise sources such as light source variation, detector noise, thermal noise, etc., and gives a practical estimate of the smallest change that can be measured. The detection limit is

$$\mathrm{DL} = \frac{3\sigma}{S},$$
(10.27)

where 3σ is three standard deviations of the observed total system noise.

Although the detection limit is more useful than the sensor figure of merit for practical comparisons of label-free sensing system implementations, it still excludes such important factors as the ability to resolve fast reaction kinetics. Recently, Hu et al. [85] proposed a figure of merit that includes the response time of the optical system, but the influence of analyte mass transport also needs consideration [86].

10.4 Fabrication

10.4.1 *Sensors*

Photonic ring resonators can be fabricated by a number of different microfabrication techniques, but their application as biosensors limits the choice of fabrication method somewhat, since biosensing includes exposure to water-based biosamples. This exposure can cause drift of the optical properties of the ring resonators, which is very difficult to compensate for. Accordingly, we focus our discussion below on fabrication methods that have successfully been applied to biosensing.

10.4.1.1 Polymer-based sensors

One of the first demonstrations of biosensing using ring resonators was performed on polymer ring resonators molded into polystyrene (PS) [34]. The molds were fabricated by electron beam (e-beam) lithography and then used to imprint a waveguide core into PS at elevated temperatures [87]. In this approach, the PS layer is spin-coated on top of a thermal silicon dioxide bottom cladding before imprinting. Furthermore, to increase the Q of the resonators, the authors employed a thermal reflow process for reducing the sidewall roughness [88, 89].

Another method for structuring polymer waveguides is by UV curing. In this case, a liquid prepolymer is cast onto a transparent mold and the structure fixed by UV polymerization. A recent example of such an approach, which has been applied to biosensing, is the UV imprinting of ring resonators into polysiloxane-liquid (PSQ-L) [90, 91]. Here, the imprint mold was made out of the elastomer polydimethylsiloxane (PDMS), and thus the authors refer to the process as soft lithography.

An interesting aspect of polymer waveguides is that they can be made porous by selectively dissolving one component of a two-component polymer. This approach has recently been suggested to increase the sensitivity of waveguide-based biosensors [92]. In this case, the core layer is imprinted into a mixture of PS and poly(methyl methacrylate) (PMMA), as above, followed by soaking in dimethylsulfoxide (DMSO), to selectively dissolve the PMMA, leaving a porous waveguide structure. So far, no biosensing experiments have been demonstrated by this approach, and the limited mass transport into the porous waveguides will likely be a practical limitation of this scheme for biosensing.

Finally, polymer-based sensors can also be fabricated by UV lithography of photopatternable core layers. For example, the UV-curable epoxy SU-8 can be used as a core material, in combination with a cladding of the fluoropolymer Cytop. A vertically coupled ring resonator fabricated by this method was recently demonstrated [93].

10.4.1.2 Thin-solid-film-based sensors

The most advanced ring resonator sensor systems fabricated to date have been made by patterning solid thin films using the fabrication methods of the semiconductor industry. The thin films can be thermally grown, such as in the case of silicon dioxide, deposited by chemical vapor deposition (CVD) or plasma-enhanced chemical vapor deposition (PECVD), or transfer-bonded.

A silicon nitride core layer on a silicon dioxide bottom cladding was one of the first material systems applied to ring resonator sensor fabrication [31]. The bottom cladding is typically grown by oxidizing a silicon substrate and the core layer deposited either by low-pressure CVD or PECVD. Patterning of the waveguide layers then proceeds with deep-UV or e-beam lithography, followed by dry plasma etching. The main benefit of this approach is the maturity of these processes, since they have been applied for decades to the fabrication of electronics, and the stability of the materials, since silicon nitride is commonly used for device passivation in electronics fabrication. The earliest example of biosensing with silicon nitride ring resonators is the work of Ksendzov and Lin [36], and some more recent examples are [47, 48, 94].

As explained in Section 10.2.2, the high refractive index of silicon allows rings to be made very small without reducing Q by bending loss and also yields a high evanescent electric field at the silicon surface. Silicon ring resonator biosensor fabrication starts with a silicon-on-insulator (SOI) substrate consisting of a silicon bulk with a layer of thermally grown silicon dioxide buried below a device layer of crystalline silicon. These substrates are made by transfer bonding techniques and are commercially available [95]. The fabrication sequence is thus very simple and consists of a single deep-UV lithography and dry-etching step to form the silicon waveguide core.

Apart from silicon and silicon nitride, successful biosensing experiments have been performed on glass resonators [35].

10.4.2 Sample-Handling System

From pipetting in [31] to a highly integrated lab-on-a-chip system in [67], the sample-handling systems for photonic ring resonators have evolved in parallel with the development of the microfluidics field.

Due to their optical transparency and mechanical properties, polymers are the most commonly used materials for integration of microfluidics with photonic ring resonators. Some examples include SU-8 [49], PDMS [96], and off-stoichiometric thiol-ene (OSTE) [67].

10.4.2.1 SU-8

SU-8 is an epoxy-based negative photoresist commonly used to make molds for soft lithography.

Due to its stiffness, SU-8 has been scarcely used for sample-handling systems on photonic ring resonators. Xu et al. [49] patterned sensing windows on top of ring resonators in a 2 μm thin layer of SU-8, followed by photolithography of a microfluidic channel in a 50 μm layer of SU-8. A PDMS layer was then placed on top of the channel to seal during testing.

10.4.2.2 PDMS

PDMS is a polymeric organosilicon compound belonging to the group of the silicones. Among polymers, PDMS is currently the most used material for microfluidic devices, due to its elasticity, ease of fabrication, and optical transparency [97]. Consequently, PDMS has been integrated with photonic ring resonator biosensors in several proofs of concept [37, 42, 48]. Figure 10.7 shows four ring resonators in PDMS microchannels.

PDMS microfluidic chips containing microchannels can be easily fabricated by soft lithography and bonded to the sensing surface by different approaches.

One of them is stamping with epoxy glue. In 2009, De Vos et al. fabricated a PDMS layer by soft lithography and bonded it by

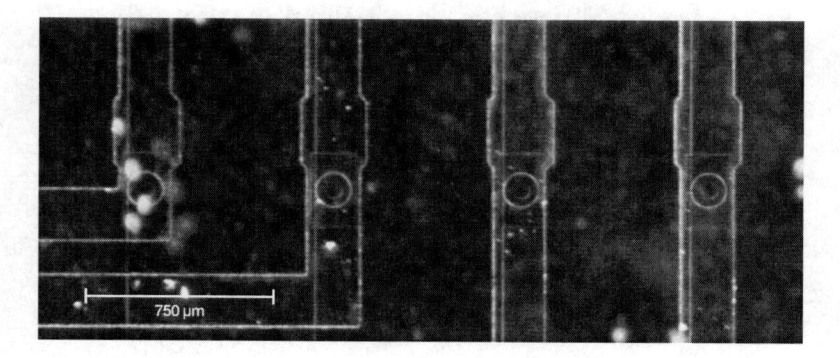

Figure 10.7 Four ring resonator sensors in PDMS microchannels. The image is a top view through the transparent PDMS.

stamping with an epoxy glue to an SOI sensor chip [37]. The authors mention an important limitation in the stamping, in which small variations in the amount of glue result in channel clogging.

Oxygen plasma activation is a second approach for PDMS bonding, used by Claes et al. to bond PDMS microfluidics to an SOI sensor surface [42]. However, the poor time stability of oxygen plasma activation, combined with the long times needed for surface biofunctionalization, makes PDMS incompatible with surface biofunctionalization at a wafer scale [72].

Due to squeeze film formation, it is difficult to create open through-holes (vias) by standard soft lithography of PDMS [98]. Hence, liquid connections to the outside world are usually created by a low-accuracy punching step after final curing of PDMS microfluidics. Consequently, the wafer layout of the photonics must include dead area to increase alignment tolerance of the bonding to the microfluidics. For example, the chip presented in [48] was 20-fold the footprint of the stand-alone photonic system, thus losing the scaling benefit of silicon photonic sensors.

Other limitations of PDMS in biological and medical applications are the high absorption and subsequent leaching of biomolecules [70] and biocompatibility issues with respect to cell culturing [71].

10.4.2.3 OSTE

In 2011, Carlborg et al., at the Royal Institute of Technology (KTH) Stockholm, Sweden, presented a novel polymer platform specifically developed for lab-on-a-chip applications. The polymer, named OSTE, is polymerized from two monomers, each with either thiol or alkene functional groups, in an off-stoichiometric ratio. By addition of a photoinitiatior, the polymerization can be triggered by UV light, enabling photopatterning of the material by the use of a photomask. By choosing the off-stoichiometry ratio, a wide range of mechanical properties can be obtained, from a highly stiff material to a highly elastic one [70].

Recently, OSTE has been shown to address the manufacturing challenge of matching the wafer footprint of microfluidics with that of photonics. In 2013, Errando-Herranz et al. combined photopatterning and molding processes in order to fabricate an OSTE microfluidic chip with vias in a single step, and then dry-bonded it to a silicon photonic interferometer sensor array chip [74] and a silicon photonic ring resonator array chip [67]. By combining OSTE microfluidics with grating coupled photonic sensors, the group fabricated a densely integrated sensor chip.

Figure 10.8 shows the fabrication process and layout of the chip in [67]. To fabricate the microfluidic layer (Fig. 10.8b), the OSTE prepolymer was sandwiched between a glass mask, acting both as a photolithography mask and a top mold, and a PDMS bottom mold. The glass mask contains chromium patterns for photolithography of openings in the microfluidic layer and SU-8

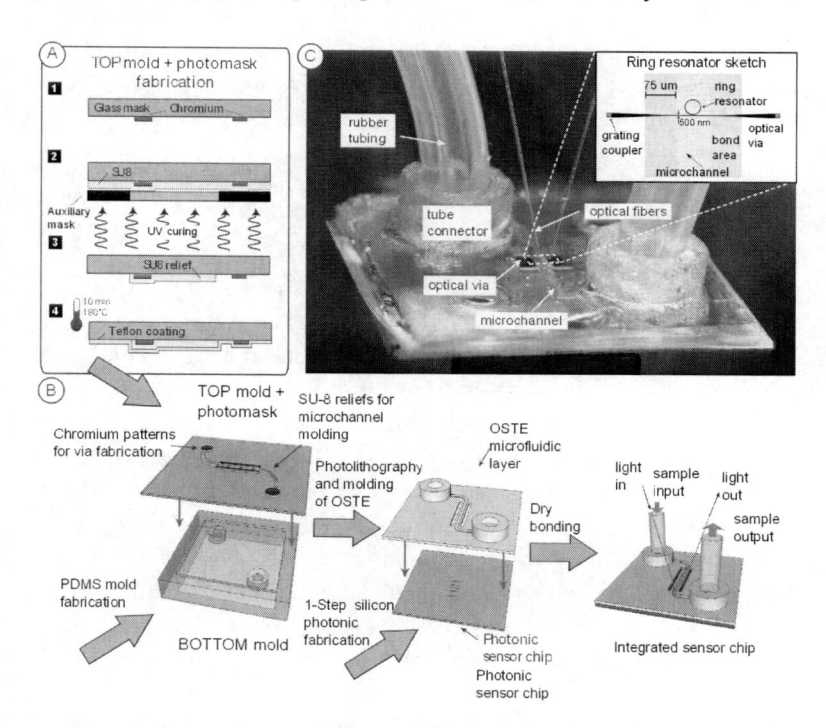

Figure 10.8 Fabrication of a ring resonator array integrated with microfluidics presented in [67]. (a) The process starts with the fabrication of the top mold and photomask. (b) The OSTE microfluidic layer is then molded and photopatterned in a single step and bonded to the silicon photonic resonator chip. (c) The final device with rubber tubing connected for sample transport.

reliefs for molding of channels (fabrication of the mask is shown in Fig. 10.8a). Photopatterning of vias in the microfluidic layer avoids the manual punching process present in PDMS fabrication, enabling patterning of tightly spaced structures in OSTE. The bottom PDMS mold defines both the outer dimensions of the microfluidic chip and the form of the integrated standard connectors, which enable a sealed connection to an external pumping system through rubber tubes.

Furthermore, OSTE features low-temperature bonding to silicon in a dry-bonding process compatible with biofunctionalization [99], and several OSTE formulations have shown good biocompatibility for cell culturing [100].

In [67], an OSTE microfluidic chip with a channel width of 100 μm separated by only 75 μm bond areas from the optical vias was integrated on a silicon photonic ring resonator sensor device with a total length (bus waveguide) of 500 μm (Fig. 10.8c).

10.5 Summary

Since the beginning of biosensing, much work has focused on the detection of analytes without the need for complicated, expensive, and time-consuming labeling techniques. Among label-free biosensors, optical planar waveguide sensors offer important advantages such as immunity to electromagnetic interference, simple guidance of light, small size, and a standardized fabrication process, which enables fabrication of dense arrays of biosensors on a small chip. In particular, photonic ring resonators are especially effective for label-free biosensing, due to the combination of a large, effective interaction length with a small size, enabling the resolution of fast reaction kinetics when integrated in lab-on-a-chip systems.

In this chapter, after introducing the concept of ring resonator biosensing, we traced the historical development of ring resonator biosensor technology, from the first steps in refractive index sensing in 1997 [31] until the development of highly integrated lab-on-a-chip devices in 2013 [48, 67]. After the historical review, the theory of ring resonator sensors was introduced, resulting in the definition of relevant figures of merit for sensing. We concluded the chapter by discussing the fabrication methods used for both the sensors and the sample-handling systems.

The features of photonic ring resonator biosensing that we have discussed show great potential for this technology for becoming an important medical and biological tool for research and diagnostics.

Acknowledgments

This work was partially supported by the Swedish Research Council (B0460801), the Göran Gustafsson Foundation, and the European Research Council (267528).

References

1. Brecht, A., and Gauglitz, G. (1995). Optical probes and transducers, *Biosens. Bioelectron.*, **10**, 9–10, pp. 923–936, doi:10.1016/0956-5663 (95)99230-i, http://dx.doi.org/10.1016/0956-5663(95)99230-i.

2. Kerman, K., Kobayashi, M., and Tamiya, E. (2004). Recent trends in electrochemical DNA biosensor technology, *Meas. Sci. Technol.*, **15**, 2, pp. R1–R11, doi:10.1088/0957-0233/15/2/r01, http://dx.doi.org/ 10.1088/0957-0233/15/2/r01.

3. Ronkainen-Matsuno, N. (2002). Electrochemical immunoassay moving into the fast lane, *TrAC Trends Anal. Chem.*, **21**, 4, pp. 213–225, doi:10.1016/s0165-9936(02)00401-6, http://dx.doi.org/ 10.1016/s0165-9936(02)00401-6.

4. Zheng, G., Patolsky, F., Cui, Y., Wang, W.U., and Lieber, C.M. (2005). Multiplexed electrical detection of cancer markers with nanowire sensor arrays, *Nat. Biotechnol.*, **23**, 10, pp. 1294–1301, doi:10.1038/ nbt1138, http://dx.doi.org/10.1038/nbt1138.

5. Lambeck, P.V. (2006). Integrated optical sensors for the chemical domain, *Meas. Sci. Technol.*, **17**, 8, pp. R93–R116, doi:10.1088/ 0957-0233/17/8/r01, http://dx.doi.org/10.1088/0957-0233/17/8/ r01.

6. Passaro, V. M.N., Dell'Olio, F., Casamassima, B., and De Leonardis, F. (2007). Guided-wave optical biosensors, *Sensors*, **7**, 4, pp. 508– 536, doi:10.3390/s7040508, http://www.mdpi.com/1424-8220/7/ 4/508.

7. Gauglitz, G., and Proll, G. (2008). Strategies for label-free optical detection, *Adv. Biochem. Eng./Biotechnol.*, **109**, pp. 395–432, doi:10.1007/10_2007_076, http://dx.doi.org/10.1007/10_2007_076.

8. Waggoner, P.S., and Craighead, H.G. (2007). Micro- and nanomechanical sensors for environmental, chemical, and biological detection, *Lab Chip*, **7**, 10, pp. 1238–1255, doi:10.1039/b707401h, http://dx.doi.org/10.1039/b707401h.

9. Vestergaard, M., Kerman, K., and Tamiya, E. (2007). An overview of label-free electrochemical protein sensors, *Sensors*, **7**, 12, pp. 3442–3458, doi:10.3390/s7123442, http://dx.doi.org/10.3390/s7123442.

10. Washburn, A.L., Gunn, L.C., and Bailey, R.C. (2009). Label-free quantitation of a cancer biomarker in complex media using silicon photonic microring resonators, *Anal. Chem.*, **81**, 22, pp. 9499–9506, doi:10.1021/ac902006p, http://dx.doi.org/10.1021/ac902006p.

11. Sun, Y.S., Landry, J.P., Fei, Y.Y., Zhu, X.D., Luo, J.T., Wang, X.B., and Lam, K.S. (2008). Effect of fluorescently labeling protein probes on kinetics of protein-ligand reactions, *Langmuir: ACS J. Surf. Colloids*, **24**, 23, pp. 13399–13405, doi:10.1021/la802097z, http://dx.doi.org/10.1021/la802097z.

12. Swann, M.J., Freeman, N.J., and Cross, G.H. (2007). *Dual Polarization Interferometry: A Real-Time Optical Technique for Measuring (Bio)Molecular Orientation, Structure and Function at the Solid/Liquid Interface*, Vol. 1, Chap. 33, pp. 549–568, Wiley, ISBN 978-0-470-01905-4.

13. Qavi, A., Washburn, A., Byeon, J.-Y., and Bailey, R. (2009). Label-free technologies for quantitative multiparameter biological analysis, *Anal. Bioanal. Chem.*, **394**, 1, pp. 121–135, doi:10.1007/s00216-009-2637-8, http://dx.doi.org/10.1007/s00216-009-2637-8.

14. Fan, X., White, I.M., Shopova, S.I., Zhu, H., Suter, J.D., and Sun, Y. (2008). Sensitive optical biosensors for unlabeled targets: a review, *Anal. Chim. Acta*, **620**, 1–2, pp. 8–26, doi:10.1016/j.aca.2008.05.022, http://dx.doi.org/10.1016/j.aca.2008.05.022.

15. Vollmer, F., and Arnold, S. (2008). Whispering-gallery-mode biosensing: label-free detection down to single molecules, *Nat. Methods*, **5**, 7, pp. 591–596, doi:10.1038/nmeth.1221, http://dx.doi.org/10.1038/nmeth.1221.

16. Jokerst, N., Royal, M., Palit, S., Luan, L., Dhar, S., and Tyler, T. (2009). Chip scale integrated microresonator sensing systems, *J. Biophotonics*, **2**, 4, pp. 212–226, doi:10.1002/jbio.200910010, http://dx.doi.org/10.1002/jbio.200910010.

17. De Feijter, J.A., Benjamins, J., and Veer, F.A. (1978). Ellipsometry as a tool to study the adsorption behavior of synthetic and biopolymers at the air-water interface, *Biopolymers*, **17**, 7, pp. 1759–1772, doi: 10.1002/bip.1978.360170711, http://dx.doi.org/10.1002/bip.1978.360170711.

18. Hu, J., Carlie, N., Feng, N.-N., Petit, L., Agarwal, A., Richardson, K., and Kimerling, L. (2008). Planar waveguide-coupled, high-index-contrast, high-Q resonators in chalcogenide glass for sensing, *Opt. Lett.*, **33**, 21, pp. 2500–2502, doi:10.1364/ol.33.002500, http://dx.doi.org/10.1364/ol.33.002500.

19. Squires, T.M., and Quake, S.R. (2005). Microfluidics: fluid physics at the nanoliter scale, *Rev. Mod. Phys.*, **77**, 3, pp. 977–1026, doi:10.1103/revmodphys.77.977, http://dx.doi.org/10.1103/revmodphys.77.977.

20. Cross, G.H., Reeves, A.A., Brand, S., Popplewell, J.F., Peel, L.L., Swann, M.J., and Freeman, N.J. (2003). A new quantitative optical biosensor for protein characterisation, *Biosens. Bioelectron.*, **19**, 4, pp. 383–390, doi:10.1016/s0956-5663(03)00203-3, http://dx.doi.org/10.1016/s0956-5663(03)00203-3.

21. Ymeti, A., Kanger, J.S., Greve, J., Lambeck, P.V., Wijn, R., and Heideman, R.G. (2003). Realization of a multichannel integrated young interferometer chemical sensor, *Appl. Opt.*, **42**, 28, pp. 5649–5660, doi:10.1364/AO.42.005649, http://dx.doi.org/10.1364/AO.42.005649.

22. Heideman, R.G., and Lambeck, P.V. (1999). Remote opto-chemical sensing with extreme sensitivity: design, fabrication and performance of a pigtailed integrated optical phase-modulated Mach-Zehnder interferometer system, *Sens. Actuators, B*, **61**, 1–3, pp. 100–127, doi:10.1016/s0925-4005(99)00283-x, http://dx.doi.org/10.1016/s0925-4005(99)00283-x.

23. Gervais, T., and Jensen, K.F. (2006). Mass transport and surface reactions in microfluidic systems, *Chem. Eng. Sci.*, **61**, 4, pp. 1102–1121, doi:10.1016/j.ces.2005.06.024, http://dx.doi.org/10.1016/j.ces.2005.06.024.

24. Tiefenthaler, K., and Lukosz, W. (1984). Integrated optical switches and gas sensors, *Opt. Lett.*, **9**, 4, pp. 137+, http://www.opticsinfobase.org/abstract.cfm?id=8209.

25. Lukosz, W., and Stamm, C. (1990). Integrated optical interferometer as relative humidity sensor and differential refractometer, *Sens. Actuators, A*, **25**, 1–3, pp. 185–188, doi:10.1016/0924-4247(90)87029-i, http://dx.doi.org/10.1016/0924-4247(90)87029-i.

26. Lukosz, W., and Tiefenthaler, K. (1983). Embossing technique for fabricating integrated optical components in hard inorganic waveguiding materials, *Opt. Lett.*, **8**, 10, pp. 537–539, doi:10.1364/ol.8.000537, http://dx.doi.org/10.1364/ol.8.000537.

27. Lukosz, W., and Tiefenthaler, K. (1988). Sensitivity of integrated optical grating and prism couplers as (bio)chemical sensors, *Sens. Actuators*, **15**, 3, pp. 273–284, doi:10.1016/0250-6874(88)87016-1, http://dx.doi.org/10.1016/0250-6874(88)87016-1.

28. Tiefenthaler, K., and Lukosz, W. (1989). Sensitivity of grating couplers as integrated optical chemical sensors, *J. Opt. Soc. Am. B: Opt. Phys.*, **6**, 2, pp. 209–220, doi:10.1364/JOSAB.6.000209, http://dx.doi.org/10.1364/JOSAB.6.000209.

29. Heideman, R.G., Kooyman, R.P.H., and Greve, J. (1993). Performance of a highly sensitive optical waveguide Mach-Zehnder interferometer immunosensor, *Sens. Actuators, B*, **10**, 3, pp. 209–217, doi:10.1016/0925-4005(93)87008-d, http://dx.doi.org/10.1016/0925-4005(93)87008-d.

30. Koster, T., and Lambeck, P. (2002). Fully integrated optical polarimeter, *Sens. Actuators, B*, **82**, 2–3, pp. 213–226, doi:10.1016/s0925-4005(01)01008-5, http://dx.doi.org/10.1016/s0925-4005(01)01008-5.

31. Sohlström, H., and Öberg, M. (1997). Refractive index measurement using integrated ring resonators. In *The 8th European Conference on Integrated Optics*, pp. 322–325.

32. Krioukov, E., Klunder, Driessen, A., Greve, J., and Otto, C. (2002). Sensor based on an integrated optical microcavity, *Opt. Lett.*, **27**, 7, pp. 512–514, doi:10.1364/ol.27.000512, http://dx.doi.org/10.1364/ol.27.000512.

33. Chao, C.Y., and Guo, L.J. (2003). Biochemical sensors based on polymer microrings with sharp asymmetrical resonance, *Appl. Phys. Lett.*, **83**, 8, pp. 1527–1529, doi:10.1063/1.1605261, http://dx.doi.org/10.1063/1.1605261.

34. Chao, C.-Y., Fung, W., and Guo, L.J. (2006). Polymer microring resonators for biochemical sensing applications, *IEEE J. Sel. Top. Quant. Electron.*, **12**, 1, pp. 134–142, doi:10.1109/jstqe.2005.862945, http://dx.doi.org/10.1109/jstqe.2005.862945.

35. Yalcin, A., Popat, K.C., Aldridge, J.C., Desai, T.A., Hryniewicz, J., Chbouki, N., Little, B.E., King, O., Van, V., Chu, S., Gill, D., Anthes-Washburn, M., Unlu, M.S., and Goldberg, B.B. (2006). Optical sensing of biomolecules using microring resonators, *IEEE J. Sel. Top. Quant. Electron.*, **12**, 1,

pp. 148–155, doi:10.1109/jstqe.2005.863003, http://dx.doi.org/10.1109/jstqe.2005.863003.

36. Ksendzov, A., and Lin, Y. (2005). Integrated optics ring-resonator sensors for protein detection, *Opt. Lett.*, **30**, 24, pp. 3344–3346, doi: 10.1364/ol.30.003344, http://dx.doi.org/10.1364/ol.30.003344.

37. De Vos, K., Girones, J., Claes, T., De Koninck, Y., Popelka, S., Schacht, E., Baets, R., and Bienstman, P. (2009). Multiplexed antibody detection with an array of silicon-on-insulator microring resonators, *IEEE Photonics J.*, **1**, 4, pp. 225–235, doi:10.1109/jphot.2009.2035433, http://dx.doi.org/10.1109/jphot.2009.2035433.

38. De Vos, K., Girones, J., Popelka, S., Schacht, E., Baets, R., and Bienstman, P. (2009). SOI optical microring resonator with poly(ethylene glycol) polymer brush for label-free biosensor applications, *Biosens. Bioelectron.*, **24**, 8, pp. 2528–2533, doi:10.1016/j.bios.2009.01.009, http://dx.doi.org/10.1016/j.bios.2009.01.009.

39. Xu, D.X., Densmore, A., Delâge, A., Waldron, P., McKinnon, R., Janz, S., Lapointe, J., Lopinski, G., Mischki, T., Post, E., Cheben, P., and Schmid, J.H. (2008). Folded cavity SOI microring sensors for highsensitivity and real time measurement ofbiomolecular binding, *Opt. Express*, **16**, 19, pp. 15137–15148, doi:10.1364/oe.16.015137, http://dx.doi.org/10.1364/oe.16.015137.

40. Fukuyama, M., Yamatogi, S., Ding, H., Nishida, M., Kawamoto, C., Amemiya, Y., Ikeda, T., Noda, T., Kawamoto, S., Ono, K., Kuroda, A., and Yokoyama, S. (2010). Selective detection of Antigen-antibody reaction using si ring optical resonators, *Jpn. J. Appl. Phys.*, **49**, 4, pp. 04DL09+, doi:10.1143/JJAP.49.04DL09, http://dx.doi.org/10.1143/JJAP.49.04DL09.

41. Kim, K.W., Song, J., Kee, J.S., Liu, Q., Lo, G.-Q., and Park, M.K. (2013). Label-free biosensor based on an electrical tracing-assisted silicon microring resonator with a low-cost broadband source, *Biosens. Bioelectron.*, **46**, pp. 15–21, doi:10.1016/j.bios.2013.02.002, http://dx.doi.org/10.1016/j.bios.2013.02.002.

42. Claes, T., Bogaerts, W., and Bienstman, P. (2010). Experimental characterization of a silicon photonic biosensor consisting of two cascaded ring resonators based on the Vernier-effect and introduction of a curve fitting method for an improved detection limit, *Opt. Express*, **18**, 22, pp. 22747–22761, doi:10.1364/oe.18.022747, http://dx.doi.org/10.1364/oe.18.022747.

43. Claes, T., Bogaerts, W., and Bienstman, P. (2011). Vernier-cascade label-free biosensor with integrated arrayed waveguide grating for wavelength interrogation with low-cost broadband source, *Opt. Lett.*, **36**, 17, pp. 3320–3322, doi:10.1364/ol.36.003320, http://dx.doi.org/ 10.1364/ol.36.003320.

44. Ramachandran, A., Wang, S., Clarke, J., Ja, S., Goad, D., Wald, L., Flood, E., Knobbe, E., Hryniewicz, J., Chu, S., Gill, D., Chen, W., King, O., and Little, B. (2008). A universal biosensing platform based on optical microring resonators, *Biosens. Bioelectron.*, **23**, 7, pp. 939–944, doi:10.1016/ j.bios.2007.09.007, http://dx.doi.org/10.1016/j.bios.2007.09.007.

45. Wang, S., Ramachandran, A., and Ja, S.-J. (2009). Integrated microring resonator biosensors for monitoring cell growth and detection of toxic chemicals in water, *Biosens. Bioelectron.*, **24**, 10, pp. 3061–3066, doi:10.1016/j.bios.2009.03.027, http://dx.doi.org/10.1016/j. bios.2009.03.027.

46. Barrios, C.A., Gylfason, K.B., Sánchez, B., Griol, A., Sohlström, H., Holgado, M., and Casquel, R. (2007). Slot-waveguide biochemical sensor, *Opt. Lett.*, **32**, 21, pp. 3080–3082, doi:10.1364/OL.32.003080, http://dx.doi.org/10.1364/OL.32.003080.

47. Barrios, C.A., Bañuls, M.J., González-Pedro, V., Gylfason, K.B., Sánchez, B., Griol, A., Maquieira, A., Sohlström, H., Holgado, M., and Casquel, R. (2008). Label-free optical biosensing with slot-waveguides, *Opt. Lett.*, **33**, 7, pp. 708–710, http://www.opticsinfobase.org/abstract.cfm?id= 156931.

48. Carlborg, C.F., Gylfason, K.B., Kazmierczak, A., Dortu, F., Polo, M. J.B., Catala, A.M., Kresbach, G.M., Sohlström, H., Moh, T., Vivien, L., Popplewell, J., Ronan, G., Barrios, C.A., Stemme, G., and Wijngaart (2010). A packaged optical slot-waveguide ring resonator sensor array for multiplex label-free assays in labs-on-chips, *Lab Chip*, **10**, 3, pp. 281–290, doi:10.1039/b914183a, http://dx.doi.org/10.1039/ b914183a.

49. Xu, D.X., Vachon, M., Densmore, A., Ma, R., Delâge, A., Janz, S., Lapointe, J., Li, Y., Lopinski, G., Zhang, D., Liu, Q.Y., Cheben, P., and Schmid, J.H. (2010). Label-free biosensor array based on silicon-on-insulator ring resonators addressed using a WDM approach, *Opt. Lett.*, **35**, 16, pp. 2771–2773, doi:10.1364/ol.35.002771, http://dx.doi.org/10. 1364/ol.35.002771.

50. Washburn, A.L., Luchansky, M.S., Bowman, A.L., and Bailey, R.C. (2010). Quantitative, label-free detection of five protein biomarkers using

multiplexed arrays of silicon photonic microring resonators, *Anal. Chem.*, **82**, 1, pp. 69–72, doi:10.1021/ac902451b, http://dx.doi.org/10.1021/ac902451b.

51. Luchansky, M.S., and Bailey, R.C. (2010). Silicon photonic microring resonators for quantitative cytokine detection and T-cell secretion analysis, *Anal. Chem.*, **82**, 5, pp. 1975–1981, doi:10.1021/ac902725q, http://dx.doi.org/10.1021/ac902725q.

52. Luchansky, M.S., and Bailey, R.C. (2011). Rapid, multiparameter profiling of cellular secretion using silicon photonic microring resonator arrays, *J. Am. Chem. Soc.*, **133**, 50, pp. 20500–20506, doi:10.1021/ja2087618, http://dx.doi.org/10.1021/ja2087618.

53. McClellan, M.S., Domier, L.L., and Bailey, R.C. (2012). Label-free virus detection using silicon photonic microring resonators, *Biosens. Bioelectron.*, **31**, 1, pp. 388–392, doi:10.1016/j.bios.2011.10.056, http://dx.doi.org/10.1016/j.bios.2011.10.056.

54. Qavi, A.J., and Bailey, R.C. (2010). Multiplexed detection and label-free quantitation of microRNAs using arrays of silicon photonic microring resonators, *Angew. Chem., Int. Ed.*, **49**, 27, pp. 4608–4611, doi:10.1002/anie.201001712, http://dx.doi.org/10.1002/anie.201001712.

55. Luchansky, M.S., Washburn, A.L., McClellan, M.S., and Bailey, R.C. (2011). Sensitive on-chip detection of a protein biomarker in human serum and plasma over an extended dynamic range using silicon photonic microring resonators and sub-micron beads, *Lab Chip*, **11**, 12, pp. 2042–2044, doi:10.1039/c1lc20231f, http://dx.doi.org/10.1039/c1lc20231f.

56. Qavi, A.J., Mysz, T.M., and Bailey, R.C. (2011). Isothermal discrimination of single-nucleotide polymorphisms via real-time kinetic desorption and label-free detection of DNA using silicon photonic microring resonator arrays, *Anal. Chem.*, **83**, 17, pp. 6827–6833, doi:10.1021/ac201659p, http://dx.doi.org/10.1021/ac201659p.

57. Qavi, A.J., Kindt, J.T., Gleeson, M.A., and Bailey, R.C. (2011). Anti-DNA:RNA antibodies and silicon photonic microring resonators: increased sensitivity for multiplexed microRNA detection, *Anal. Chem.*, **83**, 15, pp. 5949–5956, doi:10.1021/ac201340s, http://dx.doi.org/10.1021/ac201340s.

58. Kindt, J.T., and Bailey, R.C. (2012). Chaperone probes and bead-based enhancement to improve the direct detection of mRNA using silicon photonic sensor arrays, *Anal. Chem.*, **84**, 18, pp. 8067–8074, doi:10.1021/ac3019813, http://dx.doi.org/10.1021/ac3019813.

59. Kindt, J.T., Luchansky, M.S., Qavi, A.J., Lee, S.-H., and Bailey, R.C. (2013). Subpicogram per milliliter detection of interleukins using silicon photonic microring resonators and an enzymatic signal enhancement strategy, *Anal. Chem.*, **85**, 22, pp. 10653–10657, doi: 10.1021/ac402972d, http://dx.doi.org/10.1021/ac402972d.

60. Sloan, C.D., Marty, M.T., Sligar, S.G., and Bailey, R.C. (2013). Interfacing lipid bilayer nanodiscs and silicon photonic sensor arrays for multiplexed protein-lipid and protein-membrane protein interaction screening, *Anal. Chem.*, **85**, 5, pp. 2970–2976, doi:10.1021/ac3037359, http://dx.doi.org/10.1021/ac3037359.

61. Byeon, J.-Y., and Bailey, R.C. (2011). Multiplexed evaluation of capture agent binding kinetics using arrays of silicon photonic microring resonators, *Analyst*, **136**, 17, pp. 3430–3433, doi:10.1039/c0an00853b, http://dx.doi.org/10.1039/c0an00853b.

62. Park, M.K., Kee, J.S., Quah, J.Y., Netto, V., Song, J., Fang, Q., La Fosse, E.M., and Lo, G.-Q. (2013). Label-free aptamer sensor based on silicon microring resonators, *Sens. Actuators, B*, **176**, pp. 552–559, doi: 10.1016/j.snb.2012.08.078, http://dx.doi.org/10.1016/j.snb.2012.08.078.

63. Luan, L., Evans, R.D., Schwinn, D., Fair, R.B., and Jokerst, N.M. (2008). Chip scale integration of optical microresonator sensors with digital microfludics systems. In *21st Annual Meeting of the IEEE Lasers and Electro-Optics Society, 2008. LEOS 2008*, IEEE, ISBN 978-1-4244-1931-9, pp. 259–260, doi:10.1109/leos.2008.4688588, http://dx.doi.org/10.1109/leos.2008.4688588.

64. Claes, T., Molera, J.G., De Vos, K., Schacht, E., Baets, R., and Bienstman, P. (2009). Label-free biosensing with a slot-waveguide-based ring resonator in silicon on insulator, *IEEE Photonics J.*, **1**, 3, pp. 197–204, doi:10.1109/jphot.2009.2031596, http://dx.doi.org/10.1109/jphot.2009.2031596.

65. Bailey, R.C., Washburn, A.L., Qavi, A.J., Iqbal, M., Gleeson, M., Tybor, F., and Gunn, L.C. (2009). A robust silicon photonic platform for multiparameter biological analysis, *SPIE*, pp. 72200N+, doi:10.1117/12.809819, http://dx.doi.org/10.1117/12.809819.

66. Lerma Arce, C., Witters, D., Puers, R., Lammertyn, J., and Bienstman, P. (2012). Silicon photonic sensors incorporated in a digital microfluidic System, *Anal. Bioanal. Chem.*, **404**, 10, pp. 2887–2894, doi:10.1007/s00216-012-6319-6, http://dx.doi.org/10.1007/s00216-012-6319-6.

67. Errando-Herranz, C., Saharil, F., Romero, A.M., Sandström, N., Shafagh, R.Z., van der Wijngaart, W., Haraldsson, T., and Gylfason, K.B. (2013). Integration of microfluidics with grating coupled silicon photonic sensors by one-step combined photopatterning and molding of OSTE, *Opt. Express*, **21**, 18, pp. 21293–21298, doi:10.1364/oe.21.021293, http://dx.doi.org/10.1364/oe.21.021293.

68. De Vos, K., Bartolozzi, I., Schacht, E., Bienstman, P., and Baets, R. (2007). Silicon-on-insulator microring resonator for sensitive and label-free biosensing, *Opt. Express*, **15**, 12, pp. 7610–7615, doi:10.1364/OE.15.007610, http://dx.doi.org/10.1364/OE.15.007610.

69. Paguirigan, A.L., and Beebe, D.J. (2008). Microfluidics meet cell biology: bridging the gap by validation and application of microscale techniques for cell biological assays, *BioEssays: News Rev. Mol. Cell. Dev. Biol.*, **30**, 9, pp. 811–821, doi:10.1002/bies.20804, http://dx.doi.org/10.1002/bies.20804.

70. Carlborg, C.F., Haraldsson, T., Oberg, K., Malkoch, M., and van der Wijngaart, W. (2011). Beyond PDMS: off-stoichiometry thiol-ene (OSTE) based soft lithography for rapid prototyping of microfluidic devices, *Lab Chip*, **11**, 18, pp. 3136–3147, doi:10.1039/c1lc20388f, http://dx.doi.org/10.1039/c1lc20388f.

71. Paguirigan, A.L., and Beebe, D.J. (2009). From the cellular perspective: exploring differences in the cellular baseline in macroscale and microfluidic cultures, *Integr. Biol.*, **1**, 2, pp. 182–195, doi:10.1039/b814565b, http://dx.doi.org/10.1039/b814565b.

72. Luchansky, M.S., and Bailey, R.C. (2011). High-Q optical sensors for chemical and biological analysis, *Anal. Chem.*, **84**, 2, pp. 793–821, doi:10.1021/ac2029024, http://dx.doi.org/10.1021/ac2029024.

73. Waldbaur, A., Rapp, H., Lange, K., and Rapp, B.E. (2011). Let there be chip-towards rapid prototyping of microfluidic devices: one-step manufacturing processes, *Anal. Methods*, **3**, 12, pp. 2681–2716, doi:10.1039/c1ay05253e, http://dx.doi.org/10.1039/c1ay05253e.

74. Errando-Herranz, C., Saharil, F., Mola Romero, A., Sandström, N., Shafagh, R.Z., van der Wijngaart, W., Haraldsson, T., and Gylfason, K.B. (2013). Integration of polymer microfluidic channels, vias, and connectors with silicon photonic sensors by one-step combined photopatterning and molding of OSTE. In *Proceedings of the 2013 17th International Solid-State Sensors, Actuators and Microsystems Conference (Transducers)*, IEEE, pp. 1613–1616.

75. Kindt, J.T., and Bailey, R.C. (2013). Biomolecular analysis with microring resonators: applications in multiplexed diagnostics and interaction screening, *Curr. Opin. Chem. Biol.*, **17**, 5, pp. 818–826, doi: 10.1016/j.cbpa.2013.06.014, http://dx.doi.org/10.1016/j.cbpa.2013.06.014.

76. Hammer, M., Hiremath, K.R., and Stoffer, R. (2004). *Analytical Approaches to the Description of Optical Microresonator Devices*, pp. 48–71, AIP, doi:10.1063/1.1764013, http://dx.doi.org/10.1063/1.1764013.

77. Yariv, A. (2000). Universal relations for coupling of optical power between microresonators and dielectric waveguides, *Electron. Lett.*, **36**, 4, pp. 321–322, doi:10.1049/el:20000340, http://dx.doi.org/10.1049/el:20000340.

78. Amarnath, K. (2006). *Active Microring and Microdisk Optical Resonators on Indium Phosphide*, PhD thesis, University of Maryland, College Park, http://drum.lib.umd.edu/bitstream/1903/3513/1/umi-umd-3343.pdf.

79. Heebner, J.E., Grover, R., and Ibrahim, T.A. (2008). *Optical Microresonators: Theory, Fabrication, and Applications*, Springer, http://www.worldcat.org/oclc/166358155.

80. Yariv, A. (2002). Critical coupling and its control in optical waveguide-ring resonator systems, *Photonics Technol. Lett., IEEE*, **14**, 4, pp. 483–485, doi:10.1109/68.992585, http://dx.doi.org/10.1109/68.992585.

81. Rabiei, P., Steier, W.H., Zhang, C., and Dalton, L.R. (2002). Polymer micro-ring filters and modulators, *J. Lightwave Technol.*, **20**, 11, pp. 1968–1975, doi:10.1109/jlt.2002.803058, http://dx.doi.org/10.1109/jlt.2002.803058.

82. Rabus, D.G. (2007). *Integrated Ring Resonators: The Compendium*, Springer, http://www.worldcat.org/oclc/123893382.

83. Kabashin, A.V., Evans, P., Pastkovsky, S., Hendren, W., Wurtz, G.A., Atkinson, R., Pollard, R., Podolskiy, V.A., and Zayats, A.V. (2009). Plasmonic nanorod metamaterials for biosensing, *Nat. Mater.*, **8**, 11, pp. 867–871, doi:10.1038/nmat2546, http://dx.doi.org/10.1038/nmat2546.

84. White, I.M., and Fan, X. (2008). On the performance quantification of resonant refractive index sensors, *Opt. Express*, **16**, 2, pp. 1020–1028, doi:10.1364/oe.16.001020, http://dx.doi.org/10.1364/oe.16.001020.

85. Hu, J., Sun, X., Agarwal, A., and Kimerling, L.C. (2009). Design guidelines for optical resonator biochemical sensors, *J. Opt. Soc. Am. B*, **26**, 5,

pp. 1032–1041, doi:10.1364/josab.26.001032, http://dx.doi.org/10. 1364/josab.26.001032.

86. Squires, T.M., Messinger, R.J., and Manalis, S.R. (2008). Making it stick: convection, reaction and diffusion in surface-based biosensors, *Nat. Biotech.*, **26**, 4, pp. 417–426, doi:10.1038/nbt1388, http://dx.doi.org/ 10.1038/nbt1388.

87. Chao, C.-Y., and Guo, L.J. (2002). Polymer microring resonators fabricated by nanoimprint technique, *J. Vac. Sci. Technol. B: Microelectron. Nanometer Struct.*, **20**, 6, pp. 2862–2866, doi:10.1116/1.1521729, http://dx.doi.org/10.1116/1.1521729.

88. Chao, C.-Y., and Guo, L.J. (2004). Thermal-flow technique for reducing surface roughness and controlling gap size in polymer microring resonators, *Appl. Phys. Lett.*, **84**, 14, pp. 2479+, doi:10.1063/1. 1691492, http://dx.doi.org/10.1063/1.1691492.

89. Chao, C.-Y., and Guo, L.J. (2004). Reduction of surface scattering loss in polymer microrings using thermal-reflow technique, *Photonics Technol. Lett., IEEE*, **16**, 6, pp. 1498–1500, doi:10.1109/lpt.2004. 827413, http://dx.doi.org/10.1109/lpt.2004.827413.

90. Wang, L., Han, X., Gu, Y., Lv, H., Cheng, J., Teng, J., Ren, J., Wang, J., Jian, X., and Zhao, M. (2012). Optical biosensors utilizing polymer-based athermal microring resonators, pp. 842731–842731–9, doi:10.1117/ 12.922209, http://dx.doi.org/10.1117/12.922209.

91. Teng, J., Scheerlinck, S., Zhang, H., Jian, X., Morthier, G., Beats, R., Han, X., and Zhao, M. (2009). A PSQ-l polymer microring resonator fabricated by a simple UV-based Soft-Lithography process, *Photonics Technol. Lett., IEEE*, **21**, 18, pp. 1323–1325, doi:10.1109/lpt.2009.2026551, http://dx.doi.org/10.1109/lpt.2009.2026551.

92. Mancuso, M., Goddard, J.M., and Erickson, D. (2012). Nanoporous polymer ring resonators for biosensing, *Opt. Express*, **20**, 1, pp. 245–255, doi:10.1364/oe.20.000245, http://dx.doi.org/10.1364/oe. 20.000245.

93. Delezoide, C., Salsac, M., Lautru, J., Leh, H., Nogues, C., Zyss, J., Buckle, M., Ledoux-Rak, I., and Nguyen, C.T. (2012). Vertically coupled polymer microracetrack resonators for label-free biochemical sensors, *Photonics Technol. Lett., IEEE*, **24**, 4, pp. 270–272, doi:10.1109/lpt. 2011.2177518, http://dx.doi.org/10.1109/lpt.2011.2177518.

94. Ghasemi, F., Eftekhar, A.A., Gottfried, D.S., Song, X., Cummings, R.D., and Adibi, A. (2013). Self-referenced silicon nitride array microring biosensor for toxin detection using glycans at visible wavelength,

pp. 85940A–85940A–9, doi:10.1117/12.2005653, http://dx.doi.org/ 10.1117/12.2005653.

95. Bruel, M. (1995). Silicon on insulator material technology, *Electron. Lett.*, **31**, 14, pp. 1201–1202, doi:10.1049/el:19950805, http://dx.doi. org/10.1049/el:19950805.

96. Carlborg, C.F., Gylfason, K.B., Kazmierczak, A., Dortu, F., Banuls Polo, M.J., Maquieira Catala, A., Kresbach, G.M., Sohlstrom, H., Moh, T., Vivien, L., Popplewell, J., Ronan, G., Barrios, C.A., Stemme, G., and van der Wijngaart, W. (2010). A packaged optical slot-waveguide ring resonator sensor array for multiplex label-free assays in labs-on-chips, *Lab Chip*, **10**, 3, pp. 281–290, doi:10.1039/b914183a, http://dx.doi. org/10.1039/b914183a.

97. Whitesides, G.M. (2006). The origins and the future of microfluidics, *Nature*, **442**, 7101, pp. 368–373, doi:10.1038/nature05058, http://dx. doi.org/10.1038/nature05058.

98. Carlborg, C.F., Haraldsson, T., Cornaglia, M., Stemme, G., and van der Wijngaart, W. (2010). A high-yield process for 3-D large-scale integrated microfluidic networks in PDMS, *J. Microelectromech. Syst.*, **19**, 5, pp. 1050–1057, doi:10.1109/jmems.2010.2067203, http://dx.doi.org/ 10.1109/jmems.2010.2067203.

99. Saharil, F., Carlborg, C.F., Haraldsson, T., and van der Wijngaart, W. (2012). Biocompatible "click" wafer bonding for microfluidic devices, *Lab Chip*, **12**, 17, pp. 3032–3035, doi:10.1039/c2lc21098c, http://dx. doi.org/10.1039/c2lc21098c.

100. Errando-Herranz, C., Vastesson, A., Zelenina, M., Pardon, G., Bergström, G., Wijngaart, W., Haraldsson, T., Brismar, H., and Gylfason, K.B. (2013). Biocompatibility of OSTE polymers studied by cell growth experiments. In *Proceedings of the 17th International Conference on Miniaturized Systems for Chemistry and Life Sciences (microTAS)*.

101. Iqbal, M., Gleeson, M.A., Spaugh, B., Tybor, F., Gunn, W.G., Hochberg, M., Baehr-Jones, T., Bailey, R.C., and Gunn, L.C. (2010). Label-free biosensor arrays based on silicon ring resonators and high-speed optical scanning instrumentation, *IEEE J. Sel. Top. Quant. Electron.*, **16**, 3, pp. 654–661, doi:10.1109/jstqe.2009.2032510, http://dx.doi.org/ 10.1109/jstqe.2009.2032510.

Index